AF785259

The Cannabis Genome

Cannabis is undergoing a remarkable transformation—from a misunderstood plant to a cornerstone of modern medicine, industry, and agriculture. This book bridges the gap between tradition and innovation, providing a holistic exploration of cannabis genetics, breeding, genomics, and its wide-ranging applications. Authored by leading researchers, industry pioneers, and legacy breeders, it offers a unique blend of academic rigor and practical insights.

Unlike other resources, this book integrates perspectives from public and private sectors, honoring those who have advanced cannabis cultivation despite legal and social barriers. It recognizes the pivotal role of startups like MyFloraDNA, which invest heavily in groundbreaking research, shaping the future of this extraordinary plant.

Key Features:

- Comprehensive insights into cannabis origins, taxonomy, and genetic diversity.
- Advanced breeding techniques and applications of tissue culture, CRISPR, and "omics" technologies.
- Practical strategies for pest management, crop improvement, and intellectual property protection.
- Forward-looking perspectives on cannabis in personalized medicine, sustainability, and global agriculture.
- Written for scientists, growers, policymakers, and enthusiasts, this book addresses the economic, social, and scientific dimensions of cannabis. It offers actionable knowledge to drive progress in cultivation, industry, and research.

Angel Fernandez Marti, a former scientist at UC Davis and UC Berkeley, is a renowned expert in plant breeding and biotechnology. Now leading MyFloraDNA, he focuses on advancing research and innovation in cannabis genetics.

Jaime Prohens, a distinguished professor at the Universitat Politècnica de València, is a globally respected expert in plant breeding and genetics. His groundbreaking contributions have shaped modern breeding techniques, including applications for cannabis.

Depiction and Designing of Plant Genomes
Series editor: Chittaranjan Kole

The Cannabis Genome
A New Era in Genetics, Genomics, and Breeding
Edited by Angel Fernandez Marti, Jaime Prohens

For more information, please visit our website: www.routledge.com/Depiction-and-Designing-of-Plant-Genomes/book-series/DDPG

The Cannabis Genome

A New Era in Genetics, Genomics, and Breeding

Edited by
Angel Fernandez Marti and Jaime Prohens

CRC Press
Taylor & Francis Group
Boca Raton London New York

CRC Press is an imprint of the
Taylor & Francis Group, an **informa** business

Designed cover image: Shutterstock

First edition published 2025
by CRC Press
2385 NW Executive Center Drive, Suite 320, Boca Raton FL 33431

and by CRC Press
4 Park Square, Milton Park, Abingdon, Oxon, OX14 4RN

CRC Press is an imprint of Taylor & Francis Group, LLC

© 2025 selection and editorial matter, Angel Fernandez Marti and Jaime Prohens; individual chapters, the contributors

This book contains information obtained from authentic and highly regarded sources. While all reasonable efforts have been made to publish reliable data and information, neither the author[s] nor the publisher can accept any legal responsibility or liability for any errors or omissions that may be made. The publishers wish to make clear that any views or opinions expressed in this book by individual editors, authors or contributors are personal to them and do not necessarily reflect the views/opinions of the publishers. The information or guidance contained in this book is intended for use by medical, scientific or health-care professionals and is provided strictly as a supplement to the medical or other professional's own judgement, their knowledge of the patient's medical history, relevant manufacturer's instructions and the appropriate best practice guidelines. Because of the rapid advances in medical science, any information or advice on dosages, procedures or diagnoses should be independently verified. The reader is strongly urged to consult the relevant national drug formulary and the drug companies' and device or material manufacturers' printed instructions, and their websites, before administering or utilizing any of the drugs, devices or materials mentioned in this book. This book does not indicate whether a particular treatment is appropriate or suitable for a particular individual. Ultimately it is the sole responsibility of the medical professional to make his or her own professional judgements, so as to advise and treat patients appropriately. The authors and publishers have also attempted to trace the copyright holders of all material reproduced in this publication and apologize to copyright holders if permission to publish in this form has not been obtained. If any copyright material has not been acknowledged please write and let us know so we may rectify in any future reprint.

Except as permitted under U.S. Copyright Law, no part of this book may be reprinted, reproduced, transmitted, or utilized in any form by any electronic, mechanical, or other means, now known or hereafter invented, including photocopying, microfilming, and recording, or in any information storage or retrieval system, without written permission from the publishers.

For permission to photocopy or use material electronically from this work, access www.copyright.com or contact the Copyright Clearance Center, Inc. (CCC), 222 Rosewood Drive, Danvers, MA 01923, 978-750-8400. For works that are not available on CCC please contact mpkbookspermissions@tandf.co.uk

Trademark notice: Product or corporate names may be trademarks or registered trademarks and are used only for identification and explanation without intent to infringe.

ISBN: 978-1-032-47897-5 (hbk)
ISBN: 978-1-032-47900-2 (pbk)
ISBN: 978-1-003-38643-8 (ebk)

DOI: 10.1201/9781003386438

Typeset in Times
by Newgen Publishing UK

Dedication

This book series is dedicated to my brothers and sisters –

Bansi Badan Kole
Niranjan Kole
Bharati Kole Nandi
Arati Kole Kundu
Ram Ranjan Kole

Contents

Preface to the Series

Genome sequencing of the model plant Arabidopsis in 2000 followed by the crop plant rice in 2002 flagged of genome sequencing in an array of plants. As on March 05, 2024, 4604 genomes of 1482 different plant species have been sequenced, and the number has been continuing to grow exponentially. The list includes several crop plants as well as some model plants. Sequencing of the nuclear genomes was followed by sequencing of extranuclear genomes such as mitochondria and chloroplast and by several studies on functional genomics. However, detailed reports on them are scattered in the literature, including research articles and reviews. Therefore, a comprehensive compilation of deliberations on the sequences with enumeration of the sequences, annotation, assembly, gene families and comparative genomics of these useful genomes and elucidation of functional genomics including transcriptomics, metabolomics and proteomics in these higher plants is timely.

A number of higher plants provide the basic F5—food, feed, fuel, fiber and furniture—needs; and also, nutraceuticals, beverages, luxury and aesthetic items. Genetic improvement in these plants with regard to their utilities was being done in the last century mainly by traditional breeding. Advent of molecular breeding and genetic engineering in the last two decades of the last century strengthened the breeding strategies in the economic plants. Advances in plant genomics has ultimately revolutionized the concepts and strategies of crop improvement. Employment of high-throughput genotyping and phenotyping, together with other genomics assisted breeding tools like gene discovery, allele mining, genome-wide association studies, genomic selection and genome-wide prediction etc., has facilitated accelerated and precise plant breeding. Recent emergence of the genome editing technique has enormously supplemented the molecular and genomics tools, making it possible to design and tailor the genomes according to the needs. However, success stories of next-generation breeding techniques need to be compiled for easy access to the readers.

This present book series entitled, Depiction and Designing of Plant Genomes, includes a number of books each dedicated to the deliberations on genomic attributes and/or genetic improvement of a higher plant and hopefully will serve as handy compendium for readers.

The books of this series have been edited by highly reputed scientists, and I remain thankful for their invaluable academic efforts and cooperation. I am also thankful to the publisher and specifically to its Senior Commissioning Editor (Chemistry / Agricultural Sciences), Dr. Renu Upadhyay and her Editorial Assistant, Ms. Jyotsna Jangra for their outstanding and timely help and cordial cooperation since inception of this book series.

Finally, I must express my thanks to my wife Phullara and our children, Sourav and Devleena, for assisting me in editing the books of this series and also bearing with me for my channelizing the time that should have been kept for them towards editing this book series.

Kolkata, 3rd August 2024 **Chittaranjan Kole**

Special Acknowledgment to
Phullara Kole
For her outstanding assistance in editing this book series

Preface

After many years working in academia and researching various crops, from fruit trees to forest species and horticultural crops, I am thrilled to be able to contribute to the improvement of cannabis cultivation. This plant deserves special attention because, unfortunately, it has not been researched and developed like other plants. The recent legalization in many countries and its acceptance by society open new paradigms about the benefits this plant brings to human health.

Cannabis, often misunderstood and stigmatized, is now emerging as a significant player in the field of medicinal plants. The therapeutic potentials of cannabis are vast and varied, offering benefits for a range of conditions from chronic pain to epilepsy. The shift in legal status across various regions has sparked a wave of scientific interest and investment, leading to advancements that were previously unimaginable.

Unfortunately, very few universities and public research centers conduct research on cannabis, leaving much of the scientific advancements to be carried out by private companies or startups. A notable example is MyFloraDNA, a company that allocates 50% of its annual revenue to research, demonstrating its commitment to genetic and scientific advancement of this plant. This dedication highlights the critical role that private enterprises play in pioneering research and development efforts in areas that are often neglected by public institutions.

This book is different from other compendiums published on various species. While most books focus exclusively on scientific results and feature only researchers as authors, in this book I wanted to combine the perspectives of scientists from public universities, researchers from private companies, legacy breeders and pioneers of cannabis cultivation. This mixed approach honors those who have dedicated their lives to advancing this crop and those who have had their freedom violated for simply possessing and believing in its potential.

By including voices from diverse backgrounds, this book aims to provide a holistic view of cannabis cultivation and its potential. It recognizes the contributions of individuals who have worked tirelessly, often under challenging conditions, to unlock the secrets of this remarkable plant. Their work has paved the way for new therapeutic applications and has helped to break down the barriers of stigma and misinformation.

I dedicate this book to all these people and give them the recognition they deserve. I also want to acknowledge all the mentors I have had throughout my scientific career, without whom I would not have professionalized as a scientist. Two of them have been especially influential in my life: Dr. Rafael Socias, from CITA of Aragon (Spain), who rests in peace, and the geneticist Prof. Dodd from UC Berkeley (USA). Their guidance and wisdom have been instrumental in shaping my approach to scientific inquiry and research.

A special thanks to the entire MyFloraDNA family, especially to my partner Javier, for his constant support and help in leading this company dedicated to advancing cannabis genetics. Finally, I thank my family: my parents and sister, and especially my wife Flora for her great support over the years. Her name itself is a symbol of life and growth, perfectly embodying the essence of what we strive to achieve. Her support over the years has been nothing short of extraordinary. She has been my rock, my confidant, and my greatest source of strength. To my sons, Oliver and Max, you are my source of inspiration and motivation every day. Your curiosity and wonder fuel my passion to push the boundaries of what we know and to explore the unknown.

This book not only seeks to share scientific knowledge but also to pay tribute to all those who have contributed to the advancement of cannabis, a plant that, despite its historical challenges, is now beginning to be recognized for its many benefits. May this work serve as a bridge between science, industry, and the community, promoting a greater understanding and appreciation for this extraordinary plant.

In conclusion, the journey of cannabis from a stigmatized plant to a cornerstone of modern medicinal therapy is a testament to the relentless pursuit of knowledge and the courage to challenge the status quo. It is my hope that this book will inspire future researchers, growers, and enthusiasts to continue exploring and innovating in the field of cannabis science. Together, we can unlock the full potential of this remarkable plant and ensure that its benefits are accessible to all who need them.

This book is a call to action for researchers, policymakers, and the general public to view cannabis through a new lens—one that appreciates its potential to improve lives and drive scientific progress. Let us move forward with an open mind and a commitment to uncovering the truths that lie within this extraordinary plant.

Angel Fernandez Marti
Jaime Prohens

Contributors

Riaz Ahmad
MyFloraDNA, Inc., Sacramento, California, USA

Ladan Ajdanian
Institute of Integrative Biology and Systems, Université Laval, Quebec, Canada

Ajith Anand
MyFloraDNA, Inc., Sacramento, California, USA

Mehdi Babaei
Institute of Integrative Biology and Systems, Université Laval, Quebec, Canada

Oussama Badad
Trilogine Seeds, Colorado, USA

Yousoon Baek
Cornell University, School of Integrative Plant Science, Ithaca, NY

Justin Boissinot
Institute of Integrative Biology and Systems, Université Laval, Quebec, Canada

Miriam Caballero Cerveró
Institute for Integrative Systems Biology CSIC- UV

Martina Catani
Department of Chemical, Pharmaceutical, and Agricultural Sciences, University of Ferrara, Ferrara, Italy

Tatiana Chenet
Department of Environmental and Prevention Sciences, University of Ferrara, Ferrara, Italy

Natasha Damiana Spadafora
Department of Chemical, Pharmaceutical, and Agricultural Sciences, University of Ferrara, Ferrara, Italy

Julie Duong
MyFloraDNA Inc., Sacramento California, USA

Chandrakanth Emani
MyFloraDNA Inc., Sacramento California, USA

Graham Farrar
Glass House Farms, Camarillo, California, USA

Simona Felletti
Department of Environmental and Prevention Sciences, University of Ferrara, Ferrara, Italy

Angel Fernandez Marti
MyFloraDNA, Inc., Sacramento, California, USA

Matthew Gates
Zenthanol Consulting, San Diego, California, USA

Matt Haddad
Trilogine Seeds, Colorado, USA

Mohsen Hesami
Department of Plant Agriculture, University of Guelph, Guelph, Ontario, Canada

Dale Hunt
JD, Hunt IP Law, San Diego, California, USA

Athar Hussain
School of Food and Agricultural Sciences, University of Management and Technology (UMT), Lahore, Pakistan
Genomics and Informatics Lab (GIL) Ltd, Lahore, Pakistan

Hirsh Jain
Ananda Strategy, Los Angeles, California, USA

Kevin Jodrey
Wonderland Nursery, Garberville, California, United States

Andrew Maxwell Phineas Jones
Department of Plant Agriculture, University of Guelph, Guelph, Ontario, Canada

Josh Karchmer
Glass House Farms, Camarillo, California, USA

Tiziana Maria Sirangelo
ENEA-Italian National Agency for New Technologies, Energy and Sustainable Economic Development-Division Biotechnologies and Agroindustry, Rome, Italy

José Tomás Matus
Institute for Integrative Systems Biology CSIC- UV

Mahboubeh Mostafaei Dehnavi
Department of Plant Sciences, University of California, Davis, CA, USA
Department of Horticulture Sciences, College of Agriculture and Natural Resources, University of Tehran, Karaj, Iran

David Navarro-Payá
Institute for Integrative Systems Biology CSIC- UV

Arturo Ortega
MyFloraDNA Inc., Sacramento California, USA

Antonio Santiago Pajuelo
Institute for Integrative Systems Biology CSIC- UV

Marcus Parungao
MyFloraDNA Inc., Sacramento California, USA

Sergio Picazo
Sacramento, California, USA

Gastón A. Pizzio
Institute for Integrative Systems Biology CSIC- UV

Jaime Prohens
Universitat Politècnica de València, Valencia, Spain

Pablo Romero
Institute for Integrative Systems Biology CSIC- UV

Ed Rosenthal
Quick American Publishing, Piedmont, California

Anna L. Schwabe
Department of Ecology and Evolutionary Biology, University of Colorado, Boulder, Colorado

Gail Taylor
Department of Plant Sciences, University of California, Davis, CA, USA
Department of Genetics, Evolution and Environment, University College London, London, UK

Davoud Torkamaneh
Université Laval, Canada

Mireia Uranga
KU Leuven Plant Institute (LPI) and Laboratory for Plant Genetics and Crop

Daniela Vergara
Department of Ecology and Evolutionary Biology, University of Colorado Boulder, Boulder, Colorado
Harvest New York, Cornell Cooperative Extension, Geneva, NY

Chen Zhang
Institute for Integrative Systems Biology CSIC- UV

Abbreviations

2,4-D – 2,4	Dichlorophenoxyacetic Acid
ABA	Abscisic Acid
AI	Artificial Intelligence
ARS	Agricultural Research Service
BA	Benzyladenine
BC	Backcross
BAM	Binary Alignment/Map
BSHW	Big Sur Holy Weed
CAOA	Cannabis Administration and Opportunity Act
CANNUSE	Comprehensive Database Classifying Medicinal, Alimentary, Fiber, Psychoactive, and Ethnobotanical Uses of Cannabis
Cas	CRISPR-associated Protein
CBC	Cannabichromene
CBCA	Cannabichromenic Acid
CBD	Cannabidiol
CBDA	Cannabidiolic Acid
CBDAS	Cannabidiolic Acid Synthase
CBG	Cannabigerol
CBGA	Cannabigerolic Acid
CBN	Cannabinol
CDC	Centers for Disease Control and Prevention
CGRI LLC	Company Name (No Expansion Given)
CGN	Center for Genetic Resources, The Netherlands
CGRC	Cannabis Germplasm Resource Center
ChIP-Seq	Chromatin Immunoprecipitation Sequencing
CIAT	Centro Internacional de Agricultura Tropical
C. sativa	*Cannabis sativa*
CCI	Certificate of Cultivar Identity
CLTVTD	Cultivated (Genetics/Validated Curations)
CO2	Carbon Dioxide
CRISPR	Clustered Regularly Interspaced Short Palindromic Repeats
CSA	Controlled Substances Act
CTAB	Cetyltrimethylammonium Bromide
CTX	Inter-chromosomal Translocations
DAPC	Discriminant Analysis of Principal Components
DARE	Drug Abuse Resistance Education
DEA	Drug Enforcement Administration
DKW	Driver and Kuniyuki Walnut Medium
DNA	Deoxyribonucleic Acid
DOJ	Department of Justice
DUS	Distinctness, Uniformity, and Stability
E-Nose	Electronic Nose
E.U.	European Union
EI	Electron Ionization
ENA	European Nucleotide Archive
EST-SSRs	Expressed Sequence Tag-Simple Sequence Repeats
FAIR	Findable, Accessible, Interoperable, and Reusable

F1	First Filial Generation
F2	Second Filial Generation
FDA	Food and Drug Administration
FID	Flame Ionization Detection
FNA	Flora of North America
GC	Gas Chromatography / Guanine-Cytosine Content
GBS	Genotyping-By-Sequencing
GEO	Gene Expression Omnibus
GE	Genetic Engineering
GIS	Glabrous Inflorescence
GISH	Genomic In Situ Hybridization
GPPS	Geranyl Pyrophosphate Synthase
GWAS	Genome-Wide Association Study
HHS	Department of Health and Human Services
HPLC	High-Performance Liquid Chromatography
HRM	High Resolution Melting
HRMS	High Resolution Mass Spectrometry
HLVd	Hop Latent Viroid
ICNCP	International Code of Nomenclature for Cultivated Plants
InDel	Insertion-Deletion
IP	Intellectual Property
IPGRI	International Plant Genetic Resources Institute
IPP	Isopentenyl Diphosphate
IRS	Internal Revenue Service
ITX	Intra-chromosomal Translocations
JD	Juris Doctor
LC	Liquid Chromatography
LC-HRMS	Liquid Chromatography-High Resolution Mass Spectrometry
LED	Light-Emitting Diode
MADC	Male Associated DNA Coded Markers
Mb	Megabase
Mbp	Million Base Pairs
MAS	Marker-Assisted Selection
MEP	Methylerythritol Phosphate
MeJA	Methyl-Jasmonate
ML	Mildew Locus O
MORE	Marijuana Opportunity Reinvestment and Expungement (Act)
MS	Mass Spectrometry / Murashige and Skoog (Plant Tissue Culture Medium)
MTA	Material Transfer Agreement
MYB	MYB Transcription Factor Gene Family
NCBI	National Center for Biotechnology Information
NGS	Next-Generation Sequencing
NILs	Near-Isogenic Lines
NMR	Nuclear Magnetic Resonance Spectroscopy
NSGC	National Small Grains Collection
OA	Olivetolic Acid
OLS	Olivetol Synthase
P1	Parent 1
P2	Parent 2
PBR	Plant Breeders' Rights

PCA	Principal Component Analysis
PCR	Polymerase Chain Reaction
PhD	Doctor of Philosophy
PM	Powdery Mildew
PP	Plant Patent
PRJNA	Project Identifier in NCBI SRA
PT	Prenyltransferase
PVP	Plant Variety Protection
QTL	Quantitative Trait Loci
QTLs	Quantitative Trait Loci
qRT-PCR	Quantitative Reverse Transcription Polymerase Chain Reaction
R&D	Research and Development
RNA	Ribonucleic Acid
RNA-Seq	RNA Sequencing
RNAi	RNA Interference
RS	Raman Spectroscopy
SA	Salicylic Acid
SAFE	Secure and Fair Enforcement (Banking Act)
SAM	Sequence Alignment/Map
SB-34	California Senate Bill 34
SBS	Sequencing By Synthesis
SGRNA	Single Guide RNA
SNP	Single Nucleotide Polymorphism
SNPs	Single Nucleotide Polymorphisms
SNV	Single Nucleotide Variant
SPME	Solid Phase Microextraction
SRA	Sequence Read Archive
STR	Short Tandem Repeat
SVs	Structural Variants
TF	Transcription Factor
THC	Tetrahydrocannabinol / Δ9-Tetrahydrocannabinol
THCA	Tetrahydrocannabinolic Acid
THCAS	Tetrahydrocannabinolic Acid Synthase
THCV	Tetrahydrocannabivarin
TPS	Terpene Synthases
UPOV	International Union for the Protection of New Varieties of Plants
USDA	United States Department of Agriculture
USDA-ARS	United States Department of Agriculture - Agricultural Research Service
USPTO	United States Patent and Trademark Office
UV	Ultraviolet
VSUP	Variety-Specific Utility Patent
WGCNA	Weighted Gene Co-expression Network Analysis
WGS	Whole Genome Sequencing
XX	Female Sex Chromosome Pair
XY	Male Sex Chromosome Pair
XXY	Atypical Female Chromosomes
XXXY	Atypical Female Chromosomes
youPCR	A product name used for early sex determination in *Cannabis sativa*
ZFN	Zinc Finger Nucleases

1 The Debated Nomenclature of *Cannabis sativa*

Anna L. Schwabe and Daniela Vergara

1.1 INTRODUCTION

Cannabis sativa L., commonly known as marijuana or hemp, stands as one of the most captivating yet stigmatized plants globally, boasting an extensive and intertwined history with humans with evidence of cultivation dating back thousands of years (Li 1973, 1974; Kalant 2001; Russo 2007; Clarke and Merlin 2016; McPartland et al. 2019). Although humans have used cannabis for food, fuel, fiber, ceremonies, medicine, and more, it has also been vilified as an illicit drug worldwide (United Nations Conference for the Adoption of a Single Convention on Narcotic 1962) dating back as far as 1378. In the United States, Cannabis was formally criminalized under the Controlled Substances Act (CSA), passed by the 91st United States Congress, and signed into Law by President Richard Nixon in 1970 (Congress 1970).

In this chapter, we will adhere to the proper scientific name format *Cannabis sativa* or *C. sativa* when appropriate; however, when referring to the colloquial vernacular, we will use the word "cannabis".

Throughout the extensive relationship between humans and cannabis, it has been cultivated and selected for various uses. Applied uses encompass every part of the plant, including the seed, fiber, roots, oils, and unique phytochemicals from the glandular trichomes. While the aromatic flower has attracted most of the attention surrounding cannabis, historical documentation reveals a diverse variety of uses for other parts of the plant. Roots were once used as a treatment for gout and joint pain (Pliny and Ian 1854; Ryz et al. 2017), leaves have been used in the preparation of tea drinks, seeds can be processed for oils and serve as a nutrient-rich grain, and the fibrous stem is used to make rope, cloth, and building materials (Ahmed et al. 2022).

Perhaps one of the most distinctive characteristics of cannabis is the production of secondary metabolites, including terpenes and cannabinoids. These compounds are produced and stored in specialized glandular structures called trichomes, which are concentrated on the female flowers (Sirikantaramas et al. 2005; Gagne et al. 2012). The cannabinoids produced in highest quantities by the plant are Δ^9-tetrahydrocannabinolic acid (THCA), and cannabidiolic acid (CBDA), which are converted to their neutral forms Δ^9-tetrahydrocannabinol (THC) and cannabidiol (CBD) respectively (Hart et al. 2001), through heat by a process called decarboxylation. The acidic forms of both compounds are produced via an enzymatic process in which cannabigerolic acid (CBGA) serves as the precursor molecule. THC is the psychoactive compound responsible for the sensation known as the "high," and it is subject to strict governmental regulation. The plant produces more than 120 cannabinoids, depending on how these are counted (Mechoulam 2005; Brenneisen 2007; Elzinga et al. 2015; Hanuš and Appendino 2016; Hanuš et al. 2016; Radwan et al. 2021), and as many of these cannabinoids are present in small quantities, they are referred to as "minor cannabinoids".

DOI: 10.1201/9781003386438-1

Cannabis sativa is a member of the family Cannabaceae with about 170 species distributed across ten small genera, including *Cannabis, Humulus*, and *Celtis* (McPartland 2018). However, there is conflicting information regarding the current taxonomy (List 2010; Clarke and Merlin 2013; Integrated Taxonomic Information System Database 2019). The Flora of North America (FNA), the governmental repository National Center for Biotechnology Information (NCBI), and multiple genomic studies support one species (Small 1997; Sawler et al. 2015; Lynch et al. 2016; Lapierre et al. 2023), although some classical breeders and botanists support the polytypic taxonomy of the genus *Cannabis* (Lamarck 1785; Emboden 1974; Anderson 1980). However, since the species consists of distinct lineages, it remains unclear whether isolation between lineages exists. Therefore, it is uncertain if some groupings produce more fit or viable offspring when crossed within groups compared to crosses among groups.

Here, we will comment and discuss the origins and different groupings within the *Cannabis* species, the diverse nomenclature proposed both scientifically and colloquially, and the disagreements regarding these naming conventions.

1.2 ORIGINS AND CHARACTERISTICS OF THE PLANT

The genera *Cannabis* and *Humulus* (hop) both belong to the family Cannabaceae, and they appeared to have diverged from a common ancestor approximately 28 million years ago on the Tibetan Plateau (McPartland et al. 2019). It is generally accepted that there are no natural populations of cannabis remaining unadulterated from cultivated types (Small and Cronquist 1976). Wild-growing populations are thought to have emerged either by escaping cultivation (Doebley et al. 2006) becoming feral, or through introgression via pollen drift from cultivated forms. This gives rise to "wild" plants that carry traits selected by humans. However, these feral populations also undergo their own adaptive processes, and as a result may not necessarily exhibit traits that align with human preferences. Of note is that many wild-growing types may not match the quality of cultivated, hand-selected, and carefully grown modern crops. The absence of true ancestral populations indicates cannabis, both feral and cultivated, have traces of human selection rather than natural evolutionary processes (McPartland et al. 2019; Schwabe and McGlaughlin 2019). Without access to ancestral or wild-type cannabis, the evolutionary history becomes somewhat ambiguous, making it challenging to piece together the evolutionary timeline accurately. The absence of these foundational forms hinders a comprehensive understanding of the natural trajectory of the species, creating a notable gap in our knowledge of its evolutionary journey.

While the absence of ancestral or wild-type cannabis complicates our understanding of its evolutionary history, the fact remains that humans have been utilizing cannabis for thousands of years. This extended period of human interaction has led to widespread distribution, facilitated by seed dispersal and unintentional escapes from cultivation. As a result, cannabis is now found around the world and locally adapted forms are found on multiple continents and island nations (Piper 2005; Clarke and Merlin 2013; McPartland et al. 2019). Despite the global distribution, cannabis has faced legal restrictions for decades, although some countries have persisted in cultivating hemp for grain or fiber production. The recent global trend toward decriminalization and legalization has ignited public interest in the potential uses of cannabis. This shift has not only stimulated scientific research but has also led to a relaxation in restrictions, paving the way for the growth of an expansive and lucrative industry.

A continuous record of cannabis cultivation exists in China from Neolithic times through to the present (Li 1974), although the first formal description was recorded by Carl Linnaeus in 1753. Linnaeus, who was acquainted with the cultivated forms of the plant, may have chosen the term "sativa," as it translates to "cultivated". Later in 1785, Jean-Baptiste Lamarck described a second species, *C. indica*, with a unique smell with intoxicating qualities (Watts 2006). A third species has also been proposed, *C. ruderalis,* which was first described by Russian Botanist Dimitri

Janischewsky in 1924 (Schultes et al. 1975). Formally, *C. sativa* stands as the sole recognized species in the *Cannabis* genus.

Genetic and genomic research supports at least three major lineages: one is a group containing traditional hemp for fiber and grain, and there are at least two "marijuana" or "drug-type" groupings (Sawler et al. 2015; Lynch et al. 2016; Vergara et al. 2016; Dufresnes et al. 2017; Grassa et al. 2021). Scholars and enthusiasts have attempted to define and categorize cannabis in various ways, but the differentiation resulting from common vernacular or governmental connotations does not align with biological lineages, which may be the result of thousands of years of artificial breeding and selection for different uses.

The reproductive behavior of cannabis adds another layer of complexity. Cannabis most commonly exhibits separate male and female plants, a characteristic referred to as dioecious, meaning "two houses". Males, with distinct XY sex chromosomes, are heterogametic, while females, with two XX sex chromosomes, are homogametic. Interestingly, monoecious ("one house") occurrences, where both male and female flowers coexist on the same plant are also found among cannabis populations (Mandolino et al. 1999; Moliterni et al. 2004; Divashuk et al. 2014; Razumova et al. 2016). Although male and female flowers typically develop on distinct plants, certain conditions can lead to the development of male flower structures on female plants due to the labile or leaky nature of sex in cannabis (Ram and Jaiswal 1970; Ram and Jaiswal 1972; Ram and Sett 1982; Milewicz and Sawicki 2012). Notably, distinguishing between male and female plants before flowering is challenging, but it becomes crucial in cultivation, especially when aiming for flowers with high cannabinoid content. Male plants are undesirable in such scenarios because fertilized females tend to shift their energy to unwanted seed production and produce fewer cannabinoids and other phytochemicals (Meier and Mediavilla 1998; Lipson Feder et al. 2021).

While the identification of males before flowering poses a challenge, the potential for improved cultivation practices arises through advancements in genetic testing. These tests hold promise for early male identification. Nevertheless, a limitation exists, as these tests frequently struggle to discern monoecious plants, predominantly of the XX type (Punja and Holmes 2020). Additionally, genetic females exhibiting male traits, such as pollen production introduces further intricacies (Mandolino et al., 1999; Prentout et al., 2019, 2020; Toth et al., 2020). Although flower cultivation avoids the presence of males in the population at all costs, hemp cultivation for seed production requires the presence of both male and female plants or monoecious plants to facilitate fertilization and subsequent seed development. As a result, mixed or monoecious populations are commonly found among hemp varieties cultivated for grain.

1.3 HISTORY AND DIFFICULTY IN CLASSIFICATION

Various categorizations have been proposed to describe cannabis (Schultes et al. 1975; Small and Cronquist 1976; Anderson 1980; Watts 2006). These classifications may be based on phenotypic characteristics, alleged region of origin, intended uses, effects post-consumption, or governmental terminologies, described below. Employing precise language is essential for effective communication and navigation within the diverse landscape of cannabis. This is particularly crucial given the various applications and implications of different cannabis varieties. Whether it is discussing therapeutic benefits, recreational use, or industrial applications, the use of specific terminology ensures clarity and facilitates informed conversations about this versatile plant. Our main goal is to contribute to establishing clarity and avoiding generalizations or misunderstandings. Whether we use terms like "hemp" and "marijuana" or specify strain names, it's crucial to approach the topic with nuance and a well-informed perspective. We need to recognize that these terms can also have different interpretations and legal implications, depending on the jurisdiction. Therefore, aligning our language usage with specific definitions becomes crucial to ensure precision and avoid potential complications as we strive for clarity.

1.3.1 *Cannabis sativa*, *Cannabis indica*, and *Cannabis ruderalis*

Although *Cannabis sativa* is currently the only recognized species in the *Cannabis* genus, ongoing debates persist regarding the taxonomic classification on whether two (*C. sativa* and *C. indica*) or three (*C. sativa, C. indica*, and *C. ruderalis*) distinct species should be recognized, along with multiple subspecies (Hillig 2005; Clarke and Merlin 2013; Clarke and Merlin 2016; Small 2016; McPartland 2018; McPartland and Small 2020). Genetic and genomic analyses suggest that *C. sativa* and *C. indica* do not exhibit enough differentiation to be classified as distinct and separate species, especially when compared to the genetic differentiation observed in other closely related species (McPartland 2018). Such findings argue against elevating these species delineations to a taxonomic level higher than variety or subspecies. Therefore, it is generally accepted that *C. sativa* and *C. indica* should not be recognized as different species but a case to potentially recognize *C. sativa* subsp. *sativa* and *C. sativa* subsp. *indica* could be made based on these findings.

A third proposed taxon, *C. ruderalis,* described as smaller in stature (around 60 cm or 24 inches) with fewer branches and lower cannabinoid production, is thought to have originated from the Volga and Siberia regions in Russia (Schultes et al. 1975). These plants have "auto flowering" properties (Rosenthal 2001, 2009) that do not depend on the light cues for flower production (Schultes et al. 1975; Rosenthal 2001, 2009; Lapierre et al. 2023). This distinctive flowering trait in *C. ruderalis* is theorized to be a mutation that may have conferred an adaptive advantage by allowing populations to undergo flowering cycles and set seed in short growing seasons (Piper 2005; McPartland 2018; Vassilevska-Ivanova 2019; Toth et al. 2022; Kurtz et al. 2023; Leckie et al. 2023) This trait has been deliberately bred into commercial strains, leading to the development of the modern auto-flowering cannabis plants. These plants exhibit accelerated maturation and can be cultivated in smaller spaces, presenting a significant departure from photoperiod plants that flower in response to the shortening day length and light exposure during the growing cycle. For more detailed species proposals and taxonomic considerations and treatments, see Table S1.

1.3.2 "Indica" and "Sativa"

The terms "sativa" and "indica" are used concurrently to describe morphology, but also common psychoactive effects (Table 1.1), which are completely unrelated characteristics in cannabis (McPartland 2017; McPartland and Guy 2017; McPartland and Small 2020). Although some botanists, taxonomists, and enthusiasts use "indica" and "sativa" to describe a morphological type (Table 1.1), these traits are independent to the phytochemicals produced by the plant (Vergara et al. 2021a) responsible for the psychoactive effects. Additionally, phenotypic traits may be plastic and

TABLE 1.1
Different Attributes Including Phenotypic Traits, Provenance, and Effects for the Scientific and Common Groupings Using "Sativa" and "Indica" Terminology

Alleged characteristic		Sativa	Indica
Phenotype	Color	Light green	Dark green
	Height	Tall	Short
	Leaf shape	Lengthy, narrow	Broad
	Flowering time	Long time to flower	Short time to flower
	Flowering pattern	Loose	Packed
	Branching pattern	Open	Closed (bushy)
	Smell	Fruity	Earthy
Provenance	Region of origin	Tropics (warmer climates)	Northern latitudes (cooler climates)
Effect	Body v Head high (high vs. Stoned?)	Active, energizing, uplifting	Sedative (Couch-lock)

FIGURE 1.1 Mutations lead to phenotypic diversity in *Cannabis sativa.* **Top left to right**: "Freakshow" fern leaf mutation (Eddie Duckworth); variegation leading to a white and green mottling in the leaf coloration (Anna Schwabe); and faciation which is an abnormal growth pattern leading to abnormal flattened, elongated, or cresting patterns in the stems or flowers (London Niro). **Middle left to right:** Leaf buds are flowers growing in the petiole (Stephanie Beno); "Australian Bastard Cannabis" (aka "ABC") leaf mutation leading to rosette formation and smooth leaf margins (Scott John Butler); flower structure mutation leading to a simple racemic growth architecture rather than the compound raceme that cannabis typically displays (Anna Schwabe). **Bottom left to right:** Whorled phylotaxy with three leaves emerging from the same node rather than the common opposite leaf arrangement (Hayden McKenna); self-topping plant that naturally formed two main stems without any physical manipulation (Anna Schwabe); foxtailing is when the flower architecture is stacked rather than clustered (Leigh-Anne Bayer).

can change due to the environment and are not always a good tool for defining differences within or among related organisms (Schwabe et al. 2015). Moreover, heritable mutations leading to unique characteristics add another layer of complexity to the phenotypic diversity observed within the species (Figure 1.1). The terms "Sativa" and "Indica" written in quotation marks (either capitalized or not) carry different implications than *C. sativa* and *C. indica* written in italics (McPartland 2018) which is indicative of a taxonomic distinction based on morphology, geography, or genetic differentiators and recognized as a formal classification or ranking. The use of quotation marks suggests a colloquial or vernacular context, often referring to perceived effects or characteristics associated with cannabis. On the other hand, writing *C. sativa* and *C. indica* in italics indicates a taxonomic classification, emphasizing the botanical and scientific aspects of the *C. sativa* species. It's essential to distinguish between these usages to avoid confusion between colloquial and taxonomic contexts.

Because no correlation exists between the phenotypic characteristics and the phytochemicals, or between phytochemicals and these common classifications (Smith et al. 2022), confusion abounds. The phytochemicals contribute to the effects following consumption, and consumers refer to "Sativa" and "Indica" when describing effects (Table 1.1). However, effects are also not solely determined by the phytochemicals but factors such as physiology, mental state, and metabolism play a role in the consumer experience, which further contributes to the vagueness in establishing a direct correlation between plant chemotypes with specific effects (de la Fuente et al. 2020). The effects of cannabis vary between sexes (Cuttler et al. 2016), and future research may explore whether factors such as time of day, food intake, mood, and ambiance can also influence the overall experience within the same individual. It is important to recognize that the effects of cannabis are multifaceted and influenced by a complex interplay of various factors beyond strain classification or phytochemical composition (de la Fuente et al. 2020)

Further, cannabinoids and terpenes may only be responsible for a fraction of the sensory experience (Rice and Koziel 2015), and since not all cannabinoids and terpenes are typically measured, let alone other volatile chemicals present in cannabis, it is challenging to definitively assert the accuracy of an "energizing" (commonly referred to as "Sativa") or a "sedative" (commonly referred to as "Indica") effect. Because these traits are not consistently associated (Vergara et al. 2021a), morphology related with *C. indica* may produce uplifting effects. Additionally, as cannabinoid and terpene levels fluctuate throughout the plant's growth cycle (Aizpurua-Olaizola et al. 2016), the effects may vary depending on the maturity of the plant. The absence of known wild populations makes it difficult to determine whether specific compounds are associated with geographical regions.

A third common group is "Hybrid," which refers to plants that are crosses between "Sativa" and "Indica" (Vergara et al. 2016). These terms are employed to describe either the effects or the morphology of the plants, but they are not inherently correlated. A cross may exhibit a combination of morphological characteristics from *C. sativa* and *C. indica* plants, or it could reflect a blend of effects associated with the respective types "Sativa" and "Indica". The interpretation of whether a "Hybrid" is more aligned with "Sativa" or "Indica" traits (or *C. sativa* and *C. indica* morphology) relies on the specific context in which the term is utilized. "Hybrid" strains on the retail market generally only refer to effects and are frequently described using a spectrum from 100% "Sativa" to 100% "Indica" and are commonly sold as a percentage of each (70/30 "Sativa" and "Indica" respectively), and to add more complexity and confusion, this description may also be based on parentage, subjective effects, or a combination. For instance, a strain labeled as 60/40 might be considered uplifting with some pain-relieving qualities or might be a cross between a sativa-type and an indica-type landrace and exhibit morphological traits of both.

1.3.3 Cannabis

In recent years, there has been a preference for using the term "cannabis" when referring to medical or recreational flower with high cannabinoid content, particularly THC(A). This common usage tends to exclude hemp varieties cultivated for fiber and grain. However, the term "cannabis" can be confusing since hemp is also part of the genus *Cannabis*. The distinction is often made in industries, where the "cannabis industry" deals with intoxicating types for medical or recreational purposes, while the "hemp industry" focuses on plants used for fiber or grain cultivation. The legal frameworks governing these two separate industries, each regulated by different bodies, introduce additional layers of complexity. Moreover, factors such as pre-harvest and post-harvest testing contribute to the intricacies and potential discrepancies within the industry, further highlighting the challenges and nuances associated with the classification of cannabis varieties (discussed below).

1.3.4 Hemp

Hemp used for fiber and grain production forms a clear genetic lineage within the *C. sativa* species (Sawler et al. 2015; Lynch et al. 2016; Vergara et al. 2021b), however, the divergence is not enough to warrant hemp types as separate species (Sawler et al. 2015; Lynch et al. 2016; Schwabe and

McGlaughlin 2019; Vergara et al. 2021b). The USDA definition of hemp includes cannabis plants that have less than 0.3% total THC by dry weight in the mature flower. Under this definition, hemp may include plants grown for purposes such as fiber, grain, or flower. However, potency thresholds do not match biological parameters that delineate hemp from non-hemp, but this is the system currently used for many policies surrounding the cultivation of hemp worldwide, where hemp is generally defined from 0.2%–1.0% THC by dry weight. The USDA's Final Rules for the 2018 Farm Bill, which is set to be revised in 2024, require pre-harvest testing for total THC, considering both delta-9 THC and THCA.

The Controlled Substances Act (CSA) defines "hemp" as "the plant species *Cannabis sativa* L. and any part of that plant, including the seeds thereof and all derivatives, extracts, cannabinoids, isomers, acids, salts, and salts of isomers, whether growing or not, with a total delta-9 tetrahydrocannabinol concentration of not more than 0.3 percent on a dry weight basis." This broad definition encompasses plants that produce cannabinoids other than THC, such as CBD or CBG, as long as they meet the specified concentration criteria. However, post-harvest testing, governed by CSA and DEA regulations, specifically mentions only delta-9 THC. This distinction creates a situation where hemp flower with less than 0.3% delta-9 THC may still contain significant levels of THCA. Since post-harvest testing focuses on delta-9 THC alone, products with compliant delta-9 THC levels but higher THCA may still qualify as legal hemp under current definitions. Because regulatory frameworks can evolve, and interpretations may vary, individuals and businesses operating in the hemp industry should stay updated to ensure compliance with the law and should consider consulting legal professionals familiar with the specific jurisdiction and regulations for accurate guidance.

The distinction between hemp and non-hemp type cannabis plants based on THC content or via genetic analyses is relatively recent and has been facilitated by advances in analytical testing methods. Due to the growing interest of other cannabinoids with therapeutic possibilities, the cultivation of cannabis plants that produce other minor cannabinoids has gained popularity. The resurgence of hemp cultivation in recent years may also be due to the growing interest in sustainable and eco-friendly alternatives for biofuel, bioplastics, building, and insulation materials, among many other products. Hemp seed and green leaves are also utilized for their nutritional value, as the seeds are rich in essential fatty acids and amino acids and the leaves are leafy greens with essential minerals and powerful antioxidants. Therefore, the uses for cannabis extend far beyond THC and the world is now re-examining previously widespread utility of the whole plant- stalk fiber for paper, rope, and cloth; seeds and leaves for their nutritional value; and a multitude of therapeutic uses which largely center around our own endogenous cannabinoid system, which regulates homeostasis and governs the other systems in the body.

1.3.5 Marijuana

The term "marijuana" was once widely accepted in common, legal, scientific, and medical discourse. However, in contemporary times, it has become a taboo for some due to its associations with race and drug use. In response, there is a shift towards using the term "cannabis," although this introduces confusion mentioned above. Another alternative is the term "drug-type," but it presents a dual challenge: (1) the word "drug" carries negative connotations, and (2) it implies that hemp-type plants lack medicinal potential, which is inaccurate. Additional terms, such as "high-cannabinoid cannabis" for plants producing cannabinoids beyond THC, and "non-hemp type cannabis" for those not meeting hemp criteria, exist. However, these latter terms are somewhat unwieldy and may face challenges in widespread adoption.

The etymology of the term "marijuana" remains a subject of ongoing scholarly investigation and discussion. While some suggest possible origins in China or India, others argue for an African connection (Duvall 2019). The term "marijuana" holds significant historical and cultural importance, as its usage has evolved and become intertwined with various social, political, and legal contexts throughout history. The term "marijuana" has been used interchangeably with various other terms such as "mariajuana," "mariahuana," "marihuana," "Indian hemp," "*Cannabis sativa* L.," "*Cannabis indica*," and the colloquial "Cannabis".

Although the exact origins remain unclear, one of the first recorded uses of a similar term occurred in an article titled "The American Congo," in 1894 (Bourke 1894; Russel and Porter 1982; Piper 2005). The term "mariguan" was used in the article to describe a species of plant found along the banks of the Rio Grande River between Texas and Mexico, as part of the description of the local flora (Russel and Porter 1982). Theories also propose Chinese influence, suggesting that the phrase "ma ren hua" which roughly translates to "hemp seed flowers," made its way to Mexico and underwent Hispanicization to become "marijuana". Yet a further hypothesis posits that the Bantu word for Cannabis, "ma'kaña," was introduced to the Americas via Angolan slaves who arrived in Brazil, and later adopted by Spanish speakers (Duvall 2019). One of the common words to refer to the plant in Brazil is "Maconha".

A crucial aspect of the term's evolution lies in its connection to the Kimbundu people from western Central Africa, and Bantu words such as "iamba," "riamba," "liamba," and "diamba" from Central, Southern, Eastern and Southeastern Africa. Further contributions to the complexity of the term's linguistic evolution include documented usage in Brazil, Uruguay, Colombia, Jamaica, and Cuba going back as far as 1839 (Duvall 2019). The linguistic evolution from the Kimbundu word "mariamba," used by slaves during the 1700s and 1800s, to the Spanish word "marihuana," and later the commonly used term "marijuana," holds particular significance. This transformation occurred because of the involuntary migration of these slaves to various regions, including West Africa, the Caribbean, and South America. Speculatively, in Central America, the Kimbundu (Angolan) word "mariamba" may have influenced the Spanish word "marihuana," potentially contributing to the emergence of the word "marijuana." It's noteworthy that the term "marijuana" as we recognize it today did not appear until 1846 in Farmacopea Mexicana, initially spelled as "mariguana," with subsequent spellings predominantly using "marihuana."

The most notorious and contentious aspect of the word "marijuana" is its association with the United States prohibitionist era in the early 20th century. During this period, the word "marijuana" was prominently used by US authorities and anti-cannabis campaigners to create a negative perception of the plant and users who were immigrants and people of color (Herer et al. 1998), although that idea has been challenged (Campos 2012). By emphasizing the foreign-sounding verbiage, proponents of prohibition sought to stigmatize and marginalize cannabis consumption (Musto 1991; Herer et al. 1998; Patton 2020). During the era of prohibition, the term "marijuana" became intricately tied to the perceived vices and threats associated with Mexican immigrants and African Americans. Proponents of prohibition strategically linked cannabis use to these minority groups, using this association to justify discriminatory policies and perpetuate racial stereotypes (Musto 1991; Herer et al. 1998; Patton 2020). It's important to note that the term itself is not inherently racist (Campos 2012); instead, it refers specifically to a preparation of the female cannabis flower intended for smoking, and it is worth emphasizing that the term "marijuana" does not encompass other specific cannabis preparations, such as hash or bhang. Nevertheless, the historical usage of the term and the context in which it was employed were deeply rooted in racial stereotypes and discrimination (Slaughter 1987; Musto 1991; Herer et al. 1998; Patton 2020).

As societal attitudes, scientific understanding, and legal frameworks undergo continual evolution, a collective awareness is emerging, emphasizing the necessity for accurate and neutral language in the discourse surrounding cannabis. This evolving language use is part of a broader initiative to cultivate well-informed dialogue, advocate for equitable policies, and challenge the historical prejudices associated with the plant and its consumption. While it is crucial to recognize the historical context and racial implications linked to the term "marijuana," it's equally important to respect the diverse preferences people may have regarding the language used in discussions about cannabis, without perpetuating racial intolerance. The overarching goal remains to foster understanding, respect, and inclusivity in conversations centered around this multifaceted plant.

1.4 OTHER CLASSIFICATION TERMINOLOGIES

A long history of cultivation, breeding, and selection has led to the variation and the wide array of phenotypes in modern cannabis (Figure 1.2), and groupings of these cultivated selections are commonly referred to as strains, cultivars, varieties, or chemovars. These terms lack consistent definitions across different sources, leading to confusion in their usage. Ongoing disputes among botanists, taxonomists, breeders, and industry experts further contribute to this linguistic complexity. Acknowledging that language is continually evolving, it is essential to recognize that certain terms within cannabis culture, even if not universally agreed upon, have become well established over the decades.

The extensive history of underground breeding and trading, coupled with a lack of precise record-keeping, registration practices, or the preservation of type specimens, has resulted in a scenario where most cannabis available in the market originates from polyhybrid crosses rather than the traditional method of crossing two inbred lines to create new phenotypes. It is imperative to appreciate that the cannabis breeding system differs significantly from other cultivars. Many cannabis varieties lack genuine, true-to-type breeding efforts, and the technical scientific terms applied to other plant species may not accurately reflect the process by which new lines are developed and brought to market in the cannabis community and industry. The unique characteristics of the cannabis industry underscore the need for a nuanced understanding of the language and breeding practices associated with this distinctive plant.

FIGURE 1.2 *Cannabis sativa* displays an incredible array of phenotypic traits, including phyto- pigments which lend to the pink, purple, yellow, orange and even black coloration of strains like "Honey Badger" and "Blackberry Kush". (Photo credit: Anna Schwabe, FimMaster Frank, Joel Bedard).

1.4.1 Landrace

The current working definition for landrace plants is "a dynamic population(s) of a cultivated plant that has a historical origin, distinct identity, and lacks formal crop improvement, as well as often being genetically diverse, locally adapted and associated with traditional farming systems" (Villa et al. 2005; Small 2017). A broader definition of landraces includes their evolution in traditional or new agricultural environments influenced by local culture (Casañas et al. 2017). Landraces are genetically diverse, cultivated varieties adapted to specific ecological and geographical conditions. They evolve through natural and human-driven selection, and the crucial aspect in defining landraces is the intervention of humans in their cultivation at some point in the past, distinguishing them from purely wild or natural varieties. As cannabis has been moved around the globe, cultivated plants have escaped and become locally adapted to different geographical regions, becoming established landraces. Wild-growing populations of locally adapted cannabis, which are considered landraces rather than purely natural varieties, can be discovered on every continent. These strains are frequently named after the regions where they are discovered (Table 1.2).

Adapting historical landraces to current conditions using advanced breeding presents a challenging opportunity for sustainable agriculture. When landraces are taken out of their locally adapted regions, they may change in response to the new environment. Regrettably, some cannabis landraces have vanished through the systematic removal of cannabis populations fueled by The War on Drugs (Halpin-McCormick et al. 2024). However, contemporary breeding techniques offer potential avenues to preserve and cultivate these landraces, thereby sustaining their crucial agricultural significance.

Landrace varieties are highly valued by cultivators and breeders because of their historical lineages, their adaptation to regional environments, and the potential to reintroduce lost genetic traits caused by recent genetic bottlenecks and selective breeding (Halpin-McCormick et al. 2024). While modern cannabis strains may excel in phytochemical production and yield, there is growing recognition that the intense selection, particularly for THC, has led to a reduction in phytochemical diversity (Mudge et al. 2018) which could result in a decrease in flavors, smells, and effects. Currently, it seems that some desirable traits, such as the distinctive skunk and gas strains from the past have disappeared from the market despite demand, and the dessert type profiles (candy, cake, cookies, confections, cereals, and fruits) appear to be dominating the market (Cup 2023; Drowns 2023).

1.4.2 Variety

The International Union for the Protection of New Varieties of Plants (UPOV) defines a variety as a group of plants within a species that share a common set of characteristics. However, UPOV also introduced the term "plant variety" as a legal designation to officially recognize specific cultivars and provide legal protection to the breeder. The term "plant variety" used in this context should not be confused with the taxonomic rank of variety, which occur naturally and tend to remain true to type (Haynes 2008; Wheeler 2017). This means that seedlings grown from a variety will exhibit the same unique characteristics as the parent plant. However, in the case of modern cultivated cannabis, several factors are distinct from naturally occurring varieties. Firstly, the variation is primarily a result of deliberate crossing and selection rather than natural development. Additionally, human intervention is often necessary for its reproduction, such as the removal of male plants or the development of feminized seeds. Furthermore, clonal propagation of cannabis also requires human intervention. Lastly, modern cannabis does not reliably produce offspring that are true to type from seed due to high genetic variation- heterozygosity- within the species. Given the factors mentioned above, most modern cultivated cannabis does not meet the requirements to be classified as varieties according to the definitions.

1.4.3 Cultivar

The International Code of Nomenclature for Cultivated Plants (ICNCP), which establishes the rules for naming new cultivars, defines specific requirements, and a cultivar must possess the following characteristics: (a) it has been bred and selected for specific traits or a combination of traits, (b) it demonstrates distinctiveness, uniformity, and stability in those traits, and (c) when propagated using appropriate methods, the traits are stable and therefore retained. Due to the plasticity of cannabis and its ability to exhibit various phenotypes in response to environmental factors, identification based on phenotypic is unreliable. Therefore, most cannabis for recreational and medical purposes on the retail market do not meet these criteria (Small 2015; Brickell et al. 2016; Pollio 2016) and due to the absence of stabilization through back breeding and inbreeding, these cannot be considered cultivars. However, hemp for grain and fiber have a longer history of legal agriculture and an established registration pathway in some countries, and therefore have a relatively small list of registered cultivars.

Registered cultivars of plants are typically documented and disseminated through various channels, which encompass government agencies, research institutions, seed companies, plant breeders' associations, and other relevant entities. Plant Breeders Associations fulfill a pivotal role in this process by publishing catalogs, online databases, or official publications that provide comprehensive descriptions of registered cultivars within specific crop categories. These resources outline the characteristics of each cultivar, performance, and availability. Government Agencies are responsible for overseeing plant variety protection and intellectual property rights. They often maintain databases or lists of registered cultivars and make this information accessible through their websites or official publications. Notable examples include the Plant Variety Protection Office in the US and the Community Plant Variety Office in the EU. This has not yet been established or allowed with cannabis.

Research Institutions and agricultural universities significantly contribute to plant breeding and variety development. When they publish research findings, scientific papers, or technical reports, they often include detailed descriptions of newly developed cultivars. These publications encompass information specifying phenotypic traits and may include facts regarding genetic profiles and recommended uses. Seed companies and nurseries play a vital role in commercializing and distributing registered cultivars. To facilitate this, they commonly release catalogs, brochures, or online platforms containing comprehensive descriptions, availability, and performance information for the registered cultivars. Plant Genetic Resources Centers or repositories serve as custodians for collections of plant genetic material, including registered cultivars. They may publish information about the cultivars they preserve, encompassing descriptions, genetic profiles, and availability. These centers are often affiliated with research institutions, botanical gardens, or gene banks. International Organizations dedicated to plant genetic resources, such as the International Union for the Protection of New Varieties of Plants (UPOV) or the International Plant Genetic Resources Institute (IPGRI), play a significant role in providing guidelines, databases, and publications related to registered plant cultivars, aiding in their documentation and dissemination. This has also not been established or allowed with cannabis.

1.4.4 Strain

The term "strain" is commonly used in the cannabis community to describe varieties for both recreational and medical purposes. Despite reservations expressed by some academics and taxonomists, who traditionally associate the term with bacteria and viruses (Russo 2019), it is important to note that "strain" does not hold an official taxonomic ranking (McPartland and Small 2020). In botanical usage, it refers to a group of descendants from a common ancestor that share uniform morphological or physiological traits. This botanical definition includes a designated group of offspring resulting from either conventional breeding or genetic mutation (Usher 1996).

While not a recognized taxonomic rank, the term "strain" is widely employed, especially in cannabis to denote a group of descendants with consistent morphological or physiological traits. This usage is appropriate in most cases for cannabis, given that cultivated plants often result from selective breeding and shaping genetic traits and phenotypic expression through artificial selection. The International Code of Nomenclature for Cultivated Plants (ICNPC) defines "strain" as "a confused term having several meanings; in cultivated plant nomenclature, often referring to a seed-raised cryptic variety".

The use of "strain" as a botanical term extends beyond cannabis, as other economically significant crop species such as rice (*Oryza sativa)* and wheat (*Triticum aestivum*) also have strains that characterize a genetic line with shared traits (De et al. 1988; Cheng et al. 2003; Bao et al. 2005). Cannabis strains, many with unclear backgrounds, are utilized in cultivation and often originate from a chosen plant or exhibit a shared morphological character. These are frequently the result of polyhybrid lines, rather than true breeding lines, which is a relatively unique aspect in cultivation. Therefore, describing most cannabis variations (Usher 1996; Laursen 2015) as "strains" is appropriate (Pollio 2016; McPartland and Small 2020). Academics often use the terms variety, cultivar, strain, and/or chemovar interchangeably (Payment and Cvetkovska 2023) for marijuana variants (Naraine et al. 2020) which further adds to the convolution.

Strain names are also used regardless of propagation methods, which includes clones (cuttings) derived from a mother plant, seeds obtained from crossing a male and female plants, feminized seeds produced by crossing a female plant that was induced to produce pollen to another female plant, or seeds resulting from self-fertilization using a single plant. Because of the high genetic diversity in cannabis (Lynch et al. 2016; Vergara et al. 2016; Kovalchuk et al. 2020), a cross between two plants (female X male, or female X female) may result in high variation within the progeny and therefore produce many different phenotypes. This is particularly true, as most cannabis plants are not reared from inbred lines. Therefore, cloning is often the most reliable method to produce progeny that are true-to-type replicas of the parent plant, justifying the use of the same strain name for the propagated offspring (Wilson 2023). All this said, cannabis strains have the potential to be cultivars, but require the appropriate breeding work to develop an inbred genetic line. Unfortunately, the history of illegality and the current illicit status has excluded marijuana-type cannabis from being eligible for registration. Therefore, without any protections, breeders and growers that have been working clandestinely may not want to spend time and resources on the complexities of breeding.

The recognized number of cannabis strains in existence is highly disputed. Although strain names are heavily used in commercial activities and in cannabis culture, strain names are unreliable as strains with the same name may be unrelated (Lynch et al. 2016; Vergara et al. 2016; Schwabe and McGlaughlin 2019), and they don't necessarily share a chemotype (Smith et al. 2022). Evidence of misnaming may have arisen due to several reasons: (1) strain names are sometimes accidentally mislabeled during cultivation, harvest, and processing; (2) strains may also be mistaken if exhibiting similar traits; or, (3) in the case where a given name was not recorded, a cultivator might give an unknown strain a new name, creating redundant strains with different names; and (4) strain names may be purposefully renamed or relabeled to enhance their value or marketability. Despite the difficulty with strain names, these are useful to categorize and facilitate communication and this nomenclatural confusion is not unique to cannabis (ie. fish (Marko et al. 2011; Muñoz-Colmenero et al. 2016; Willette et al. 2017; Hu et al. 2018), wine (Tomić et al. 2013; Lin and Salcido-Keamo 2021), olive (Lopes et al. 2004) among other examples (Lopes et al. 2004; Scarano et al. 2015)).

Names of strains are influenced by a wide range of factors that add character, differentiation, and appeal in the commercial cannabis industry, allowing consumers to identify and choose based on their preferences and desired experiences (Table 1.2).

TABLE 1.2
Factors Influencing Some Popular Strain Names

Influencing factor	Strain name examples
Flavor or aroma	'Banana Kush', 'Blueberry', 'Cheese', 'Skunk'
Regionality	'Santa Marta Gold', 'Durban Poison' 'Afghan Kush', 'Hawaiian', 'Maui Waui' 'Lebanese'
People	'Neville's Haze', 'Jack Herer', 'Barack Obama', 'Dr. Grinspoon', 'Ringo's Gift'
Music	'AC/DC', 'Just Like Heaven', 'Rapper's Delight'
Fictional characters	'Bruce Banner' (The Hulk), 'Old Toby' (The Lord of the Rings), 'Pennywise' (It)
Phenotypic traits	'Gorilla Glue', 'Freakshow', 'Purple Urkle' 'Greaseball',
Effects	'Lucid Blue', 'Euphoria Cookies', 'Cosmic Bomb' 'Headband'
Pedigree	'Berry White': a cross between 'Blueberry' and 'White Widow'
	'Trainwreck': a cross between 'Lohan' and 'LA Confidential'
	'Tangerine Haze': a cross between 'Tangerine Dream' and 'Super Lemon Haze'
	'Blueberry DJ': a cross between 'Blueberry' and 'DJ Shorts'

TABLE 1.3
Five Chemotypic Descriptions With Types of Cannabis. Those With Higher Levels of THC Utilized for Medical or Recreational Consumption Would Fall Under Type I or II. Other Types Grown for CBD, CBG, and Minor Cannabinoids (Other Than THC) and Terpenes, as Well as for Fiber, Seed, Oil, and Other Industrial Uses Fall Into Types III, IV and V

Group	Phenotype
Type I	THC dominant (THC/CBD > 1)
Type II	Balanced THC:CBD, produces moderate levels of both cannabinoids (THC/CBD ≈ 1)
Type III	CBD dominant (THC/CBD < 1)
Type IV	CBG dominant
Type V	Little to no cannabinoids

Source: Ratios from Jin et al. 2021

1.4.5 Chemotypes I-V

Another common way of grouping cannabis plants is based on the ratios of cannabinoid levels of three major acidic cannabinoids – tetrahydrocannabinolic (THCA), cannabidiolic (CBDA), and cannabigerolic acid (CBGA) (Fournier et al. 1987; Mandolino and Carboni 2004; Pacifico et al. 2006; Small 2016; Wenger et al. 2020). Cannabigerolic acid (CBGA) is the precursor to both THCA and CBDA which are formed through the enzymatic action of THCA Synthase and CBDA Synthase respectively (Taura et al. 1996; Fellermeier and Zenk 1998). It was thought that these synthases were produced by one locus with two codominant alleles (de Meijer et al. 2003; de Meijer and Hammond 2005), but now it is known that the genes coding these enzymes are many (Weiblen et al. 2015; Vergara et al. 2019; van Velzen and Schranz 2021), they are in close proximity in the genome (Laverty et al. 2019; Grassa et al. 2021; Innes and Vergara 2023), and multiple potential alleles for each of these genes could exist (Innes and Vergara 2023). Additionally, the enzymes responsible for the production of these cannabinoids are sloppy (Auldridge et al. 2006; Franco 2011; Chakraborty et al. 2013; Vergara et al. 2020), promiscuous (Onofri et al. 2015; Vergara et al. 2020), and can produce at least eight compounds in-vitro (Zirpel et al. 2018). However, the phenotype they produce, and the description (Table 1.3) is useful to understand what cannabinoids a particular strain will produce.

While THC and CBD content is often used as a general indicator of the potential effects, they are not the sole determining factor as cannabis plants contain numerous other compounds. The overall effects can be influenced by their unique phytochemical makeup, which includes other cannabinoids, terpenes, thiols, and more. These compounds work together in the body, acting on receptors to create a variety of effects (McPartland and Russo 2001; Adams and Taylor 2010; Spindle et al. 2024). As such, relying solely on THC and CBD content almost certainly does not provide a complete understanding of the effects a strain may elicit in the body. Therefore, while THC and CBD categorization can serve certain purposes within research, taxonomy, and regulation, it is essential to consider a broader range of factors when seeking specific effects from cannabis. This can include the presence of specific terpenes (Spindle et al. 2024) or other cannabinoids, understanding the genotype of particular strains, and how cultivation methods, harvesting and curing processes can impact the final chemotype of the final product (Vergara et al. 2023). Additionally, personal tolerance and preferences also significantly impact the overall experience. Classification of these five types based on major cannabinoids (Table 1.2), while useful in some circumstances, is not universally recognized or accepted by the cannabis or scientific community. Researchers, breeders, and industry professionals may use alternative classification systems or have variations in their understanding of the types. Furthermore, the chemical composition of the plants can vary significantly within each type, as there are numerous strains with distinct characteristics.

1.4.6 Chemovar

The term "chemovar" has relatively recently gained popularity, emphasizing biochemical and pharmacological distinctions between cannabis accessions (Russo 2019). It is a chemotype variant within a group, strain, or cultivar, emphasizing the phytochemical distinctions between accessions. But this classification ignores that not all cannabis is grown for phytochemical production as well as the relationships among and within genetic lines. Adopting this way of classifying may either simplify or complicate classification treatments. It could simplify classification schemes as it may reduce the number of groups to include only major cannabinoids and terpenes (Hazekamp and Fischedick 2012; Henry et al. 2018; Lewis et al. 2018; Reimann-Philipp et al. 2020). However, it ignores minor components that we are now beginning to understand play a major role in the sensory, psychotropic, or physiological effects. Our understanding and testing of minor cannabinoids, terpenes, and other phytochemicals such as flavonoids, thiols, and alkaloids is limited and many of these overlooked constituents may contribute to the effects. Additionally, the current testing methods only measure major cannabinoids and terpenes while new cannabinoids are being discovered (Citti et al. 2019), and others are not measured (Vergara et al. 2020), suggesting that our current measurements provide an incomplete chemotypic profile, and that perhaps more compounds are awaiting discovery. Of great importance is that abundance of a particular component may not be indicative of potency, and to accurately describe a chemical variant all compounds would need to be measured to provide a complete chemical profile for each plant. In this context, chemovar classification would greatly complicate matters.

The "ensemble effect" (aka "entourage" or "synergistic" effect) suggests that all the compounds found in the plant interact to produce a particular effect (McPartland and Russo 2001; Adams and Taylor 2010). However, assumptions about the role of terpenes and their contribution to the ensemble effect may not fully translate to smoked cannabis, as terpene composition can change over time due to storage conditions and during the heating process involved in vaping or smoking, rendering an analysis of terpenes inaccurate in terms of what actually gets into the body upon consumption. Therefore, more research is required to fully understand the role of terpenes and other constituents, what is entering the body, and the mechanisms of action (Spindle et al. 2024). However, some phenotypic (Smith et al. 2022) and genetic (Watts et al. 2021) studies on terpenes suggest that these may be a good option for grouping cannabis strains compared to other classification methods.

While assessing chemotypes is valuable for therapeutic and medical applications, it may not necessarily translate effectively to cultivation, categorization, or identification purposes. Hence, careful consideration is warranted in this regard. Furthermore, relying on chemovars to describe variants could complicate cannabis classification, as individuals with the same genetic identity grown under different conditions can vary in their chemotype (Schwabe 2019). This complexity underscores the importance of initiatives like "The Grow Off" (www.gronska.org/), where participants receive clones of the same stock and employ individual cultivation techniques, allowing for a comprehensive assessment of plant characteristics, and the phenotypic differences achieved by each growers' cultivation expertise.

1.5 OTHER TERMS FOR RECREATIONAL-TYPE CANNABIS

1.5.1 Weed

The term "weed" originates from the Old English word "wēod," which originally meant "grass" or "herb" (From American Heritage Dictionary of the English Language, 5th Edition (2000)). In the early 20th century, during the prohibition era and the subsequent demonization of cannabis, this term became associated with the plant. Alongside "weed," other slang terms like "pot" and "grass" also gained popularity as informal references to cannabis. The choice of "weed" may have been influenced by rapid growth characteristics and its ability to thrive in diverse environments. Like other plants labeled as weeds, cannabis plants have resilient and vigorous growth, often flourishing without intentional cultivation (Bridges 1994; 2000). Of note is that the term "weed" is colloquial and informal, and its usage can vary based on the context and regional differences. This informal designation reflects both historical associations and the plant's tenacious growth, highlighting the adaptability and robust nature of cannabis.

1.5.2 Ganja

The term "ganja" is thought to have originated in the Indian subcontinent and has been used for centuries to refer to cannabis (Ganjapreneur n.d.c). According to online sources, the word "ganja" is derived from the Sanskrit language, an ancient Indo-Aryan language used in India. In Sanskrit, the term "gāñjā" (pronounced "ganja") refers to hemp plants and their resinous flowers. Hemp has been cultivated in the Indian subcontinent for various purposes, including religious, medicinal, and recreational use, for thousands of years. Ganja holds cultural and religious significance in India, particularly within certain sects of Hinduism and is considered one of the five sacred plants mentioned in ancient Hindu scriptures. Lord Shiva is often depicted as an ascetic who consumed cannabis as part of his spiritual practices (Godlaski 2012). The term "ganja" gained more widespread recognition in the Western world during the 20th century because of the cultural influence of Rastafarianism, a religious and social movement that originated in Jamaica. Rastafarians consider cannabis, referred to as "ganja," as a sacrament and integral to their spiritual practices (Virdi 2021). Popular musical artists such as Peter Tosh, Cypress Hill with Damian Marley, and Eminem, have referred to "ganja" in their music.

1.5.3 Sticky Icky

The term "sticky icky" is a slang term commonly used to refer to high-quality, potent cannabis buds that are resinous and sticky to the touch. The exact origin of the term is uncertain, but apparently gained popularity and recognition through the influence of hip-hop and rap music (Ganjapreneur n.d.a, b). Musicians known for their affinity for cannabis culture have used the term "sticky icky" in their songs and interviews, contributing to its widespread usage (Pitbull "Sticky Icky" 2007, Chingy "Holidae Inn" 2009).

1.6 CONCLUSIONS

The way people commonly refer to the plant, or the way that laws and regulations refer to it, does not necessarily reflect the distinct genetic lineages uncovered through genomic research. The categorization, taxonomy, and grouping of cannabis continues to be a subject of contention, but a distinct naming system has already been established within the cannabis culture. Dispensaries have products labeled based on the traditional classification of "Sativa" and "Indica," with each type associated with unique physiological or psychoactive effects. "Budtenders," the customer service representatives at these dispensaries, often guide consumers using this classification system for products that align with their desired effects. Perhaps in the future we can provide suggestions that can be adopted by the industry that provides information for consumers, caregivers, and particularly medical patients. Academic institutions in their state of privileged seclusion judge and condemn these popular terminologies. Therefore, academics tend to use scientific terminology, whereas the legacy market employs colloquial or regional names. This disconnect between the academic and legacy perspectives intensifies the divide between the University Ivory tower circles and the general public.

The surging popularity of cannabis for both consumption and industrial applications has brought the industry to a point of growing demand for education from academic institutions, and where the scientific and academic communities are finally acknowledging the importance of instruction on the plant. Despite the long-standing expertise within the cannabis community, academic institutions have been slow to join the conversation. Instead of embracing and building upon the rich knowledge, most academic institutions have kept their doors closed to cannabis research, primarily driven by baseless fears linked to the plant's legal federal status. This delayed integration is a departure from the principle of academic freedom, which should be the cornerstone of scholarly pursuits. Rather than proactively engaging with the wealth of insights amassed by the cannabis community, academic institutions have, at times, adopted a dismissive stance. This approach appears to mimic historical colonizing behaviors, wherein institutions impose their own "truths" without genuine collaboration or understanding. It is crucial for academic institutions to recognize the existing expertise within the cannabis community, pull up seats to join the ongoing conversation, and foster an environment that appreciates and incorporates the decades of knowledge already present. This shift can lead to more meaningful collaboration, where scientific rigor combines with established wisdom for the betterment of both academia and the cannabis community.

Strategies to foster relationships and bonds between academia and the cannabis community include the effective communication of evidence-based knowledge using common terminology that the public understands. Therefore, translating scientific research into accessible language and disseminating it through various channels, including media, websites, and educational initiatives are crucial. These strategies will promote an open dialogue of mutual respect and will hopefully build long-lasting collaborations between academics and the cannabis community to advance public health, legislation, and the multiple promising companies that are arising in this nascent industry.

The distrust the cannabis community has towards the government is understandable given the illegality of the plant for so many decades. Additionally, the American Federal Government that has managed a monopoly for Cannabis production for medical and scientific research has produced plants that are not reflective of those grown by the industry (Vergara et al. 2017; Schwabe et al. 2019; Vergara et al. 2021b), increasing the mistrust towards governmental institutions. Currently, several states that have legalized marijuana consumption are trying to include the legacy population, but many of their nonsensical laws and regulations prevent the entrance of small farmers and the cannabis community into legal markets. These challenges underscore the need for more thoughtful and inclusive regulatory frameworks that address the historical context and nuances of the cannabis industry, ensuring fair opportunities for all stakeholders.

Nevertheless, the cannabis community has opened the doors to a new prejudice-free profitable industry with diverse mindsets. Many science communicators, policymakers, and healthcare professionals that see the benefit of this thriving industry are striving to make amends, to bring scientific knowledge mainstream, and to build an amicable environment for all.

Acknowledgments: The authors would like to thank Ángel Fernandez for kindly accepting this contribution.

Author contributions: ALS and DV wrote the manuscript.

Competing financial interests: ALS and DV are board members of the non-profit Agricultural Genomics Foundation; DV is the sole owner of the company CGRI, LLC.

Table S1/Appendix. Alternative nomenclature suggestions for variations of cannabis, including both hemp and drug-type variants.

Reference	Proposed nomenclature
Clarke and Merlin, 2013	*Cannabis indica* Lamarck; *Cannabis sativa* var. *sativa*; *Cannabis sativa* subsp. *sativa* var. *sativa*; *Cannabis sativa* subsp. *indica* var. *indica*; *Cannabis sativa* subsp. *sativa* var. *spontanea*; *Cannabis sativa* subsp. *indica* var.; kafiristanica; *Cannabis sativa* subsp. *spontanea* + *Cannabis ruderalis*; *Cannabis sativa* subsp. *spontanea* "narrow leaf hemp ancestor"; *Cannabis sativa* subsp. *kafiristanica*; *Cannabis sativa* var. indica; *Cannabis sativa* "hemp biotype"; *Cannabis sativa* "hemp biotype"; *Cannabis sativa* "feral biotype"; *Cannabis ruderalis* + *Cannabis sativa* "feral biotype"; *Cannabis sativa* subsp. *sativa*; *Cannabis sativa* subsp. *sativa* "narrow leaf hemp"; *Cannabis sativa* subsp. *chinesensis*; *Cannabis sativa* subsp. *chinesensis* "broad leaf hemp"; *Cannabis sativa* subsp. *indica*; *Cannabis sativa* subsp. *afghanica*
Lapierre et al. 2023	*Cannabis sativa*, *Cannabis sativa*, *Cannabis ruderalis*
theplantlist.org	*Cannabis sativa* Lamarck; *Cannabis sativa* f. afghanica Vavilov; *Cannabis sativa* f. chinensis (Delile) A.DC.; *Cannabis sativa* var. gigantea (Delile ex Vilm.) Alef.; *Cannabis sativa* var. indica (Lam.) Wehmer; *Cannabis sativa* subsp. indica (Lam.) E. Small & Cronquist; *Cannabis sativa* var. indica (Lam.) E. Small & Cronquist; *Cannabis sativa* subsp. intersita (Soják) Soják; *Cannabis sativa* var. kafiristanica (Vavilov) E. Small & Cronquist; *Cannabis sativa* var. kif A.DC.; *Cannabis sativa* var. macrosperma (Stokes) Asch. & Graebn.; *Cannabis sativa* var. monoica Hol.; *Cannabis sativa* f. pedemontana A.DC.; *Cannabis sativa* var. praecox Serebr.; *Cannabis sativa* var. ruderalis Janisch.; *Cannabis sativa* var. ruderalis (Janisch.) S.Z. Liou; *Cannabis sativa* var. spontanea Vavilov; *Cannabis americana* Pharm. ex Wehmer; *Cannabis chinensis* Delile; *Cannabis erratica* Siev.; *Cannabis foetens* Gilib.; *Cannabis generalis* E.H.L. Krause; *Cannabis gigantea* Crevost; *Cannabis sativa* f. afghanica (Vavilov) Vavilov; *Cannabis sativa* var. kafiristanica Vavilov; *Cannabis kafiristanica* (Vavilov) Chrtek; *Cannabis intersita* Soják; *Cannabis lupulus* Scop.; *Cannabis macrosperma* Stokes; *Cannabis ruderalis* Janischewsky
https://npgsweb.ars-grin.gov/gringlobal/taxon/taxonomysearch.aspx	*Cannabis sativa* subsp. sativa; *Cannabis sativa* subsp. indica (Lam.) E. Small & Cronquist
www.ipni.org	*Cannabis sativa* subsp. indica (Lam.) E. Small & Cronquist; *Cannabis sativa* subsp. intersita (Soják) Soják; *Cannabis sativa* var. kafiristanica (Vavilov) E. Small & Cronquist; *Cannabis sativa* var. ruderalis (Janisch.) S.Z. Liou; *Cannabis sativa* var. indica (Lam.) Wehmer

REFERENCES

Adams, T. B. and S. V. Taylor. 2010. Safety evaluation of essential oils: A constituent-based approach. CRC Press.

Ahmed, A. F., M. Z. Islam, M. S. Mahmud, M. E. Sarker, and M. R. Islam. 2022. Hemp as a potential raw material toward a sustainable world: A review. Heliyon 8(1):e08753.

Aizpurua-Olaizola, O., U. Soydaner, E. Öztürk, D. Schibano, Y. Simsir, P. Navarro, N. Etxebarria, and A. Usobiaga. 2016. Evolution of the cannabinoid and terpene content during the growth of *Cannabis sativa* plants from different chemotypes. Journal of Natural Products 79 :324–331.

Anderson, L. C. 1980. Leaf variation among *Cannabis* species from a controlled garden. Botanical Museum Leaflets, Harvard University 28(1):61–69.

Auldridge, M. E., D. R. McCarty, and H. J. Klee. 2006. Plant carotenoid cleavage oxygenases and their apocarotenoid products. Current Opinion in Plant Biology 9:315–321.

Bao, J., S. Lee, C. Chen, X. Zhang, Y. Zhang, S. Liu, T. Clark, J. Wang, M. Cao, and H. Yang. 2005. Serial analysis of gene expression study of a hybrid rice strain (LYP9) and its parental cultivars. Plant Physiology 138:1216–1231.

Bourke, J. G. 1894. the American Congo. Scribner's Magazine 15:1–21.

Brenneisen, R. 2007. Chemistry and analysis of phytocannabinoids and other *Cannabis* constituents. Pp. 17–49. Marijuana and the cannabinoids. Springer.

Brickell, C., C. Alexander, J. Cubey, J. David, M. Hoffman, A. Leslie, V. Malécot, and X. Jin. 2016. International code of nomenclature for cultivated plants (ICNCP or cultivated plant code) Scripta horticulturae 18. ISHS.

Bridges, D. C. 1994. Impact of weeds on human endeavors. Weed Technology 8:392–395.

Campos, I. 2012. Home grown: Marijuana and the origins of Mexico's war on drugs. Univ of North Carolina Press.

Casañas, F., J. Simó, J. Casals, and J. Prohens. 2017. Toward an evolved concept of landrace. Frontiers in Plant Science 8:145.

Chakraborty, S., R. Minda, L. Salaye, A. M. Dandekar, S. K. Bhattacharjee, and B. J. Rao. 2013. Promiscuity-based enzyme selection for rational directed evolution experiments. Enzyme Engineering: Methods and Protocols:205–216.

Cheng, C., R. Motohashi, S. Tsuchimoto, Y. Fukuta, H. Ohtsubo, and E. Ohtsubo. 2003. Polyphyletic origin of cultivated rice: Based on the interspersion pattern of SINEs. Molecular Biology and Evolution 20:67–75.

Citti, C., P. Linciano, F. Russo, L. Luongo, M. Iannotta, S. Maione, A. Laganà, A. L. Capriotti, F. Forni, and M. A. Vandelli. 2019. A novel phytocannabinoid isolated from *Cannabis sativa* L. with an in vivo cannabimimetic activity higher than Δ 9-tetrahydrocannabinol: Δ 9-Tetrahydrocannabiphorol. Scientific Reports 9:1–13.

Clarke, R. and M. Merlin. 2013. Cannabis: Evolution and ethnobotany. University of California Press.

Clarke, R. C. and M. D. Merlin. 2016. Cannabis domestication, breeding history, present-day genetic diversity, and future prospects. Critical Reviews in Plant Sciences 35:293–327.

Cup, T. E. 2023. The 19th annual Emerald cup competition and awards will once again shape tastes in the marketplace. https://theemeraldcup.com/tag/emerald-cup-awards/

Cuttler, C., L. K. Mischley, and M. Sexton. 2016. Sex differences in cannabis use and effects: A cross-sectional survey of cannabis users. Cannabis and Cannabinoid Research 1:166–175.

de la Fuente, A., F. Zamberlan, A. S. Ferrán, F. Carrillo, E. Tagliazucchi, and C. Pallavicini. 2020. Relationship among subjective responses, flavor, and chemical composition across more than 800 commercial cannabis varieties. Journal of Cannabis Research 2:1–18.

de Meijer, E. P. M., M. Bagatta, A. Carboni, P. Crucitti, V. M. C. Moliterni, P. Ranalli, and G. Mandolino. 2003. The inheritance of chemical phenotype in *Cannabis sativa* L. Genetics 163:335–346.

de Meijer, E. P. M. and K. M. Hammond. 2005. The inheritance of chemical phenotype in *Cannabis sativa* L. (II): Cannabigerol predominant plants. Euphytica 145:189–198.

De, R., R. Seetharaman, M. Sinha, and S. Banerjee. 1988. Genetic divergence in rice. Indian Journal of Genetics and Plant Breeding 48:189–194.

Divashuk, M. G., O. S. Alexandrov, O. V. Razumova, I. V. Kirov, and G. I. Karlov. 2014. Molecular cytogenetic characterization of the dioecious *Cannabis sativa* with an XY chromosome sex determination system. PloS One 9:e85118.

Doebley, J. F., B. S. Gaut, and B. D. Smith. 2006. The molecular genetics of crop domestication. Cell 127:1309–1321.

Drowns, D. 2023. Leafly Buzz: 13 top cannabis strains of '23.

Dufresnes, C., C. Jan, F. Bienert, J. Goudet, and L. Fumagalli. 2017. Broad-scale genetic diversity of *Cannabis* for forensic applications. PloS One 12:e0170522.

Duvall, C. S. 2019. The African roots of marijuana. Duke University Press, 360 pp.

Elzinga, S., J. Fischedick, R. Podkolinski, and J. Raber. 2015. Cannabinoids and terpenes as chemotaxonomic markers in cannabis. Natural Products & Chemistry & Research 3:2.

Emboden, W. A. 1974. Cannabis: A polytypic genus. Economic Botany 28:304–310.

Fellermeier, M. and M. H. Zenk. 1998. Prenylation of olivetolate by a hemp transferase yields cannabigerolic acid, the precursor of tetrahydrocannabinol. FEBS Letters 427:283–285.

Fournier, G., C. Richez-Dumanois, J. Duvezin, J.-P. Mathieu, and M. Paris. 1987. Identification of a new chemotype in *Cannabis sativa*: Cannabigerol-dominant plants, biogenetic and agronomic prospects. Planta Medica 53:277–280.

Franco, O. L. 2011. Peptide promiscuity: An evolutionary concept for plant defense. FEBS Letters 585:995–1000.

Gagne, S. J., J. M. Stout, E. Liu, Z. Boubakir, S. M. Clark, and J. E. Page. 2012. Identification of olivetolic acid cyclase from *Cannabis sativa* reveals a unique catalytic route to plant polyketides. Proceedings of the National Academy of Sciences 109(31):12811–12816.

Ganjapreneur. (n.d.a). Ganja. www.ganjapreneur.com/vocab/ganja/#content.

Ganjapreneur. (n.d.b). Names for weed. www.ganjapreneur.com/vocab-category/names-for-weed/.

Ganjapreneur. (n.d.c). Sticky icky. www.ganjapreneur.com/vocab/sticky-icky/.

Godlaski, T. M. 2012. Shiva, lord of bhang. Substance Use & Misuse 47:1067–1072.

Grassa, C. J., G. D. Weiblen, J. P. Wenger, C. Dabney, S. G. Poplawski, S., Timothy Motley, T. P. Michael, and C. Schwartz. 2021. A new *Cannabis* genome assembly associates elevated cannabidiol (CBD) with hemp introgressed into marijuana. New Phytologist 230:1665–1679.

Halpin-McCormick, A., K. Heyduk, M. B. Kantar, N. L. Batora, R. R. Masalia, K. B. Law, and E. J. Kuntz. 2024. Examining population structure across multiple collections of *Cannabis*. Genetic Resources and Crop Evolution 71(8):1–18.

Hanuš, L. O. and G. Appendino. 2016. Natural product reports. Natural Product Reports 33:1357–1392.

Hanuš, L. O., S. M. Meyer, E. Muñoz, O. Taglialatela-Scafati, and G. Appendino. 2016. Phytocannabinoids: A unified critical inventory. Natural Product Reports 33:1357–1392.

Hart, C. L., W. Van Gorp, M. Haney, R. W. Foltin, and M. W. Fischman. 2001. Effects of acute smoked marijuana on complex cognitive performance. Neuropsychopharmacology 25:757–765.

Haynes, C. 2008. Cultivar versus variety. https://yardandgarden.extension.iastate.edu/faq/what-difference-between-cultivar-and-variety

Hazekamp, A. and J. T. Fischedick. 2012. Cannabis-from cultivar to chemovar. Drug Testing and Analysis 4:660–667.

Henry, P., A. Hilyard, S. Johnson, and C. Orser. 2018. Predicting chemovar cluster and variety verification in vegetative cannabis accessions using targeted single nucleotide polymorphisms. PeerJ Preprints 6:e27442v27441.

Herer, J., J. Meyers, and L. Cabarga. 1998. The emperor wears no clothes. Ah Ha Pub.

Hillig, K. W. 2005. Genetic evidence for speciation in *Cannabis* (Cannabaceae). Genetic Resources and Crop Evolution 52:161–180.

Hu, Y., S. Y. Huang, R. Hanner, J. Levin, and X. Lu. 2018. Study of fish products in Metro Vancouver using DNA barcoding methods reveals fraudulent labeling. Food Control 94:38–47.

Innes, P. A. and D. Vergara. 2023. Genomic description of critical cannabinoid biosynthesis genes. Botany.

Integrated Taxonomic Information System Database. 2019. Retrieved [January, 22, 2019], from the Integrated Taxonomic Information System (ITIS), www.itis.gov.

Jin, D., P. Henry, J. Shan, and J. Chen. 2021. Identification of chemotypic markers in three chemotype categories of *Cannabis* using secondary metabolites profiled in inflorescences, leaves, stem bark, and roots. Frontiers in Plant Science 12:699530.

Kalant, H. 2001. Medicinal use of cannabis: History and current status. Pain Research & Management: The Journal of the Canadian Pain Society = journal de la societe canadienne pour le traitement de la douleur 6:80–91.

Kovalchuk, I., M. Pellino, P. Rigault, R. van Velzen, J. Ebersbach, J. R. Ashnest, M. Mau, M. Schranz, J. Alcorn, and R. Laprairie. 2020. The genomics of *Cannabis* and its close relatives. Annual Review of Plant Biology 71.

Kurtz, L. E., M. H. Brand, and J. D. Lubell-Brand. 2023. Gene dosage at the autoflowering locus effects flowering timing and plant height in triploid *Cannabis*. Journal of the American Society for Horticultural Science 148:83–88.

Lamarck, J. B. P. A. d. M. d. 1785. Encyclopedie Methodique: Botanique. 1. Paris,Liège, Panckoucke;Plomteux, 1783–1808.

Lapierre, É., A. S. Monthony, and D. Torkamaneh. 2023. Genomics-based taxonomy to clarify cannabis classification. Genome 66:202–211.

Laursen, L. 2015. Botany: The cultivation of weed. Nature 525:S4–S5.

Laverty, K. U., J. M. Stout, M. J. Sullivan, H. Shah, N. Gill, L. Holbrook, G. Deikus, R. Sebra, T. R. Hughes, and J. E. Page. 2019. A physical and genetic map of *Cannabis sativa* identifies extensive rearrangements at the THC/CBD acid synthase loci. Genome Research 29:146–156.

Leckie, K. M., J. Sawler, P. Kapos, J. O. Mackenzie, I. Giles, K. Baynes, J. Lo, J. M. Celedon, and G. J. Baute. 2023. Loss of daylength sensitivity by splice site mutation in cannabis. bioRxiv. https://doi.org/10.1101/2023.03.10.532103.

Lewis, M. A., E. B. Russo, and K. M. Smith. 2018. Pharmacological foundations of cannabis chemovars. Planta medica 84:225–233.

Li, H. L. 1973. An archaeological and historical account of cannabis in China. Economic Botany 28:437–448.

Li, H. L. 1974. Origin and use of *Cannabis* in Eastern Asia; Linguistic-cultural implications. Economic Botany 28:293–301.

Lin, S. and S. Salcido-Keamo. 2021. Fraud in wine and other alcoholic beverages. Pp. 233–247. Food fraud. Elsevier.

Lipson Feder, C., O. Cohen, A. Shapira, I. Katzir, R. Peer, O. Guberman, S. Procaccia, P. Berman, M. Flaishman, and D. Meiri. 2021. Fertilization following pollination predominantly decreases phytocannabinoids accumulation and alters the accumulation of terpenoids in cannabis inflorescences. Frontiers in Plant Science 12:753847.

List, T. P. 2010. Version 1. Published on the Internet; www.theplantlist.org/ (accessed 22 January 2019). Version 1.

Lopes, M. S., D. Mendonça, K. M. Sefc, F. S. Gil, and A. da Câmara Machado. 2004. Genetic evidence of intra-cultivar variability within Iberian olive cultivars. HortScience 39:1562–1565.

Lynch, R. C., D. Vergara, S. Tittes, K. White, C. J. Schwartz, M. J. Gibbs, T. C. Ruthenburg, K. deCesare, D. P. Land, and N. C. Kane. 2016. Genomic and chemical diversity in *Cannabis*. Critical Reviews in Plant Sciences 35:349–363.

Mandolino, G. and A. Carboni. 2004. Potential of marker-assisted selection in hemp genetic improvement. Euphytica 140:107–120.

Mandolino, G., A. Carboni, S. Forapani, V. Faeti, and P. Ranalli. 1999. Identification of DNA markers linked to the male sex in dioecious hemp (*Cannabis sativa* L.). Theoretical and Applied Genetics 98:86–92.

Marko, P. B., H. A. Nance, and K. D. Guynn. 2011. Genetic detection of mislabeled fish from a certified sustainable fishery. Current Biology 21:R621–R622.

McPartland, J. M. 2017. *Cannabis sativa* and *Cannabis indica* versus "Sativa" and "Indica". Pp. 101–121. *Cannabis sativa* L.-botany and biotechnology. Springer.

McPartland, J. M. 2018. Cannabis systematics at the levels of family, genus, and species. Cannabis and Cannabinoid Research 3:203–212.

McPartland, J. M. and G. W. Guy. 2017. Models of cannabis taxonomy, cultural bias, and conflicts between scientific and vernacular names. The Botanical Review 83:327–381.

McPartland, J. M., W. Hegman, and T. Long. 2019. Cannabis in Asia: Its center of origin and early cultivation, based on a synthesis of subfossil pollen and archaeobotanical studies. Vegetation History and Archaeobotany:1–12.

McPartland, J. M. and E. B. Russo. 2001. Cannabis and cannabis extracts: Greater than the sum of their parts? Journal of Cannabis Therapeutics 1:103–132.

McPartland, J. M. and E. Small. 2020. A classification of endangered high-THC cannabis (*Cannabis sativa* subsp. *indica*) domesticates and their wild relatives. PhytoKeys 144:81.

Mechoulam, R. 2005. Plant cannabinoids: A neglected pharmacological treasure trove. British Journal of Pharmacology 146:913–915.

Meier, C. and V. Mediavilla. 1998. Factors influencing the yield and the quality of hemp (*Cannabis sativa* L.) essential oil. Journal of the International Hemp Association Archives 5:16–20.

Milewicz, M. and J. Sawicki. 2012. Mechanisms of sex determination in plants. Acta Musei Silesiae, Scientiae Naturales 61:123–129.

Moliterni, V. M. C., L. Cattivelli, P. Ranalli, and G. Mandolino. 2004. The sexual differentiation of *Cannabis sativa* L.: A morphological and molecular study. Euphytica 140:95–106.

Mudge, E., S. Murch, and P. Brown. 2018. Chemometric analysis of cannabinoids: Chemotaxonomy and domestication Syndrome. Scientific Reports 8:13090.

Muñoz-Colmenero, M., O. Blanco, V. Arias, J. L. Martinez, and E. Garcia-Vazquez. 2016. DNA authentication of fish products reveals mislabeling associated with seafood processing. Fisheries 41:128–138.

Musto, D. F. 1991. Opium, cocaine and marijuana in American history. Scientific American 265:40–47.

Naraine, S. G., E. Small, A. E. Laursen, and L. G. Campbell. 2020. A multivariate analysis of morphological divergence of "seeds"(achenes) among ruderal, fibre, oilseed, dioecious/monoecious and marijuana variants of *Cannabis sativa* L. Genetic Resources and Crop Evolution 67:703–714.

Onofri, C., E. P. M. de Meijer, and G. Mandolino. 2015. Sequence heterogeneity of cannabidiolic-and tetrahydrocannabinolic acid-synthase in *Cannabis sativa* L. and its relationship with chemical phenotype. Phytochemistry 116:57–68.

Pacifico, D., F. Miselli, M. Micheler, A. Carboni, P. Ranalli, and G. Mandolino. 2006. Genetics and marker-assisted selection of the chemotype in *Cannabis sativa* L. Molecular Breeding 17:257–268.

Patton, D. V. 2020. A history of United States cannabis law. Jl & HealTH 34:1.

Payment, J. and M. Cvetkovska. 2023. The responses of *Cannabis sativa* to environmental stress: A balancing act. Botany 101:318–332.

Piper, A. 2005. The mysterious origins of the word 'marihuana' (No. 153). Department of East Asian Languages and Civilizations, University of Pennsylvania.

Pliny, C. and L. Ian. 1854. Naturalis historiae libri XXXVII. Sumptibus et Typis BG Teubneri.

Pollio, A. 2016. The name of *Cannabis*: A short guide for nonbotanists. Cannabis and Cannabinoid Research 1:234–238.

Prentout, D., O. Razumova, B. Rhoné, H. Badouin, H. Henri, C. Feng, J. Käfer, G. Karlov, and G.A. Marais. 2020. An efficient RNA-seq-based segregation analysis identifies the sex chromosomes of *Cannabis sativa*. Genome Research 30(2):164-172.

Punja, Z. K. and J. E. Holmes. 2020. Hermaphroditism in marijuana (*Cannabis sativa* L.) inflorescences–impact on floral morphology, seed formation, progeny sex ratios, and genetic variation. Frontiers in Plant Science 11:718.

Radwan, M. M., S. Chandra, S. Gul, and M. A. ElSohly. 2021. Cannabinoids, phenolics, terpenes and alkaloids of cannabis. Molecules 26:2774.

Ram, H. M. and V. Jaiswal. 1972. Induction of male flowers on female plants of *Cannabis sativa* by gibberellins and its inhibition by abscisic acid. Planta 105:263–266.

Ram, H. Y. M. and V. S. Jaiswal. 1970. Induction of female flowers on male plants of *Cannabis sativa* L. by 2-chloroethanephos-phonic acid. Cellular and Molecular Life Sciences 26:214–216.

Ram, H. Y. M. and R. Sett. 1982. Induction of fertile male flowers in genetically female *Cannabis sativa* plants by silver nitrate and silver thiosulphate anionic complex. Theoretical and Applied Genetics 62:369–375.

Razumova, O. V., O. S. Alexandrov, M. G. Divashuk, T. I. Sukhorada, and G. I. Karlov. 2016. Molecular cytogenetic analysis of monoecious hemp (*Cannabis sativa* L.) cultivars reveals its karyotype variations and sex chromosomes constitution. Protoplasma 253:895–901.

Reimann-Philipp, U., M. Speck, C. Orser, S. Johnson, A. Hilyard, H. Turner, A. J. Stokes, and A. L. Small-Howard. 2020. Cannabis chemovar nomenclature misrepresents chemical and genetic diversity; survey of variations in chemical profiles and genetic markers in Nevada medical cannabis samples. Cannabis and Cannabinoid Research 5(3):215–230.

Rice, S. and J. A. Koziel. 2015. Characterizing the smell of marijuana by odor impact of volatile compounds: An application of simultaneous chemical and sensory analysis. PLoS One 10:e0144160.

Rosenthal, E. 2001. The big book of buds: Marijuana varieties from the world's great seed breeders. Ed Rosenthal.

Rosenthal, E. 2009. Marijuana grower's handbook: Your complete guide for medical and personal marijuana cultivation. Ed Rosenthal.

Russel, W. I. and M. G. Porter. 1982. Among the new words. American Speech 57:270–276.

Russo, E. B. 2007. History of cannabis and its preparations in saga, science, and sobriquet. Chemistry & Biodiversity 4:1614–1648.

Russo, E. B. 2019. The case for the entourage effect and conventional breeding of clinical *Cannabis*: No "strain," no gain. Frontiers in Plant Science 9:434025.

Ryz, N. R., D. J. Remillard, and E. B. Russo. 2017. *Cannabis* roots: A traditional therapy with future potential for treating inflammation and pain. Cannabis and Cannabinoid Research 2:210–216.

Sawler, J., J. M. Stout, K. M. Gardner, D. Hudson, J. Vidmar, L. Butler, J. E. Page, and S. Myles. 2015. The genetic structure of marijuana and hemp. PloS One 10:e0133292.

Scarano, D., R. Rao, P. Masi, and G. Corrado. 2015. SSR fingerprint reveals mislabeling in commercial processed tomato products. Food Control 51:397–401.

Schultes, R. E., W. M. Klein, T. Plowman, and T. E. Lockwood. 1975. Cannabis: An example of taxonomic neglect. Cannabis and culture:21–38. www.degruyter.com/document/doi/10.1515/9783110812060/pdf?licenseType=restricted#page=35

Schwabe, A. L. 2019. A multifaceted approach to address variation in *Cannabis sativa*. Dissertations 554. https://digscholarship.unco.edu/dissertations/554/

Schwabe, A. L., C. J. Hansen, R. M. Hyslop, and M. E. McGlaughlin. 2019. Research grade marijuana supplied by the National Institute on Drug Abuse is genetically divergent from commercially available *Cannabis*. bioRxiv:592725.

Schwabe, A. L. and M. E. McGlaughlin. 2019. Genetic tools weed out misconceptions of strain reliability in *Cannabis sativa*: Implications for a budding industry. Journal of Cannabis Research 1:3.

Schwabe, A. L., J. R. Neale, and M. E. McGlaughlin. 2015. Examining the genetic integrity of a rare endemic Colorado cactus (*Sclerocactus glaucus*) in the face of hybridization threats from a close and widespread congener (*Sclerocactus parviflorus*). Conservation Genetics 16:443–457.

Sirikantaramas, S., F. Taura, Y. Tanaka, Y. Ishikawa, S. Morimoto, and Y. Shoyama. 2005. Tetrahydrocannabinolic acid synthase, the enzyme controlling marijuana psychoactivity, is secreted into the storage cavity of the glandular trichomes. Plant and Cell Physiology 46(9):1578-82. doi: 10.1093/pcp/pci166. Epub 2005 Jul 15. PMID: 16024552

Slaughter, J. B. 1987. Marijuana prohibition in the United States: History and analysis of a failed policy. Columbia Journal of Law & Social Problems 21:417.

Small, E. 1997. Cannabaceae. Pp. 381–387. In: Flora of North America Editorial Committee (eds.) Flora of North America North of Mexico, Vol. 3. Magnoliophyta: Magnoliidae and Hamamelidae..

Small, E. 2015. Evolution and classification of *Cannabis sativa* (marijuana, hemp) in relation to human utilization. The Botanical Review 81:189–294.

Small, E. 2016. Cannabis: A complete guide. CRC Press.

Small, E. 2017. Classification of *Cannabis sativa* L. in relation to agricultural, biotechnological, medical and recreational utilization. Pp. 1–62. *Cannabis sativa* L.-Botany and biotechnology. Springer.

Small, E. and A. Cronquist. 1976. A practical and natural taxonomy for *Cannabis*. Taxon 25(4):405–435. https://doi.org/10.2307/1220524

Smith, C. J., D. Vergara, B. Keegan, and N. Jikomes. 2022. The phytochemical diversity of commercial cannabis in the United States. PLoS one 17:e0267498.

Spindle, T. R., C. A. Zamarripa, E. Russo, L. Pollak, G. Bigelow, A. M. Ward, B. Tompson, C. Sempio, T. Shokati, J. Klawitter, U. Christians, and R. Vandrey. 2024. Vaporized D-limonene selectively mitigates the acute anxiogenic effects of Δ9-tetrahydrocannabinol in healthy adults who intermittently use cannabis. Drug and Alcohol Dependence 257:111267.

Taura, F., S. Morimoto, and Y. Shoyama. 1996. Purification and characterization of cannabidiolic-acid synthase from *Cannabis sativa* L. Biochemical analysis of a novel enzyme that catalyzes the oxidocyclization of cannabigerolic acid to cannabidiolic acid. Journal of Biological Chemistry 271:17411–17416.

Tomić, L., N. Štajner, and B. Javornik. 2013. Characterization of grapevines by the use of genetic markers. Pp. 1–25. The Mediterranean Genetic Code-Grapevine and Olive. Intech Open Science.

Toth, J. A., G. M. Stack, A. R. Cala, C. H. Carlson, R. L. Wilk, J. L. Crawford, ... and L. B. Smart. (2020). Development and validation of genetic markers for sex and cannabinoid chemotype in *Cannabis sativa* L. Gcb Bioenergy 12(3):213–222.

Toth, J. A., G. M. Stack, C. H. Carlson, and L. B. Smart. 2022. Identification and mapping of major-effect flowering time loci Autoflower1 and Early1 in *Cannabis sativa* L. Frontiers in Plant Science 13:991680.

United Nations Conference for the Adoption of a Single Convention on Narcotic, D. 1962. Single convention on narcotic drugs, 1961. United Nations New York.

U.S. Congress. 1970. Comprehensive drug abuse prevention and control act of 1970 United States Congress. House Committee on Interstate Foreign Commerce.

Usher, G. 1996. The Wordsworth dictionary of botany. Wordsworth Reference.

van Velzen, R. and M. E. Schranz. 2021. Origin and evolution of the cannabinoid oxidocyclase gene family. Genome Biology and Evolution 13:evab130.

Vassilevska-Ivanova, R. 2019. Biology and ecology of genus *Cannabis*: Genetic origin and biodiversity. vitro.

Vergara, D., H. Baker, K. Clancy, K. G. Keepers, J. P. Mendieta, C. S. Pauli, S. B. Tittes, K. H. White, and N. C. Kane. 2016. Genetic and genomic tools for *Cannabis sativa*. Critical Reviews in Plant Sciences 35:364–377.

Vergara, D., L. C. Bidwell, R. Gaudino, A. Torres, G. Du, T. C. Ruthenburg, K. deCesare, D. P. Land, K. E. Hutchison, and N. C. Kane. 2017. Compromised external validity: Federally produced *Cannabis* does not reflect legal markets. Scientific Reports 7:46528.

Vergara, D., C. Feathers, E. L. Huscher, B. Holmes, J. A. Haas, and N. C. Kane. 2021a. Widely assumed phenotypic associations in *Cannabis sativa* lack a shared genetic basis. PeerJ 9:e10672.

Vergara, D., R. Gaudino, T. Blank, and B. Keegan. 2020. Modeling cannabinoids from a large-scale sample of *Cannabis sativa* chemotypes. PloS One 15:e0236878.

Vergara, D., E. L. Huscher, K. G. Keepers, R. M. Givens, C. G. Cizek, A. Torres, R. Gaudino, and N. C. Kane. 2019. Gene copy number is associated with phytochemistry in *Cannabis sativa*. AoB Plants 11:plz074.

Vergara, D., E. L. Huscher, K. G. Keepers, R. Pisupati, A. L. Schwabe, M. E. McGlaughlin, and N. C. Kane. 2021b. Genomic evidence that governmentally produced *Cannabis sativa* poorly represents genetic variation available in state markets. Frontiers in Plant Science 12:668315.

Vergara, D., S. Shelnutt, H. Grab, A. Demetri, S. Rice, A. BarracoIII, G. Boudinot, M. Bridgen, A. Cala, J. Crawford, C. Devine, J. Hanchar, N. Mattson, P. McMullen, L. A. Monserrate-Oyola, A. Ochterski, L. Pashow, A. Reid, J. Reid, C. Smart, L. Smart, Z. Smith, S. Snyder, L. Sosnoskie, F. A. Sparks, G. Stack, Z. Stansell, M. Ullrich, L. Vignale, B. Young, B. Zahl, and M. Zuefle. 2023. New York State *Cannabis sativa* L. Production Manual.

Villa, T. C. C., N. Maxted, M. Scholten, and B. Ford-Lloyd. 2005. Defining and identifying crop landraces. Plant Genetic Resources 3:373–384.

Virdi, J. 2021. Understanding the history of Rastafari and the sacramental use of ganja with Jahlani Niaah. https://chacruna.net/rastafari_sacramental_ganja/

Watts, G. 2006. Science commentary: *Cannabis* confusions. BMJ: British Medical Journal 332:175.

Watts, S., M. McElroy, Z. Migicovsky, H. Maassen, R. van Velzen, and S. Myles. (2021). Cannabis labelling is associated with genetic variation in terpene synthase genes. Nature Plants 7(10):1330–1334.

Weiblen, G. D., J. P. Wenger, K. J. Craft, M. A. ElSohly, Z. Mehmedic, E. L. Treiber, and M. D. Marks. 2015. Gene duplication and divergence affecting drug content in *Cannabis sativa*. New Phytologist 208(4):1241–1250.

Wenger, J. P., C. J. Dabney III, M. A. ElSohly, S. Chandra, M. M. Radwan, C. G. Majumdar, and G. D. Weiblen. 2020. Validating a predictive model of cannabinoid inheritance with feral, clinical, and industrial *Cannabis sativa*. American Journal of Botany 107:1423–1432.

Wheeler, J. 2017. Picking plants for pollinators: The cultivar conundrum. www.xerces.org/blog/cultivar-conundrum

Willette, D. A., S. E. Simmonds, S. H. Cheng, S. Esteves, T. L. Kane, H. Nuetzel, N. Pilaud, R. Rachmawati, and P. H. Barber. 2017. Using DNA barcoding to track seafood mislabeling in Los Angeles restaurants. Conservation Biology 31:1076–1085.

Wilson, J. 2023. Curious about *Cannabis* (3rd Edition): A scientific introduction to a controversial plant. Natural Learning Enterprises, LLC.

Zirpel, B., O. Kayser, and F. Stehle. 2018. Elucidation of structure-function relationship of THCA and CBDA synthase from *Cannabis sativa* L. Journal of Biotechnology 284:17–26.

2 Overcoming Genetic Obstacles to the Industrialization of Cannabinoid Farming

Ed Rosenthal

2.1 INTRODUCTION

It is generally accepted that progenitor hops/cannabis originated in the Himalayan foothills about 28 mya and then split from hops about 20 mya. It has been a turbulent time over the last 30 mya or so. The cannabis plant has survived and evolved in different geologic climates along the foothills, as well as changing climates as turbulent plates clashed, producing the Himalayas.

An additional stress it has faced is the variability of the weather from year to year. Cannabis has developed several strategies to cope with these conditions and they all directly affect its industrialization.

With variable weather, a heterogeneous feral population or landrace is likely to be more successful than a homogeneous gene pool. Since the varied alleles result in some plants in the population that thrive best or are least negatively affected by weather conditions. Cannabis employs several strategies to assure preservation, reproduction, and expansion of range.

In the past few thousand years, its usefulness to humans has been its main road to range expansion and adaptation. Much like dogs, it travels wherever humans go "domestically" but because of its previously developed survival strategies it often uses relocation as a springboard to become feral. The swing from wild or feral to domesticated to feral has also had a profound effect on the genome. In the U.S., Midwest hemp was cultivated through the 1940s. Now it's a common weed in the areas where it was farmed.

2.2 BREEDING STRATEGIES

Here are some of the other survival strategies the plant uses to assure success.

Unlike most seed species, it is one of the five percent that are monoecious. It is even rarer among annuals, which leads to the possibility that it was originally a perennial like its sister plant, hops, which is also monoecious, but which is a perennial with dieback to the roots each fall. In a Mediterranean climate, some varieties of cannabis (Figure 2.1) can flower and regrow vegetatively to produce a second crop. Further, small commercial growers sometimes grow a crop indoors and then place the plants outdoors to produce another harvest.

To make seed, a monoecious plant must combine its genes with another plant's pollen. Cannabis is wind pollinated (Figure 2.2). Most seed is produced from close by pollen of neighboring plants. However, some of it is carried by the wind for long distances and can affect foreign populations.

 DOI: 10.1201/9781003386438-2

FIGURE 2.1A A harvested plant in midst of vegetative growth under a short dark period.

FIGURE 2.1B A swirled leaf is often the first growth as plants revert to a vegetative state.

This both increases the chances of pollination of a solo female plant from a distance or for foreign pollen that may contain novel alleles, to enter a population. A new trait can also lead to diversity or, if it is favorable to adaptation, become "standard" within the population.

The complements of alleles act as a sort of insurance for uncertain weather. Traits associated with various alleles will perform better or worse depending on the weather. With more information stored in heterogeneous plants, the population will be able to be more fit for the environment (Figure 2.3).

The cannabis achene also presents an avenue for territorial expansion. Hemp seed is regularly used in wild bird seed mixes since many bird and small mammal species feast on it. Wild hemp seed is a part of many animals' diets. It's healthy because it contains oils and protein, and its large size allows for a faster fill-up requiring less energy to harvest. Not every seed gets digested. A poop in a strategic location can be the start of a new population (Figure 2.4).

FIGURE 2.2A Male flowers about to open into pale yellow 5-petaled flowers releasing pollen.

FIGURE 2.2B Opening flowers.

Modern cannabis evolves very quickly under human guidance, to the market's demand for new varieties with different tastes and effects. This has been going on for thousands of years, but as far as THC bearing plants, the intense evolution by human intervention began in the 1960s when growers began breeding using landrace varieties as the first progenitors (Figure 2.5).

FIGURE 2.3 Legal ganja field in Khandwa, Madya Pradesh, India, 1981. The plants in this landrace have quite a bit of morphological diversity.

FIGURE 2.4A Male plants being collected in Morocco to feed goats which enjoy feasting on cannabis.

FIGURE 2.4B Many animals love to feast on cannabis seed.

As the seed companies started in the 1970s, breeding and hybridization also became a passion for many growers, resulting in the profusion of varieties available today. This hybridizing of hybrids has resulted in varieties having a salad bowl of different alleles, so many "cultivars" show variation in many aspects including ripening time, morphology, and most importantly, in the bud chemistry. Modern varieties today have more genetic mix-up than landraces that were formed for the most part by isolated populations with occasional genetic contributions.

2.3 GOALS FOR CANNABIS BREEDING

When you think about the theory of agriculture, whether intentionally philosophizing or involved in its use, you realize that it separates parts of the plant from its use by nature. For instance, we are not really interested in apple trees. Our goal is their fruit. We are interested in that tree as a producer of the fruit and not holistically for the tree. This is true of the plants that produce fruits or seeds or flowers that we eat. But we may be interested in only a portion of a plant part. For instance, extractors are interested primarily in certain oils, rather than the whole bud.

Just as tobacco is just a carrier for nicotine (and many other constituents) and so it is with cannabis. Currently, the market favors the flowers, which are rich in cannabinoids, terpenes, and other unique chemicals and really have little interest in the rest of the plant. It's actually a small portion of the flower that interests us. (At some point researchers may take more interest in the root chemistry, too.)

The market for raw flowers is increasing at a much smaller rate than for refined products. Some of the desirable qualities looked for in raw flowers are of little importance to refiners. These include the look and shape of the bud and the extraneous vegetation that may be mixed with it. Other bud qualities such as morphology and yield in hash/concentrate production, chemistry and ripening time become more important.

The important issue in breeding cannabis for content is how to develop a uniform crop that performs the way you want it to, which includes many factors as mentioned previously. Using AI DNA readers, this will soon be accomplished quickly and economically, saving tens of thousands of hours of researcher time. Using advanced algorithms, breeders will be guided to making crosses or series of crosses to develop scientifically adapted plants, producing amazing combinations of cannabinoids and their entourages without having to resort to GE.

FIGURE 2.5 Skunk #1, bred by Dave Watson and Distributed by Sensi-Seed Co., was the first superstar variety bred from landraces.

Modern cannabis farming will industrialize by designing both the plants and the tools for harvesting and processing them as a singular endeavor. At the same time, crop turnover, which is producing higher yields by growing the plants for a shorter time before harvest, becomes a serious consideration.

Using the tools that are soon going to be available, both breeders and farmers will be able to consider genetic factors directly when choosing cultivars, and determining the best cultivation and harvest methods for them (Figure 2.6).

When a large acreage of cannabis is grown it is best if the plants are homogeneous, with uniform morphology, ripening time and chemistry. This is where there is a conflict between the plant's

FIGURE 2.6A Hemp harvester.

FIGURE 2.6B Hemp harvester in action.

survival strategy and agriculture. This is taken very seriously by commercial growers who often use clones or tissue culture to assure uniformity of the crop. However, this is not feasible on a large scale.

With new tools for determining genetic profile, the breeder's ability to create a more uniform crop from seed is made far easier. The problem here is that most varieties have genetic profiles that include a large number of genes with differing alleles. Maintaining uniformity when that includes large numbers of these gene pairs cannot be done with typical breeding techniques and will include algorithmic programming of combinative breeding.

FIGURE 2.7 Moroccan landrace, circa 1995. These were single stem plants that exhibited very little branching. Modern Moroccan is mostly a hybrid of the Moroccan landrace and modern European varieties introduced by hash dealers carrying the genetics from Holland.

2.4 TETRAPLOID CULTIVARS

One set of possibilities is to use tetraploid cultivars. The DNA would consist of two pairs of each chromosome. This creates a more dynamic, more three-dimensional breeding field with many more possible combinations. It could also proportionalize a group of alleles for instance, instead of the standard choices of AA,Aa,aa, with a tetraploid the choices would be AAAA,AAAa,AAaa,Aaaa,aaaa. There could also be totally new kinds of crosses used with third and fourth alleles of the gene. Using crosses like these to make triploids increases the possible steering possibilities geometrically.

Two factors that I have experimented with regarding industrialization are carbon allocation and ease of harvest, which have some correlation. The most expensive input for plants to gather is CO_2, which must be mined from the air. The use of the gas by the plant and ultimately its yield goes to the efficiency of the system, because when growing vegetatively the plant is using time, space, and energy growing material not being used to produce the desired product. Most outdoor marijuana farmers grow plants using an area between 10 and 75 ft^2. To fill that space, the plants spend 10–15 weeks growing vegetative material that is discarded. If the plants instead spent less time growing infrastructure and instead flowered earlier by limiting vegetative growth and increasing density, the number of crops per year would increase. When growing using this method, the field looks more like a cereal field, and its harvesting and processing is much simpler and faster (Figure 2.7).

3 Cannabis Genetic Resources

Mehdi Babaei, Justin Boissinot, Ladan Ajdanian and Davoud Torkamaneh

3.1 INTRODUCTION

Cannabis (*Cannabis sativa* L.), a member of the *Cannabaceae* family, is an annual flowering dioecious plant and one of the earliest domesticated plants in Est/Central Asia, dating back approximately 12,000 years (Appendino et al., 2011; Ren et al., 2021). Throughout millennia, humans have cultivated and utilized cannabis for various purposes, including the production of fiber, oil, and traditional medicine. This historical usage is closely tied to the plant's rich composition of essential medicinal compounds, such as cannabinoids, terpenoids, flavonoids, and alkaloids (Appendino et al., 2011; Ross & ElSohly, 1996; Vanhoenacker et al., 2002). Despite cannabis's capacity to produce hundreds of active secondary metabolites, the species is often divided and regulated based on the level of psychoactive cannabinoids that are produced. In most countries (e.g., Canada, the USA, the EU), plants that produce below 0.3% Δ^9-tetrahydrocannabinol (THC) are regulated as hemp and plants with 0.3% or greater THC are classified as drug-type. Although hemp is legal in most regions, drug-type cannabis remains categorized as a Schedule I drug in various jurisdictions.

With a karyotype genome of 2n = 18 + XX for females and 2n = 18 + XY for males, featuring nine autosomes and two sex chromosomes (X and Y), cannabis exhibits distinct genetic characteristics. The haploid genome size was estimated at 818 Mb for females (XX) and 843 Mb for (XY) (Braich et al., 2020). Although cannabis plants are predominantly dioecious, meaning male and female flowers are on separate plants, monoecious cultivars of *C. sativa* have been bred, especially to produce seeds and oil used in industrial hemp. Additionally, dioecious cannabis cultivars have been known to display hermaphroditism (Monthony et al., 2024). In the cannabis industry, unfertilized female plants are preferred due to their higher cannabinoid production. Despite being an outcrossing species with significant within-population variability, clonal propagation methods have been well established, making them the primary approach for large-scale production (Jones & Monthony, 2022). This involves maintaining specific plants with desirable traits, known as mother plants, for extended periods in a vegetative stage. However, this is a time-consuming and costly operation, often occupying 10–15% of production space in commercial settings (Monthony et al., 2021).

In contrast to major crops such as wheat (*Triticum* spp.), rice (*Oryza sativa*) and sorghum (*Sorghum bicolor*), which experienced significant yield increases during the Green Revolution (Evenson & Gollin, 2003), cannabis has faced limitations in genetic improvement, due to its illegal status in many countries. The evolving landscape of political and legal reforms, particularly acceptance of medical cannabis in some regions, has resulted in an increasing demand for cannabis-related products (Hammond et al., 2022). Cannabis germplasm collections are valuable resources for potential breeding, providing new genes for important traits such as increased seed productivity, improved fiber quality, and resilience to environmental challenges. Despite the increasing importance of

 DOI: 10.1201/9781003386438-3

cannabis germplasm resources, there remains a notable gap in our understanding of cannabis evolution. The diversity, population structure, and relationships between wild and cultivated populations are largely unknown (McPartland & Hegman, 2018). A recent study on cannabis germplasm exploration has revealed significant phenotypic diversity in growth and phytochemical parameters (Babaei et al., 2024), indicating the potential importance of identifying and harnessing this diversity for development of new varieties tailored to specific end-uses.

3.2 HISTORICAL CLASSIFICATION AND CONTEMPORARY CONTROVERSIES

Taxonomic classification has posed a long-standing challenge for cannabis cultivators, compounded by clandestine operations and legal prohibitions limiting formal breeding efforts (Peng & Shahidi, 2021). Traditionally, cannabis plants have been classified based on morphological, environmental, and chemical characteristics, leading to three main frameworks proposed by Lapierre, Monthony et al. (2023). The first framework groups three subspecies under a single species, while the second recognizes several distinct species (*sativa*, *indica*, and *ruderalis*). Alternatively, the third framework suggests a single, phenotypically diverse species, *Cannabis sativa*.

Historically, *sativa*-type plants have been cultivated as industrial plants for fiber, oil, and animal feed in the Western world, characterized by tall stature, sparse branching, and long, slender leaves (Cronquist, 1981). Conversely, the *indica* type, originating in South Asia, is known for its shorter stature, broader leaves, and faster maturation (Small & Cronquist, 1976). The *ruderalis*, often considered a distinct subspecies, exhibits smaller plants with origins in Russia. In the traditional (legacy) market, *sativa* plants are often associated with uplifting and energizing effects, while *indica* varieties are perceived as relaxant and stress relievers, reflecting their historical usage patterns (Holland, 2010). The terms *sativa* and *indica* have historically been used to differentiate between drug and non-drug plants based on variations in the THC/CBD ratio (Lapierre, de Ronne et al., 2023). However, this classification has been fraught with ambiguity and contradiction, with interpretations varying widely across different contexts and perspectives (McPartland & Guy, 2017). Controversy surrounding the classification of cannabis is evident not only in common parlance but also in scientific literature within the genus *Cannabis* (McPartland & Hegman, 2018). Beyond taxonomic classification, cannabis plants are further categorized based on their chemical compositions, known as chemovars (Grlic, 1968). This classification is primarily based on the THC levels, differentiating medicinal and recreational from fiber varieties (De Meijer et al., 2009; Turner et al., 1979).

Even with attempts to classify cannabis based on morphology and chemistry, the ongoing debate underscores the complexity and nuance inherent in defining cannabis varieties, highlighting the need for a more comprehensive understanding that integrates genetic, chemical, and ecological perspectives. Despite its importance, the genetic diversity of cannabis, particularly in medicinal varieties, is severely limited in seed banks due to challenges surrounding legalization and disappearing native varieties (Bahji & Stephenson, 2019). With the ongoing global decline in cannabis diversity, the importance of genetic preservation becomes increasingly evident, emphasizing the urgent need for comprehensive germplasm preservation efforts for the species.

3.3 DIVERSITY OF *CANNABIS* GENETIC POOLS AND ENVIRONMENTAL INFLUENCES

Cannabis rich genetic diversity is a product of its adaptation to diverse environmental conditions, leading to unique and distinct genetic clusters across geographical regions. Studies on Southeastern Asia hemp landraces (Hillig, 2005) revealed their closer relation to Asian drug-type varieties compared to European hemp counterparts. However, a more recent study in China highlights distinct clustering aligned with climate zones, with Northern Chinese hemp displaying greater similarity to European varieties (Gao et al., 2014). In another study, landraces of South Asian and Central Asian heritage were compared to modern varieties and add historical depth to the germplasm exploration (McPartland & Small, 2020). Additionally, studies on Iranian landraces indicate closer alignment with drug-type cannabis, reflecting unique regional genetic traits (Soorni et al., 2017).

While some trends differentiating landraces persist in current *indica* and *sativa* varieties, such as THC/CBD ratios and distinct terpenoid profiles, the emergence of intermediate forms (i.e., hybrids) presents cannabis's dynamic evolution in recent years (McPartland & Small, 2020). However, challenges persist in understanding cannabis genetic diversity, compounded by uncertainties surrounding its native ranges and the rarity of pure wild populations (Lynch et al., 2016). The presence of infraspecific taxa in landrace germplasms underscores the vulnerability of small populations with unique genetic characteristics to extinction, emphasizing the need for targeted conservation efforts (Coates et al., 2018). In essence, unraveling cannabis genetic diversity across regions sheds light on the complex interplay between environmental influences and genetic adaptation, shaping the plant's evolutionary trajectory.

3.4 GERMPLASM RESOURCES

The practice of collecting, preserving, and exchanging seeds has evolved over time to human needs in the realm of modern agriculture. Early historical seed samples, discovered in places such as Egypt and Babylon, show the long-standing importance of germplasm preservation (Hyland, 1977). The effort to establish germplasm banks, dedicated to collecting, preserving, conserving, and distributing seeds that represent the genetic diversity of species, remains a crucial endeavor. Numerous scientists and researchers actively engage in this preservation effort, recognizing the pivotal role of germplasm resources in breeding programs to improve plant characteristics and providing a foundation for applied research, offering valuable insights into plant biology (Johnson, 2008).

In the 1920s and 1940s, Nikolai Vavilov recognized the potential of plant relatives as a source of genetic diversity for improving crop quality. Vavilov emerged a key advocate for germplasm banks, focusing on collection and preservation of seeds from plant relatives and specific varieties closely related to cultivated crops (Vavilov, 1940). Initiated in 1894 by the Department of Applied Botany in Russia, seed collection efforts expanded under Vavilov's leadership, leading to the establishment of the N.I. (Nikolai Ivanovic) (Damania, 2008). Additionally, in 1946, four stations for introducing regional plants were established under the purview of marketing and research laws, with the first station officially started its activities two years later (Wilson et al., 1985). There are currently over 100 germplasm banks globally, with notable concentration in developing countries (Plucknett & Horne, 1992). These banks play a pivotal role in the preservation of plant genetic diversity, contributing to the global effort to ensure sustainable agriculture and food security.

In broad terms, germplasm can be categorized into two main types: general and specialized (Gabriel, 1992). Each type holds unique value for breeders and researchers in the realm of agriculture and genetic preservation. General germplasm is valued for its inherent, often undiscovered, diversity. This type encompasses a range of genetic resources that contribute to the preservation of natural genetic diversity. General germplasm includes inbred lines, wild and landrace cultivars, and open-pollinated cultivars. These resources serve as a reservoir for preserving the diversity inherent in plant populations. Specialized germplasm, on the other hand, is distinguished by its focus on maintaining specific genetic stocks. This category includes alleles of specific genes, insertions of transgenic genes utilized for research purposes (e.g., RNAi lines), and plants exhibiting alternative ploidy compared to normal plants, such as tetraploids. Additionally, specialized germplasm may include plants with a deviation in chromosome numbers, such as monosomic or trisomic variations. The emphasis in specialized germplasms is on precision and targeted maintenance of particular genetic attributes (Sachs, 2009). Understanding the distinctions between general and specialized germplasm is crucial for optimizing the utilization of these resources in breeding programs, research initiatives, and the overall conservation of plant genetic diversity.

3.4.1 Cannabis Germplasm Resources

Although cannabis gains legal recognition and acceptance as an agricultural commodity in many regions around the world, a considerable deficiency persists in the establishment of public germplasm repositories within the cannabis industry (Small, 2018). While some rare seeds are safeguarded in national gene banks, the reluctance to share these valuable resources with researchers and producers in different countries poses a significant challenge. Overall, a total of seven research collections are dedicated to maintaining cannabis germplasm, mostly hemp varieties (Table 3.1). These collections encompass a variety of seed types, including fiber, drug, and feral cannabis varieties.

The most extensive and comprehensive collection is currently housed at the Vavilov Research Institute (VIR) in St. Petersburg, Russia. However, the genetic diversity and the purity of the VIR collection have faced compromises due to political, technical, and financial challenges in Russia during the late 1980s. The drug-type varieties in this collection originate from Afghanistan, Kazakhstan, Syria, Turkey, and Uzbekistan. The remaining collection includes hemp type and wild cannabis from Armenia, Bulgaria, Chile, China, Czechoslovakia, Estonia, France, Germany, Hungary, Italy, Latvia, Moldova, Poland, Portugal, Romania, Russia, Spain, Sweden, Ukraine, the United States, and former Yugoslavia (Coogan, 2016).

It is indeed intriguing that while seed banks like the Millennium Seed Bank (https://apps.kew.org/seedlist/SeedlistServlet) and the Svalbard Seed Bank (https://svalbardi.com/blogs/news/what-seeds-are-in-the-svalbard-global-seed-vault) maintain extensive collections of various plant species, cannabis germplasm remains relatively underrepresented. The Millennium Seed Bank, for instance, with its global reach and focus on biodiversity conservation, holds only a single cannabis sample. Similarly, the Svalbard Seed Bank, known for its role in safeguarding crop diversity in the face of

TABLE 3.1
List of Cannabis Germplasm Collections With the Number of Accessions and Their Types

Country	Collection	Number of Accessions		References
Russia	Vavilov Institute of Plant Research (VIR)	499	6 accessions of northern hemp 56 of Central Russian hemp 322 southern accessions 94 samples of monoecious hemp 21 wild accessions	www.oats2016.org/hemp/hemp1.htm
Netherlands	Wageningen University	over 156	fiber, drug, wild or naturalized populations	(Gilmore et al., 2007)
Poland	Institute of Natural Fibers and Medicinal Plants	139	ND	(Mankowska & Silska, 2015)
China	Yunnan Academy of Social Sciences	350	ND	(Salentijn et al., 2015)
Australia	Ecofibre Global Germplasm		Eurasian accessions	(Welling, Liu, et al., 2016)
Norway	Svalbard Global Seed Vault	43	Drug and hemp type cannabis	https://svalbardi.com/blogs/news/what-seeds-are-in-the-svalbard-global-seed-vault
United Kingdom	Millennium Seed Bank	1	Wild	https://apps.kew.org/seedlist/SeedlistServlet

environmental and political challenges, houses approximately 43 cannabis accessions from diverse geographical origins. The Ecofibre collection stands out for its specialization in preserving fiber-type cannabis seeds from specific regions (Welling, Liu, et al., 2016), offering valuable resources for research and breeding programs. Yet, despite these efforts, detailed information about cannabis germplasm collections in repositories like those in Poland, China, and the University of Wageningen remains scarce (Gilmore et al., 2007; Mankowska & Silska, 2015; Salentijn et al., 2015; Welling, Shapter, et al., 2016).

The preservation of populations within seed banks is a critical aspect of germplasm conservation, with the primary goal being the safeguarding of the genetic diversity present in a species. This ensures the inclusion of a diverse array of alleles in the gene pools of the preserved seeds. For monoecious species, maintaining a minimum of 1,000 plants is considered reliable, while dioecious species such as cannabis typically require 2,000 plants to replicate 99% of alleles effectively (Crossa et al., 1993). Regrettably, many seed banks globally fall short of possessing even 1,000 viable cannabis seeds. This situation is disconcerting, given that the insufficient seed quantity poses challenges to preserving genetic diversity and establishing reserves for future production and broad distribution (Clarke & Merlin, 2016). A compelling point of comparison is the National Small Grains Collection (NSGC) maintained by the United States Department of Agriculture—Agricultural Research Service (USDA-ARS), which houses a staggering 58,392 wheat accessions (Bonman et al., 2015). This stark contrast highlights the underrepresentation of cannabis in global germplasm collections, emphasizing the need for greater attention and investment in preserving the genetic diversity of this economically and medicinally significant plant species.

3.4.2 Strategies for Germplasm Conservation

Preserving plant genetic resources is crucial for maintaining biodiversity and ensuring future agricultural sustainability. For crops like cannabis, which primarily rely on clonal propagation for propagation, it is essential to consider alternative preservation methods beyond traditional seed banks. Two proposed approaches include *in vitro* active gene banks and *in vitro* base gene banks (Agrawal et al., 2019). In the former, cultures are maintained under slow growth conditions, similar to what has been implemented for crops like cassava at the Centro Internacional de Agricultura Tropical (CIAT) (Chavarriaga-Aguirre et al., 2016). Alternatively, in the latter approach, cultures are cryopreserved, offering a long-term and safe storage solution (Hunt, 2019).

Given that clonal propagation is the predominant method for cannabis propagation, it is vital to preserve rare accessions in vegetative form. Techniques for achieving slow growth include adjusting culture temperature, adding osmotic agents to the medium, using growth inhibitors, or employing outgrowth stimulants (Lata et al., 2009). However, it is important to ensure that the chemical profile of cannabis, especially those containing pharmaceutical compounds, remains stable during these processes. Recently, cryopreservation has emerged as a promising method for long-term cannabis germplasm preservation. However, it remains cost-extensive method (Downey et al., 2021).

Considering the diverse needs and challenges of cannabis conservation, it is advisable to preserve germplasm in all three forms: seeds, clones, and cryopreserved clones. This comprehensive approach ensures the preservation of genetic diversity and facilitates future research and breeding efforts aimed at improving cannabis varieties for various applications.

3.5 PRESERVATION AND USAGE OF GENETIC RESOURCES

Investigating the population genomic parameters of natural populations is a complex yet crucial task, providing insights not only into rich sources of genetic diversity but also sheds light on the nuanced patterns of domestication that have shaped cannabis over time (Soorni et al., 2017). Compounding this concern is the absence of dedicated public seed or clonal gene banks specifically for drug-type cannabis, while existing public hemp collections are limited in scope (Torkamaneh & Jones, 2022). With the ever-increasing demand for cannabis varieties in rapidly evolving markets, especially in regions like Canada and the USA, gene banks play a critical role. Beyond safeguarding genetic diversity, these repositories contribute significantly to the adaptation and improvement of new cannabis varieties. The imminent threat to landraces, the product of millennia of selective breeding, highlights the urgency of preservation effort. This is particularly important in the face of market trends that favor newer, commercially attractive cultivars. In this context, the imperative for gene banks and researchers to proactively collect, protect, and preserve the full spectrum of available cannabis genetics becomes evident, serving as a pivotal measure against their irreversible loss (Torkamaneh & Jones, 2022; Welling, Shapter, et al., 2016).

3.5.1 Complexities in *Cannabis* Genetic Variation and Germplasm Management

Numerous studies have deepened our understanding of cannabis genetic diversity, shedding light on the complexities within drug-type and hemp-type varieties. Sawler et al. (2015) have reported the genome-wide distribution of genetic differences between drug-type and hemp. Although these two types of cannabis largely share a common pool of genetic variation, distinctions between drug- and hemp-type cannabis are distributed across the genome and potentially impacting multiple genic regions. In a separate investigation, higher levels of heterozygosity have been reported in drug-type varieties, potentially a consequence of widespread hybridization during North America's shift to indoor cultivation (Lynch et al., 2016). However, caution is warranted when comparing these findings, as their methodologies were not consistent, highlighting our current lack of definitive evidence on which conservation efforts should be based. The imperfect reflection of ancestry and geographical heritage in contemporary varieties further adds layers of complexity to germplasm management (Sawler et al., 2015). To add to this, significant genetic variation has been reported within cultivars, with individuals being associated with distinct genetic clusters and considered genetically unstable, complicating further their regeneration and conservation in germplasms (Soler et al., 2017). Despite these challenges, certain gene banks, such as the Institute of Natural Fibers and Medicinal Plants (Poland), classify their varieties into climatic zones based on adaptative potentials, therefore providing a valuable collection of genetic resources without the immediate need for a complete understanding of cannabis genetic structure (Mankowska & Silska, 2015). In the intricate mosaic of cannabis genetics, this dynamic interplay of regional influences, genetic clusters, and historical legacies underscores the importance of preserving and understanding cannabis genetic resources.

3.5.2 Unlocking the Potential of *Cannabis* Genetic Resources for Novel Traits

Unlocking the expansive potential within cannabis genetic resources requires a holistic approach that transcends legal boundaries and delves into the intricate genetic landscape of the plant (Belzile et al., 2020). The pursuit of novel cannabis cultivars is hindered by a significant challenge—a lack of access to diverse, high-quality genetic material (Nutt et al., 2013). Nevertheless, accurate early screening of germplasm emerges as a powerful tool, potentially accelerating the selection of both chemical phenotypes (chemotypes) for pharmacological exploitation and agronomic traits for cultivation (Welling, Liu, et al., 2016).

Hemp, historically valued for crucial agronomic traits related to fiber and seed oil, serves as a reservoir for innovations (Lynch et al., 2016). The genetic differences between hemp- and drug-type cannabis offer a wealth of possibilities beyond THC-related traits, offering a plethora of possibilities distributed across the genome. This suggests that a multitude of traits could be improved or developed by synergistically utilizing both cannabis types (Sawler et al., 2015). To fully exploit this potential, comprehensive chemical and genetic characterization of the cannabis gene pool is needed and could significantly facilitate the metabolic engineering of diverse cannabinoid chemotypes. However, realizing this potential faces hurdles, including the lack of chemical data for most cannabis samples, leaving the chemical diversity poorly understood within designated genetic pools (Lynch et al., 2016). To add to this, cannabinoid composition has only been characterized in a limited subset of the cannabis gene pool, complicating the utilization of genetic resources in variety development (Welling et al., 2016). Nonetheless, the vast variation in phenotypes observed within cultivars represents untapped potential for selection (Soler et al., 2017).

As uniformity of cannabis products becomes imperative, especially in the medical industry, the need for well-characterized genetic resources intensifies (Soler et al., 2017). Recent efforts to address this include the establishment of a comprehensive database (i.e., CANNUSE) classifying medicinal, alimentary, fiber, psychoactive, and ethnobotanical uses of cannabis (Balant et al., 2021). This comprehensive resource has proven crucial in detecting connections between different parts of plant used for treating illnesses in different cultures, guiding researchers toward potential traits of interest found in landraces and genetic pools globally (Balant et al., 2021).

Exploiting the vast phenotypic diversity within cannabis presents an avenue for developing novel cultivars tailored for specific end-use applications, with examples showcasing improvements in cannabinoid content and fiber quality (Welling et al., 2016). However, the most limiting step in cultivar development is access to well-represented and characterized cannabis genetic resources (Lapierre, Monthony et al., 2023). Despite recognized potential, comprehensive published data detailing the extent of genetic resources within different collections remains lacking. In the absence of contemporary, systematically characterized data, along with detailed metadata, estimates of the global allelic diversity of cannabis within genetic resource collections will remain incomplete (Welling et al., 2016). Addressing these challenges is crucial to fully unlock the potential of cannabis genetic resources for the development of novel traits and varieties.

3.6 CONCLUSION

The value of genetic resources for cannabis cannot be overstated, particularly in the face of evolving challenges such as diseases, pests, and climate variability. These resources not only provide essential genetic diversity but also serve as indispensable tools for scientific research and innovation in agriculture and biotechnology. As documented (Bretting, 2018), traits sourced from gene banks have the potential to yield significant economic benefits, underscoring the importance of conserving and utilizing cannabis genetic resources. However, the full potential of cannabis genetic resources can only be realized with the removal of legal barriers for genetic preservation, especially in regions where cannabis legalization is still evolving. Precise planning for the conservation of cannabis germplasm requires addressing these legal challenges to ensure comprehensive access to diverse genetic materials. To advance research in cannabis plant genetics, it is imperative for researchers to have access to a wide range of genetic resources, particularly landrace varieties, from various nations. This access will enable a deeper understanding of cannabis genetics and facilitate the development of improved varieties tailored to meet evolving agricultural and societal needs. Ultimately, the conservation and utilization of cannabis genetic resources not only benefit farmers and the agricultural industry but also contribute to broader goals of biodiversity preservation and habitat sustainability.

REFERENCES

Agrawal, A., Singh, S., Malhotra, E. V., Meena, D., & Tyagi, R. (2019). In vitro conservation and cryopreservation of clonally propagated horticultural species. *Conservation and utilization of horticultural genetic resources* (pp. 529–578). Springer.

Appendino, G., Chianese, G., & Taglialatela-Scafati, O. (2011). Cannabinoids: occurrence and medicinal chemistry. *Current Medicinal Chemistry*, *18*(7), 1085–1099.

Babaei, M., Nemati, H., Arouiee, H., & Torkamaneh, D. (2024). Characterization of Indigenous populations of cannabis in Iran: a morphological and phenological study. *BMC Plant Biology*, *24*(1), 151.
Bahji, A., & Stephenson, C. (2019). International perspectives on the implications of cannabis legalization: a systematic review & thematic analysis. *International Journal of Environmental Research and Public Health*, *16*(17), 3095.
Balant, M., Gras, A., Gálvez, F., Garnatje, T., Vallès, J., & Vitales, D. (2021). CANNUSE, a database of traditional *Cannabis* uses—an opportunity for new research. *Database*, *2021*, baab024.
Belzile, F., Abed, A., & Torkamaneh, D. (2020). Time for a paradigm shift in the use of plant genetic resources. *Genome*, *63*(3), 189–194.
Bonman, J. M., Babiker, E. M., Cuesta-Marcos, A., Esvelt-Klos, K., Brown-Guedira, G., Chao, S., See, D., Chen, J., Akhunov, E., & Zhang, J. (2015). Genetic diversity among wheat accessions from the USDA National Small Grains Collection. *Crop Science*, *55*(3), 1243–1253.
Braich, S., Baillie, R. C., Spangenberg, G. C., & Cogan, N. O. (2020). A new and improved genome sequence of *Cannabis sativa*. *Gigabyte*, *2020*.
Bretting, P. K. (2018). 2017 Frank Meyer Medal for plant genetic resources lecture: stewards of our agricultural future. *Crop Science*, *58*(6), 2233–2240.
Chavarriaga-Aguirre, P., Brand, A., Medina, A., Prías, M., Escobar, R., Martinez, J., Díaz, P., López, C., Roca, W. M., & Tohme, J. (2016). The potential of using biotechnology to improve cassava: a review. *In Vitro Cellular & Developmental Biology-Plant*, *52*, 461–478.
Clarke, R. C., & Merlin, M. D. (2016). Cannabis domestication, breeding history, present-day genetic diversity, and future prospects. *Critical Reviews in Plant Sciences*, *35*(5–6), 293–327.
Coates, D. J., Byrne, M., & Moritz, C. (2018). Genetic diversity and conservation units: dealing with the species-population continuum in the age of genomics. *Frontiers in Ecology and Evolution*, *6*, 165.
Coogan, C. J. (2016). *The South African hemp story: saviour crop or business as usual*. Stellenbosch University.
Cronquist, A. (1981). *An integrated system of classification of flowering plants*. Columbia University Press.
Crossa, J., Hernandez, C., Bretting, P., Eberhart, S., & Taba, S. (1993). Statistical genetic considerations for maintaining germ plasm collections. *Theoretical and Applied Genetics*, *86*, 673–678.
Damania, A. B. (2008). History, achievements, and current status of genetic resources conservation: previously published in Agron. J. 100: 9–21 (2008). *Agronomy Journal*, *100*, S-27–S-39.
De Meijer, E., Hammond, K., & Sutton, A. (2009). The inheritance of chemical phenotype in *Cannabis sativa* L.(IV): cannabinoid-free plants. *Euphytica*, *168*, 95–112.
Downey, C. D., Golenia, G., Boudko, E. A., & Jones, A. M. P. (2021). Cryopreservation of 13 commercial *Cannabis sativa* genotypes using in vitro nodal explants. *Plants*, *10*(9), 1794.
Evenson, R. E., & Gollin, D. (2003). Assessing the impact of the Green Revolution, 1960 to 2000. *Science*, *300*(5620), 758–762.
Gabriel, C. J. (1992). Managing global genetic resources: the US national plant germplasm system. *BioScience*, *42*(3), 201–203.
Gao, C., Xin, P., Cheng, C., Tang, Q., Chen, P., Wang, C., Zang, G., & Zhao, L. (2014). Diversity analysis in *Cannabis sativa* based on large-scale development of expressed sequence tag-derived simple sequence repeat markers. *PLoS One*, *9*(10), e110638.
Gilmore, S., Peakall, R., & Robertson, J. (2007). Organelle DNA haplotypes reflect crop-use characteristics and geographic origins of *Cannabis sativa*. *Forensic Science International*, *172*(2–3), 179–190.
Grlic, L. (1968). A combined spectrophotometric differentiation of samples of cannabis. *Bulletin on Narcotics*, *20*, 25–29.
Hammond, D., Goodman, S., Wadsworth, E., Freeman, T. P., Kilmer, B., Schauer, G., Pacula, R. L., & Hall, W. (2022). Trends in the use of cannabis products in Canada and the USA, 2018–2020: findings from the International Cannabis Policy Study. *International Journal of Drug Policy*, *105*, 103716.
Hillig, K. W. (2005). *A systematic investigation of cannabis*. Indiana University.
Holland, J. (2010). *The pot book: a complete guide to cannabis*. Simon and Schuster.
Hunt, C. J. (2019). Technical considerations in the freezing, low-temperature storage and thawing of stem cells for cellular therapies. *Transfusion Medicine and Hemotherapy*, *46*(3), 134–150.
Hyland, H. L. (1977). History of US plant introduction. *Environmental Review*, *2*(4), 26–32.
Johnson, R. (2008). Gene banks pay big dividends to agriculture, the environment, and human welfare. *PLoS Biology*, *6*(6), e148.

Jones, M., & Monthony, A. S. (2022). Cannabis propagation. In *Handbook of cannabis production in controlled environments* (pp. 91–121). CRC Press.

Lapierre, É., Monthony, A. S., & Torkamaneh, D. (2023). Genomics-based taxonomy to clarify cannabis classification. *Genome*, *66*(08), 202–211.

Lapierre, É., de Ronne, M., Boulanger, R., Torkamaneh, D. (2023). Comprehensive phenotypic characterization of diverse drug-type *Cannabis* varieties from the Canadian legal market. *Plants (Basel)*, *12*(21), 3756.

Lata, H., Chandra, S., Khan, I. A., & ElSohly, M. A. (2009). Propagation through alginate encapsulation of axillary buds of *Cannabis sativa* L.—an important medicinal plant. *Physiology and Molecular Biology of Plants*, *15*, 79–86.

Lynch, R. C., Vergara, D., Tittes, S., White, K., Schwartz, C., Gibbs, M. J., Ruthenburg, T. C., DeCesare, K., Land, D. P., & Kane, N. C. (2016). Genomic and chemical diversity in *Cannabis*. *Critical Reviews in Plant Sciences*, *35*(5–6), 349–363.

Mankowska, G., & Silska, G. (2015). Genetic resources of *Cannabis sativa* L. in the collection of the Gene Bank at INF&MP in Poznan. *Journal of Natural Fibers*, *12*(4), 332–340.

McPartland, J. M., & Guy, G. W. (2017). Models of *Cannabis* taxonomy, cultural bias, and conflicts between scientific and vernacular names. *The Botanical Review*, *83*, 327–381.

McPartland, J. M., & Hegman, W. (2018). Cannabis utilization and diffusion patterns in prehistoric Europe: a critical analysis of archaeological evidence. *Vegetation History and Archaeobotany*, *27*, 627–634.

McPartland, J. M., & Small, E. (2020). A classification of endangered high-THC cannabis (*Cannabis sativa* subsp. *indica*) domesticates and their wild relatives. *PhytoKeys*, *144*, 81.

Monthony, A. S., Page, S. R., Hesami, M., & Jones, A. M. P. (2021). The past, present and future of *Cannabis sativa* tissue culture. *Plants*, *10*(1), 185.

Monthony, A. S., de Ronne, M., Torkamaneh, D. (2024). Exploring ethylene-related genes in *Cannabis sativa*: implications for sexual plasticity. *Plant Reproduction*. doi: 10.1007/s00497-023-00492-5.

Nutt, D. J., King, L. A., & Nichols, D. E. (2013). Effects of Schedule I drug laws on neuroscience research and treatment innovation. *Nature Reviews Neuroscience*, *14*(8), 577–585.

Peng, H., & Shahidi, F. (2021). Cannabis and cannabis edibles: a review. *Journal of Agricultural and Food Chemistry*, *69*(6), 1751–1774.

Plucknett, D. L., & Horne, M. E. (1992). Conservation of genetic resources. *Agriculture, Ecosystems & Environment*, *42*(1-2), 75–92.

Ren, G., Zhang, X., Li, Y., Ridout, K., Serrano-Serrano, M. L., Yang, Y., Liu, A., Ravikanth, G., Nawaz, M. A., & Mumtaz, A. S. (2021). Large-scale whole-genome resequencing unravels the domestication history of *Cannabis sativa*. *Science Advances*, *7*(29), eabg2286.

Ross, S. A., & ElSohly, M. A. (1996). The volatile oil composition of fresh and air-dried buds of *Cannabis sativa*. *Journal of Natural Products*, *59*(1), 49–51.

Sachs, M. M. (2009). Maize genetic resources. In *Molecular genetic approaches to maize improvement* (pp. 197–209). Springer.

Salentijn, E. M., Zhang, Q., Amaducci, S., Yang, M., & Trindade, L. M. (2015). New developments in fiber hemp (*Cannabis sativa* L.) breeding. *Industrial Crops and Products*, *68*, 32–41.

Sawler, J., Stout, J. M., Gardner, K. M., Hudson, D., Vidmar, J., Butler, L., Page, J. E., & Myles, S. (2015). The genetic structure of marijuana and hemp. *PLoS One*, *10*(8), e0133292.

Small, E. (2018). Dwarf germplasm: the key to giant *Cannabis* hempseed and cannabinoid crops. *Genetic Resources and Crop Evolution*, *65*(4), 1071–1107.

Small, E., & Cronquist, A. (1976). A practical and natural taxonomy for *Cannabis*. *Taxon*, 405–435.

Soler, S., Gramazio, P., Figàs, M. R., Vilanova, S., Rosa, E., Llosa, E. R., Borràs, D., Plazas, M., & Prohens, J. (2017). Genetic structure of *Cannabis sativa* var. *indica* cultivars based on genomic SSR (gSSR) markers: implications for breeding and germplasm management. *Industrial Crops and Products*, *104*, 171–178.

Soorni, A., Fatahi, R., Haak, D. C., Salami, S. A., & Bombarely, A. (2017). Assessment of genetic diversity and population structure in Iranian cannabis germplasm. *Scientific Reports*, *7*(1), 15668.

Torkamaneh, D., & Jones, A. M. P. (2022). Cannabis, the multibillion dollar plant that no genebank wanted. *Genome*, *65*(1), 1–5.

Turner, C. E., Elsohly, M. A., Cheng, P. C., & Lewis, G. (1979). Constituents of *Cannabis sativa* L., XIV: intrinsic problems in classifying cannabis based on a single cannabinoid analysis. *Journal of Natural Products*, *42*(3), 317–319.

Vanhoenacker, G., Van Rompaey, P., De Keukeleire, D., & Sandra, P. (2002). Chemotaxonomic features associated with flavonoids of cannabinoid-free cannabis (*Cannabis sativa* subsp. *sativa* L.) in relation to hops (*Humulus lupulus* L.). *Natural Product Letters*, *16*(1), 57–63.

Vavilov, N. I. (1940). The new systematics of cultivated plants. In J. Huxley (Ed), *The new systematics* (pp. 549–566).Clarendon Press.

Welling, M. T., Liu, L., Shapter, T., Raymond, C. A., & King, G. J. (2016). Characterisation of cannabinoid composition in a diverse *Cannabis sativa* L. germplasm collection. *Euphytica*, *208*, 463–475.

Welling, M. T., Shapter, T., Rose, T. J., Liu, L., Stanger, R., & King, G. J. (2016). A belated green revolution for *Cannabis*: virtual genetic resources to fast-track cultivar development. *Frontiers in Plant Science*, *7*, 1113.

Wilson, R., Clark, R., & Widrelechner, M. (1985). A brief history of the North Central Regional Plant Introduction Station and a list of genera maintained. *Proceedings of the Iowa Academy of Science*, *92*(2), 63–66.

4 California's Cannabis Legacy
How It Became the Global Leader

Kevin Jodrey

4.1 INTRODUCTION

California has storied relationships in regard to cannabis, serving not only as a modern trendsetter in cultural acceptance, but also for her approach to medical cannabis regulation and the indelible stamp she put on modern cannabis genetic development. How did a state, in a country with draconian and highly punitive drug laws, with no indigenous cannabis plants and a culture of casual consumption, create the most powerful and diverse cannabis economy in the world over the last 50 years?

4.2 GEOGRAPHY AND CLIMATE

California is located between the 32.5 degrees and 42 degrees north lines of latitude, with over 1,300 miles of coastline bordering the Pacific Ocean. The state has six major mountain ranges: the Klamath in the far north, the Sierra Nevada in the easternmost portion, the Coastal range follows the coast southward, farther south and running west to east we have the transverse range, and lastly, in the far south, we have the Peninsular ranges.

The center of the state contains a fertile valley called the Central Valley. It runs north-south roughly 450 miles, and the width east to west is between 40-60 miles. The southernmost third of the 450-mile valley is called the Sacramento Valley, and the remaining ⅔ is called the San Joaquin Valley. Once an inland sea and combined with sediment from the erosion of surrounding mountains, it created one of the richest soils on earth in a balanced temperature climate with over 300 sunny days per year.

4.3 GLOBAL CONNECTIONS

When we look at a map of the world, we can see which countries and regions share a common or close latitude range as California. A few of these would be Afghanistan, Pakistan, Iran, India, Nepal, South China, northern Mexico, and a swath of Africa. That list contains countries and regions that are known as historical cannabis producing regions. Places such as Pakistan, Afghanistan, and India, the three formerly collectively known as Hindustan, have a cannabis culture of cultivation and consumption of over four thousand years.

These countries have incredible diversity in climate and cannabis types, but ironically, California, with its incredible length spanning the latitudes it does and varied topography, possesses an environment that is ideal for growing world-class cannabis from almost all of the regions sharing similar environments.

DOI: 10.1201/9781003386438-4

4.4 COUNTERCULTURE AND CANNABIS

The beatnik era began in the late 1950s and stressed artistic self-expression and rejection of conventional mores. The conformity of the 1950s, coupled with the growing awareness of social injustice, created a spark that lit a global fire in disenfranchised youth. The beatniks traveled the world seeking self-expression and a connection to other cultures. Their movement inspired new youth to embrace that ideology and eventually evolved into what is known as the hippy trail.

The hippy trail was a movement and pilgrimage that started in Europe and West Asia and then headed south into Afghanistan, Pakistan, India, Nepal, Sri Lanka, Bangladesh, and Thailand. The desire of the disenfranchised to emulate the beatniks, explore new regions, and adopt cultural practices resulted in cannabis being exposed to many in a way never experienced before. These hippies and travelers collected seed stock along the way, and when they returned to North America, those seeds came back with them.

4.5 ECONOMIC AND LEGAL TRANSFORMATION

From 1910 through the 1930s, laws were enacted against marijuana as a way to attack hemp production. One couldn't argue against an industrial hemp crop, but if it was tied to cannabis the drug and then demonized through racism, it could and did achieve that goal.

Mexican immigrants and the American black population were directly associated with cannabis to create fear in the mostly white voting populations. By using the whites' fear of losing jobs to people of color and the rhetoric that cannabis made white women attracted to men of color, and that consumption of the drug made minorities feel equal to whites, it was extremely effective.

White Americans believed the propaganda, and the subsequent drug laws put into place became the cornerstones of the modern-day drug war. Hemp became prohibited and demonized, thereby protecting the industrialists' chokehold over lumber, petroleum, and chemical products.

5 Harnessing the Genetic and Chemical Diversity of *Cannabis sativa* L. in Iran

Unveiling the Potential for Breeding and Industrial Applications

Mahboubeh Mostafaei Dehnavi and Gail Taylor

5.1 INTRODUCTION

Cannabis sativa L., a globally important plant, drives a multi-billion-dollar industry, fueling the rapid evolution of the cannabis market. As the demand for cannabis and its derivatives continues to rise, breeding programs require a foundation of germplasm with wide genetic variation. Despite its long history of cultivation and recent market growth, much remains undiscovered about the genetic diversity and chemical signatures within *C. sativa*. The plant's utility, ranging from pharmaceutical uses to industrial purposes such as fiber and oil production, is largely influenced by the concentrations of THC and CBD, distinguishing it as industrial hemp or marijuana. This chapter quantifies the THC and CBD profiles and explores morphology and genetic diversity in natural cannabis populations sourced from various locations in Iran, obtained from CGRC (www.medcannabase.org) leveraging natural populations and landrace genetics to illuminate historical patterns of population structure and genetic variation. In addition, molecular markers associated with chemotype, sex, time flowering and other important traits were developed, alongside optimizing callus formation as an initial step to enhance metabolite production and genetic engineering in cannabis. Additionally, drug and hemp accessions from the CGN and IPK gene banks were analyzed to identify distinguishing SNPs within the primary cannabinoid synthase genes using high resolution melting (HRM) analysis. This project provides foundational data crucial for future cannabis breeding programs and sheds light on the diverse chemical and genomic landscape of Iranian cannabis. It emphasizes the significance of natural populations as reservoirs of genetic variation for breeding and selection programs.

5.2 HISTORICAL CONTEXT

The history of cannabis is intertwined with human civilization, with its origins traced back to central Asia or western China. Ancient Persia, enriched by the movements of Aryan and Scythian tribes, played a pivotal role in the early spread of cannabis. Archaeological evidence suggests that the Scythians, who inhabited regions of northwest Iran from the 7th century BCE to the 4th century CE, embraced cannabis for recreational and spiritual purposes. This early dissemination of cannabis laid the foundation for Iran to become a hub of genetic diversity among cannabis populations (Soorni et al., 2017).

 DOI: 10.1201/9781003386438-5

5.3 PHENOTYPIC VARIATION

Iran stands as a repository of natural cannabis populations, offering insights into historical population structures. With over 100 landraces housed in the CGRC gene bank, collected from diverse climatic zones across the country, Iran presents a promising landscape for characterizing genetic diversity within cannabis (Soorni et al., 2017). Phenotypic variations observed in these landraces encompass a wide array of traits, including fiber characteristics, oil characteristics, morphophysiology, and biochemical properties, underscoring the richness of genetic resources within Iranian cannabis populations and representing foundational elements for breeding and selection programs. The study on morphological traits in Iranian cannabis by Riahi et al. (2016) indicated significant differences among accessions for most traits, except for the number of lateral inflorescences. In terms of plant height, they reported that the shortest plants were found in Ser-01 and Sqz-01 (81.2 cm and 78.8 cm, respectively), while the tallest were in Dshm-01, Rmhz-01, Dez-02, and Bsh-01 (203 cm, 222.4 cm, 198.4 cm, and 195.3 cm, respectively). About flowering time, early flowering occurred in Ard-01, Ser-01, Krsh-01, Sqz-01, and Saad-01 (89 to 91 days), while Qzv-01, Rmhz-01, and Shrz-01 exhibited the latest flowering (108.8 to 114.3 days).

The trait correlations showed that significant positive correlations were observed, such as between the height of the first alternative node (indicating the beginning of the reproductive phase) and traits like stem diameter, number of nodes, and mean internode length. Coefficient of variation indicated that the number of days to seed formation had the lowest coefficient of variation (2.27%), whereas the length of the main inflorescence had the highest (29.31%).

5.4 PROFILING THC AND CBD IN IRANIAN CANNABIS

Mostafaei et al. (2022) aimed to measure Δ9-THC and CBD levels in 20 natural cannabis plants from various regions of Iran using High-Performance Liquid Chromatography (HPLC). They found significant variations in cannabinoid content, with THC levels ranging from 0.02% to 15.66% and CBD from 0.03% to 15.89% (% DW). The greatest THC variation was observed in plants from Dez-02 and Dez-01, while the highest CBD variation was in Rmhz-01. Notably, Sam-01 had a THC level ≤ 0.3%, aiding law enforcement efforts in Iran.

The findings categorized the plants into three distinct chemotypes based on THC/CBD ratios:

- Type I: THC-predominant – characterized by high THC content and low CBD levels. The populations of this collection with a prevalence of THC and THC/CBD content well beyond 1.0, including Pir-01, San-01, Sqz-01, Zah-01, Qzv-01, Naq-01, Abh-01, Krmn-01, Krsh-01, Bsh-01, Bam-01, Dez-01, and Dez-02, are defined as Type I.
- Type II: Intermediate (balanced) – exhibiting roughly equal proportions of THC and CBD. Ard-01 was defined as Type II and it failed to classify into either Type I or Type III.
- Type III: CBD-predominant – showcasing high CBD content and low THC levels. Six populations, including Nhv-01, Ban-01, Sir-01, Rmhz-01, Ark-01, and Sam-01, with a prevalence of CBD, hereafter, THC/CBD well below 1.0, were considered CBD- predominant populations.

Additionally, Mehrani et al. (2019) highlighted that Ban-01 and Nhv-01, with high bast and woody core fiber content, alongside Rmhz-01 and Ark-01, are promising candidates for breeding programs focused on fiber production.

5.5 ASSESSMENT OF IRANIAN CANNABIS POPULATIONS REVEALS HIGH-POTENTIAL FIBER CANDIDATES FOR DIVERSE INDUSTRIAL APPLICATIONS

Mehrani et al. (2019) evaluated several Iranian cannabis populations based on wood and fiber anatomy and stem biometry. They measured vessel and fiber biometry characteristics, identifying five populations (Frs-01, San-02, Mahb-01, Zah-01, and Mahl-01) as promising fiber candidates due to their low vessel density and porosity, indicating higher woody core fiber content. Additionally, three populations (Ban-01, Bsh-01, and Frs-01) showed superior bast fiber length, while Kash-01, Bsh-01, and Nhv-01 excelled in woody core fibers. These traits suggest these populations are well-suited for enhancing textile and paper strength, making them valuable for paper making, fiberboard, and textile production, as well as for selective breeding. Despite Bsh-01 and Zah-01 being previously classified as Type I for THC prevalence, they also exhibit favorable fiber anatomy, allowing for diverse product extraction from the same crops (Mostafaei Dehnavi et al., 2022).

5.6 ENVIRONMENTAL AND GEOGRAPHICAL INFLUENCES

- The correlation between cannabinoid content and various environmental and geographical factors was explored by Mostafaei Dehnavi et al. (2022), yielding important insights: Temperature Correlation: THC and CBD levels showed positive correlations with mean, minimum, and maximum annual temperatures, suggesting that warmer climates may enhance cannabinoid production.
- Negative Correlations: Factors such as latitude, elevation, and humidity exhibited negative correlations with cannabinoid levels, indicating that higher altitudes and more humid conditions might suppress cannabinoid synthesis.
- THC vs. CBD: A negative correlation was observed between THC and CBD concentrations, implying an inverse relationship between the two cannabinoids.

These insights provide valuable guidance for understanding how different environmental conditions influence the chemical profiles of cannabis plants, informing future cultivation and breeding strategies (Mostafaei Dehnavi et al., 2022).

5.7 MORPHOLOGICAL FEATURES AND CHEMOTYPES

In addition to chemical analysis, significant variations in morphological features were noted among the cannabis populations. These included differences in plant height (tall, mid, or dwarf), growth patterns (thin or squat growth), phyllotaxy, number of nodes, inflorescence structure such as compact or non-compact inflorescence and number of lateral inflorescences, internode length, flowering time (fast, mid, or late), seeding time (fast, mid, or late), and seed feature (Mostafaei Dehnavi et al., 2022). Notably, certain morphological traits were associated with specific chemotypes:

- Dwarf and early flowering populations: with thinner stems, more branches, and a higher density of floral tissues were observed in Type I (THC-predominant) populations such as Pir-01, Sqz-01, and Naq-01.
- Unique morphologies: The Ard-01 population, characterized by distinct inflorescence structure (broom-shaped) and seed characteristics (early seeds mature ad drop), tall and few leaves, fell into the Type II category, indicating balanced THC and CBD levels. This population, distinct both morphologically and chemically, also exhibits genetic uniqueness. Previous genetic diversity research, as reported by Soorni et al. (2017), showed that Ard-01 did not cluster with either marijuana or hemp populations. This finding is intriguing and holds potential significance for commercialization, particularly in therapeutic applications requiring a balanced THC/CBD ratio, such as in medications like Sativex.

These associations between morphology and chemical composition offer valuable markers for selecting plants with desired traits in breeding programs.

5.8 SALICYLIC ACID (SA) AND Γ-AMINOBUTYRIC ACID (GABA) AS KEY MODULATORS OF CANNABINOID PRODUCTION

Understanding gene expression helps pinpoint where and when genes are active, shedding light on how plants respond to environmental stress. Salicylic acid (SA) has gained attention for its role in activating plant defense responses and stimulating secondary metabolite pathways. It's an economical and readily available option for enhancing valuable secondary metabolites in cannabis.

Jalali et al. (2019) focused on female *Cannabis sativa* L. plants of an Iranian population (Sqz-01). They investigated how different treatments of salicylic acid (SA) and γ-aminobutyric acid (GABA) affected gene expression patterns, particularly in relation to the cannabinoid pathway enzymes, once the majority of the plants reached the flowering stage. Various concentrations of SA and GABA were sprayed onto the plants, with distilled water serving as a control. The leaves were thoroughly soaked during spraying. Sampling took place with three replications for each treatment, 72 hours after the treatments, specifically from the inflorescence. The researchers then conducted RNA extraction, cDNA synthesis, and designed specific primers for qRT-PCR analysis targeting genes such as *THCAS*, *CBDAS*, *OLS*, *PT*, and *rbcL* (used as a housekeeping gene). Additionally, HPLC analysis was performed to quantify THC and CBD metabolites in response to the SA and GABA treatments (Jalali et al., 2019).

Signaling compounds like SA and GABA were found to modulate key genes in the cannabinoid pathway, influencing the production of THC and CBD in cannabis. SA enhanced THC levels, with peak production at 1 mM, while significantly reducing CBD content compared to controls. GABA, at 0.1 mM, maximized THC production and minimized CBD content. These findings underscore SA and GABA as effective elicitors for modulating cannabinoid pathway gene expression.

Considering the pharmaceutical importance of THC and CBD, understanding how elicitors like SA and GABA affect their production is crucial. The research conducted by Jalali et al. (2019) and their findings highlight the potential of SA and GABA as regulators of cannabinoid production in cannabis, with implications for pharmaceutical applications.

5.9 GENOMIC INVESTIGATION AND BREEDING IMPLICATIONS

Genetic investigations of natural cannabis populations offer invaluable insights into the underlying genetic architecture governing desirable traits. Leveraging genomic data, researchers can unravel the genetic information of natural cannabis populations, facilitating breeding programs for diverse industrial and medical purposes. By characterizing historical patterns of population structure and genetic diversity, Iran emerges as a rich resource for identifying new varieties and initiating a pipeline of novel discoveries toward commercialization (Mostafaei Dehnavi et al., 2024).

5.9.1 High Resolution Melting (HRM) Curve Analysis for Discriminating Drug and Non-Drug Cannabis Types

Distinguishing drug-type from non-drug-type cannabis during the early vegetative stage is challenging using traditional morphological and chemical analyses. Such differentiation is crucial for cannabis breeding programs. DNA polymorphism-based PCR markers in key cannabinoid genes are essential tools for identifying genetic variations in cannabis.

The THC/CBD ratio is a key marker for differentiating fiber from drug-type cannabis. Accurate chemotype characterization is vital for developing *Cannabis sativa* cultivars for pharmacological, fiber, or seed use. High Resolution Melting (HRM) curve analysis effectively detects single nucleotide polymorphisms (SNPs) in plants. Borna et al. (2017) developed an HRM protocol for this purpose, targeting polymorphisms in the *THCAS* and *CBDAS* genes. Sequencing HRM-PCR products revealed that SNPs in the *THCAS* gene accurately differentiate drug-type cannabis, aiding legal and illegal cultivation control.

Their study confirmed that the *THCAS* gene enables distinction between drug and fiber types among 87 accessions from the IPK (The Leibniz Institute of Plant Genetics and Crop Plant Research, Germany) and CGN (Center for Genetic Resources, The Netherlands) gene banks using an SNP-based HRM assay. The HRM method demonstrated high sensitivity and specificity for identifying drug-type genotypes. Conversely, SNPs in the *CBDAS* gene were insufficient for discriminating the accessions under study (Borna et al., 2017).

5.9.2 Genotyping-By-Sequencing (GBS) Analysis of Cannabis Populations

The study by Soorni et al. (2017) employed Genotyping-By-Sequencing (GBS) to analyze the genetic structure of cannabis populations. A total of 98 samples, including 70 from 35 locations in Iran, 2 from Afghanistan, and 26 from CGN and IPK, were genotyped. Both male and female plants were sampled from each location of Iran. Using *C. sativa* "Purple Kush" and "Finola" reference genomes for mapping, the study applied *ApekI* for library preparation. This process identified 24,710 high-quality single nucleotide polymorphisms (SNPs). The GBS data for this study is available at NCBI BioProject PRJNA419020.

Estimation of genetic differentiation among marijuana and hemp accessions and Iranian samples showed larger FST values for hemp (0.086) compared to marijuana (0.039) and indicated that Iranian collections are more genetically similar to marijuana than hemp.

5.9.3 Gender, Drug, and Non-Drug Differentiation

To identify DNA markers for gender determination, Soorni et al. (2017) analyzed allele frequency differences between female and male samples, but no high-frequency sex-specific alleles outside the sex-determining region were found. Previous analyses showed marijuana and fiber types differ across the genome, but their approach failed to identify significant SNPs for type differentiation. Their reanalysis identified SNPs with allele frequencies that differentiate hemp from marijuana, including 9 SNPs with an allele frequency of 1 in hemp and 0 in marijuana, and 92 SNPs with an allele frequency of 0 in hemp and 1 in marijuana (Soorni et al., 2017).

5.9.4 Population Structure

Understanding the population genomics of natural cannabis populations can reveal genetic diversity sources and domestication patterns for this widely utilized plant. Soorni et al. (2017) performed a population structure analysis using PCA on Iranian cannabis collections, CGN, IPK, and earlier GBS data of hemp and marijuana populations (Sawler et al., 2015). This analysis confirmed two main genetic clusters, supporting the differentiation between marijuana and hemp, with Iranian collections clustering closely with marijuana. Discriminant Analysis of Principal Components (DAPC) further identified four distinct genetic clusters, highlighting the unique genetic diversity of Iranian cannabis. Their findings suggest that Iranian cannabis is more closely related to marijuana, yet distinct from marijuana and contains unique genetic diversity, potentially representing a distinct marijuana lineage. This genetic diversity is crucial for breeding programs but is at risk due to strict legal penalties against cannabis cultivation in Iran.

Further fine structure analysis identified distinct genetic populations within Iranian cannabis, revealing significant genetic diversity despite evidence of admixture likely due to breeding. The wind-pollinated and highly heterozygous nature of cannabis, combined with clonal propagation practices for high THC marijuana, complicates population structure. While not a subspecies, Iranian cannabis represents a unique genetic variety of marijuana, offering valuable material for cultivar development (Soorni et al., 2017).

Mostafaei et al. (2024) conducted another GBS study on Iranian cannabis populations, which is available as a preprint online (bioRxiv). The sequencing data for this study can be accessed at NCBI SRA under BioProject PRJNA1076947. They identified 23,266 high-quality single nucleotide polymorphisms (SNPs), offering a comprehensive dataset for genotype-phenotype association analysis. Principal Component Analysis (PCA), admixture analyses, and population structure analysis using GBS data aligned with the *C. sativa* "CBDRx" reference assembly (cs10 v.1.0) revealed that Iranian cannabis populations are genetically distinct from other global hemp and marijuana populations prepared by Sawler et al. (2015) and obtained from the NCBI SRA BioProject PRJNA285813. This highlights the potential of Iranian cannabis as a valuable genetic resource for breeding programs to enhance specific traits.

5.9.5 Genotype-Phenotype Associations

Furthermore, Mostafaei Dehnavi and colleagues carried out an association analysis to identify significant SNP markers associated with essential traits such as flowering time, cannabinoid content, sex determination, inflorescence characteristics, and morphophysiological characteristics. This comprehensive analysis provided insights into the genetic loci influencing key traits and the molecular markers distributed across the cannabis genome which offer valuable insights into the genetic basis of trait variation and provide molecular targets for marker-assisted breeding programs (Mostafaei Dehnavi et al., 2024).

5.9.6 Implications for Breeding Programs

The GBS studies on natural Iranian populations have expanded the genetic resources of cannabis, helping to uncover more of the plant's genetic secrets and fill existing genome gaps. Understanding the genetic basis of important traits in cannabis enables more precise and effective breeding strategies. Identifying SNP markers associated with desirable attributes allows breeders to select parent plants with favorable genetic profiles, thus accelerating the development of superior cannabis varieties. This approach is particularly beneficial for traits such as early sex determination, flowering time, and cannabinoid content, which are critical for both medical and industrial cannabis production. Mostafaei Dehnavi et al. (2024) have identified novel genetic loci not previously discovered, that contribute to sex determination, yield, and chemotype traits in cannabis through a genome-wide association study (GWAS).

Moreover, the genetic differentiation observed among Iranian cannabis populations, especially the distinctiveness of the northeast population, highlights the importance of incorporating diverse genetic resources into breeding programs. These populations provide a rich genetic foundation for developing new varieties with enhanced traits, contributing to the overall improvement of cannabis quality and productivity.

5.10 OPTIMIZING CALLUS FORMATION IN CANNABIS FOR ENHANCED CANNABINOID PRODUCTION: A COMPARATIVE STUDY OF CULTURE MEDIA AND GROWTH REGULATORS

Given the need to enhance the production of valuable cannabinoids in hemp for industrial, food, and especially pharmaceutical applications, traditional cultivation methods fall short due to their low yield and lengthy production times for secondary metabolites. Efficient biotechnological approaches are thus essential for the rapid and large-scale production of these valuable medicinal compounds. Eskandari et al. (2023) addressed this by optimizing callus formation in medicinal cannabis to identify the best culture media and plant growth regulator combinations.

They cultured leaf explants on MS and DKW basal media, supplemented with B5 vitamins, 30 g/L sucrose, and 7 g/L agar, and tested various concentrations of 2,4-D (2, 5, and 8 µmol/L) and BA (0.5 and 2.5 µmol/L). Callus formation was observed 5–7 days after establishment across different media and hormone levels. The calli varied in size, color, and texture. In DKW medium, calli were most compact with low cytokinin levels at any auxin concentration. In MS medium, calli were most friable with high cytokinin levels at any auxin concentration. The optimal medium for friable callus was MS with 2 µmol/L 2,4-D and 2.5 µmol/L BA, while the least effective was DKW with 5 µmol/L 2,4-D and 2.5 µmol/L BA (Eskandari et al., 2023).

Eskandari et al. (2023) concluded that friable calli, ideal for somatic embryogenesis and suspension culture, require a balanced auxin-cytokinin combination, whereas compact calli, better suited for regeneration, result from higher auxin and lower cytokinin levels. MS medium with 2 µmol/L 2,4-D and 2.5 µmol/L BA is recommended for producing friable calli.

5.11 FUTURE DIRECTIONS

Future research should aim to characterize better the genetic and phenotypic diversity of global cannabis populations. Expanding GBS studies to more diverse germplasm collections will improve our understanding of key genetic traits. Integrating advanced technologies like GWAS and CRISPR-based gene editing can further accelerate the discovery and modification of critical genes, leading to more targeted breeding programs and accelerated trait development and commercialization.
Several key avenues for future research and application are highlighted below:

1. **Enhancing breeding strategies**: Utilizing SNP markers identified GWAS, breeders can expedite the development of cannabis varieties optimized for specific traits such as cannabinoid content, flowering time, and sex. Future studies should focus on validating and refining these markers across diverse cannabis populations to enhance breeding precision.
2. **Optimizing cultivation practices**: Continued investigation into environmental influences on cannabinoid production will refine cultivation techniques, enabling growers to maximize yields and cannabinoid profiles under varying climatic conditions. Integration of biotechnological approaches, such as callus optimization for enhanced cannabinoid production, holds promise for scaling up production in a controlled and efficient manner.
3. **Exploring therapeutic potential**: Further exploration of cannabinoids beyond THC and CBD, including minor cannabinoids and terpenes, could unveil novel therapeutic applications. Understanding how specific chemotypes influence medicinal properties will inform the development of cannabis-based medications optimized to diverse health needs.
4. **Conservation and legal frameworks**: Protecting the genetic diversity within natural cannabis populations is essential, especially in the context of changing legal landscapes. Collaborative efforts to conserve landraces and wild cannabis relatives will ensure future access to valuable genetic resources for sustainable cultivation and scientific research.
5. **Global collaboration and knowledge sharing**: Establishing international collaborations and platforms for sharing genetic data and research findings will accelerate global cannabis research efforts. Open access to genomic datasets and standardized methodologies will foster innovation and informed decision-making across the cannabis industry.

5.12 CONCLUSION

Cannabis sativa L., a cornerstone of the global economy, continues to shape the multi-billion-dollar cannabis industry through its diverse applications in medicine, industry, and recreation. As demand for cannabis and its derivatives increases, understanding the plant's genetic diversity and chemical profiles becomes increasingly crucial for advancing breeding programs and optimizing cultivation practices.

This chapter has underscored Iran's fundamental role as a repository of natural cannabis populations, providing a source of genetic variation. Through comprehensive studies on THC and CBD profiles, morphology, and genetic diversity, Iranian cannabis has revealed distinct chemotypes and morphological adaptations suited for various industrial and pharmaceutical applications. These findings highlight the intricate interplay between genetic structure, environmental factors, genotype-phenotype associations, and cannabinoid production, elucidating critical insights for future breeding efforts.

ACKNOWLEDGMENT

We thank Dr. Salami's lab at the University of Tehran, Iran, for their significant research on cannabis, which has enriched this chapter.

REFERENCES

Borna T, Salami SA, Shokrpour M. 2017. High resolution melting curve analysis revealed SNPs in major cannabinoid genes associated with drug and non-drug types of cannabis. *Biotechnology & Biotechnological Equipment*, **31**, 4: 839–845. https://doi.org/10.1080/13102818.2017.1333456

Eskandari N, Ebadi A, Salami S. 2023. Optimization of callus formation in *Cannabis sativa* L. *Cell and Tissue Journal* **14**: 337–352. https://doi.org/10.61186/JCT.14.4.337

Jalali S, Salami SA, Sharifi M, Sohrabi S. 2019. Signaling compounds elicit expression of key genes in cannabinoid pathway and related metabolites in cannabis. *Industrial Crops and Products* **133**: 105–110. https://doi.org/10.1016/j.indcrop.2019.03.004

Mostafaei Dehnavi M, Ebadi A, Peirovi A, Taylor G, Salami SA. 2022. THC and CBD fingerprinting of an elite cannabis collection from Iran: Quantifying diversity to underpin future cannabis breeding. *Plants* **11**: 129. https://doi.org/10.3390/plants11010129

Mostafaei Dehnavi MM, Damerum A, Taheri S, Ebadi A, Panahi S, Hodgin G, Brandley B, Salami SA, Taylor G. 2024. Population genomics of a natural *Cannabis sativa* L. collection from Iran identifies novel genetic loci for flowering time, morphology, sex and chemotyping. https://doi.org/10.1101/2024.05.07.593022

Sawler J, Stout JM, Gardner KM, Hudson D, Vidmar J, Butler L, Page JE, Myles S. 2015. The genetic structure of marijuana and hemp (NA Tinker, Ed.). *PLOS ONE* **10**: e0133292. https://doi.org/10.1371/journal.pone.0133292

Soorni A, Fatahi R, Haak DC, Salami SA, Bombarely A. 2017. Assessment of genetic diversity and population structure in Iranian cannabis germplasm. *Scientific Reports* **7**: 15668. https://doi.org/10.1038/s41598-017-15816-5

Mehrani S, Azarnivand H, Pourtahmasi K, Salami SA. 2019. Comprehensive evaluation of native hemp populations using wood anatomy and stem fiber biometry characteristics towards producing fibers. *Journal of Forest and Wood Products* **72**: 135–145. DOI: 10.22059/jfwp.2019.271104.982

Riahi L, Shokrpour M, Salami SA, Khandan A. 2016. Some quantitative traits associations and their application in characterization of Iranian accessions of Cannabis (*Cannabis sativa* L.). *Iranian Journal of Field Crop Science* **46**, 4: 629–639 ref. 18. DOI: 10.22059/ijfcs.2015.56812

6 Designing Cannabis Tissue Culture Laboratory

Essential Laboratory Facilities for Precision and Reproducibility

Mohsen Hesami and Andrew Maxwell Phineas Jones

6.1 INTRODUCTION

As the cannabis industry continues to gain momentum, the utility of advanced horticultural methods such as tissue culture has become increasingly apparent (Hesami et al. 2020). Perhaps the most immediate application of plant tissue culture for the cannabis industry is to establish a clean plant system to provide growers with certified insect/disease/virus free planting material. These systems rely on the sterile nature of plant tissue culture to multiply verified clean plants to provide to growers so that they can begin cultivation with healthy plants. Clean plant programs have been established in a wide range of vegetatively propagated crops including potato, strawberry, and citrus. The implementation of these programs has helped to limit the spread of various insects, diseases, and viruses, and has even eradicated them from production systems in some cases.

In the case of cannabis, hop latent viroid (HPLV) has spread throughout the industry. Recent surveys have reported that 90% of Californian and about 50% of Canadian producers tested positive for HPLV, causing billions in economic losses and demonstrating the widespread occurrence and challenge represented by HPLV (Warren et al. 2019). While HPLV is among the most challenging diseases faced by cannabis producers, there are many others that can be transmitted through stem cuttings and, in some cases, seed (i.e., HPLV). As the industry matures, it will become increasingly important to implement a tissue culture based clean plant program to help reduce the spread and impact of a variety of pests and diseases domestically and internationally.

In addition to clean plant production, plant tissue culture and micropropagation provide a variety of other benefits for the cannabis industry. Among these is for long-term genetic preservation of plants. Traditionally, mother plants are maintained as vegetative plants in the growth facility under long-day conditions. The amount of space required to do this varies based on the scale of operation and the number of genotypes that are being maintained, but can represent over 25% of growth space in some cases. Using plant tissue culture, genetics can be maintained in a much smaller space, freeing up growth facilities for cultivation. Further, plant tissue culture techniques such as low temperature storage and cryopreservation can reduce labor costs and genetic mutations. Overall, plant tissue culture offers a number of advantages for maintaining cannabis genetics.

Several studies have reported micropropagation methods for cannabis and can be adopted to establish a cannabis micropropagation program, but efficient implementation requires additional practical considerations. We previously conducted a comprehensive review of various methodologies for cannabis micropropagation (Monthony et al. 2021; Hesami et al. 2021) and cannabinoid

 DOI: 10.1201/9781003386438-6

production via *in vitro* culture (Hesami et al. 2023). This chapter aims to highlight specific points that must be carefully considered when choosing the site and building out a plant tissue culture laboratory for cannabis micropropagation. While specific decisions will depend on the scale and focus of the facility, this is meant to provide a general guide upon which to build.

6.2 SITE SELECTION

A pivotal requirement for the success of cannabis tissue culture lies not only in the intricacies of the techniques employed, but also in the establishment of a meticulously designed laboratory equipped with essential facilities. Indeed, the site selection, layout, and setup of the laboratory holds paramount importance in creating an environment conducive to the delicate processes involved in plant tissue culture (Bhatia and Dahiya 2015). This all starts with the site selection to ensure it is suitable for the intended application.

A fundamental consideration in site selection is the environmental quality of the chosen area. The selected site should be relatively free from contaminants such as spores, mold, smoke, dust, and chemicals (Bhatia and Dahiya 2015). Maintaining a pristine environment is essential to maintain the aseptic conditions required for successful plant tissue culture. Contaminant-free surroundings not only safeguard the integrity of cultures but also contribute to the overall success and reproducibility of experiments (Reddy 2024). Often, micropropagation labs are setup ad-hoc in-production facilities without due consideration, resulting in increased contamination rates. While this may be unavoidable due to spaces available, it should be avoided when possible.

In addition to environmental considerations, the proximity of the laboratory to other facilities plays a crucial role in site selection. It is important that the plant tissue culture laboratory be situated far away from a laboratory of plant pathology, even though these facilities share a lot of common equipment (Bhatia and Dahiya 2015). Furthermore, establishing access to the laboratory through an outer room hall serves as an additional protective measure, minimizing the risk of cross-contamination between different types of laboratories and ensuring a controlled entry point for personnel (Reddy 2024).

Beyond these considerations, the physical surroundings of the chosen location are equally significant. Optimal site selection involves choosing a location with a well-maintained lawn and disease-free shrubs, gravel, or other environment that will contribute to reducing spore count. This landscaping contributes to the overall cleanliness and biosecurity of the environment, reducing the potential for external contamination (Bhatia and Dahiya 2015).

6.3 KEY ZONES IN CANNABIS TISSUE CULTURE LABORATORY

Once a suitable location is identified to establish the tissue culture lab, the floor plan and layout must be determined. While the specific requirements of a tissue culture lab will vary based on its scale and focus, they all share a selection of essential areas (e.g., washing room, media preparation room, sterilization room, and inoculation room). While it is not an absolutely essential component of a tissue culture lab, significant emphasis is placed on the necessity of a quarantine room for cannabis to protect the sterility of the lab.

An example of a cannabis tissue culture floor plant demonstrating the organization of critical zones and a list of key equipment is presented in Figure 6.1 and Table 6.1, respectively. Each room serves a unique function in the tissue culture process, from initial preparation to the nurturing of delicate cultures. By examining the purpose and requirements of these rooms, cultivators and laboratory designers can gain valuable insights into creating environments that foster robust growth, minimize contamination risks, and uphold stringent quality standards.

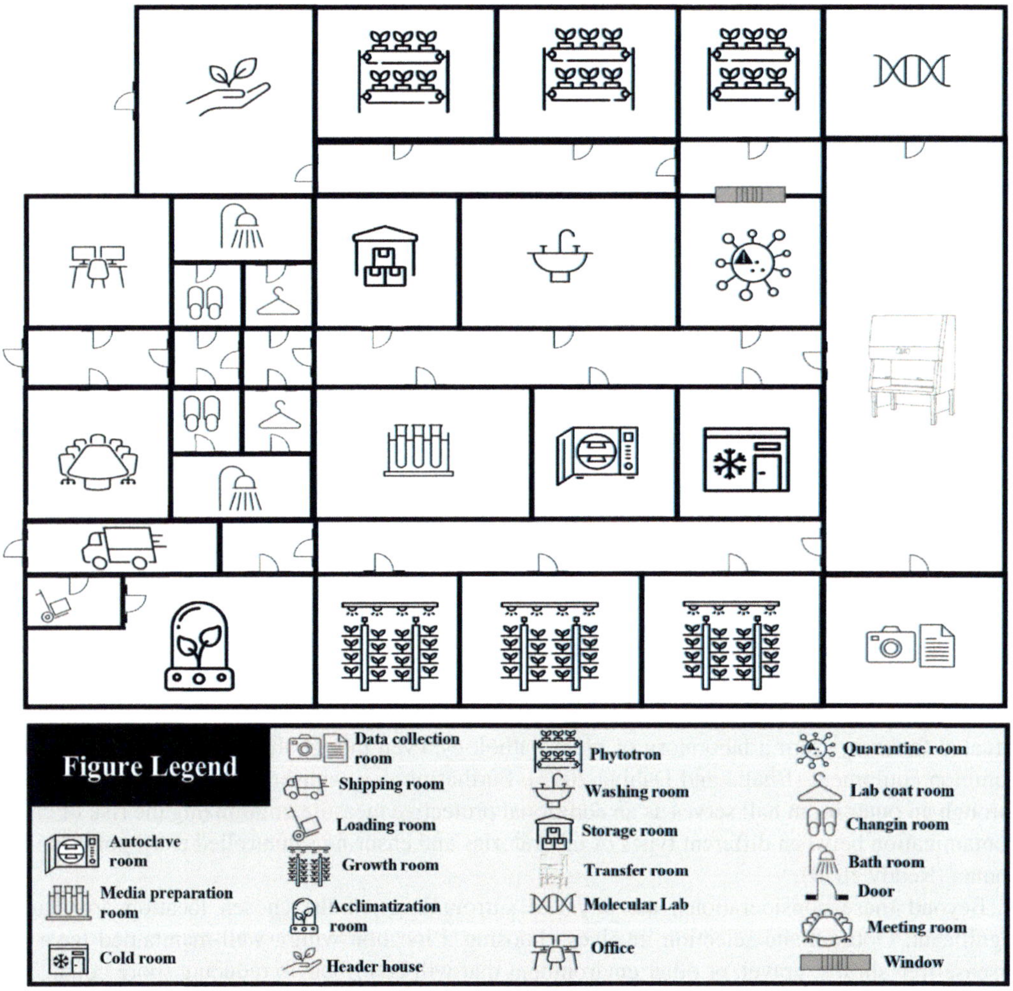

FIGURE 6.1 A sample layout for the design and organization of a cannabis tissue culture laboratory.

6.3.1 Office Room

The inclusion of an office within a cannabis tissue culture laboratory can serve as a centralized hub for managerial oversight and administrative functions. A dedicated office space allows the laboratory manager to efficiently coordinate and manage daily operations, ensuring smooth workflow and adherence to protocols. Furthermore, the office serves as a strategic location for the manager to organize and maintain inventory, facilitating timely procurement of essential materials and equipment required for tissue culture experiments. This enhances operational efficiency and contributes to the overall success and sustainability of the laboratory by ensuring that resources are effectively utilized and replenished as needed. Thus, the integration of an office space within the laboratory setting plays a pivotal role in facilitating effective management and optimizing the functionality of a cannabis tissue culture lab.

6.3.2 Meeting Room

The incorporation of a meeting room within a cannabis tissue culture laboratory holds significant importance, serving as a multifunctional area for communication, collaboration, and

TABLE 6.1
Key Equipment for Cannabis Tissue Culture and Their Functions

Equipment	Function
Water distillation/deionization system	Distillation and/or deionization of water for preparing media to ensure reproducible results
Electronic Balances (g, mg, and μg readability)	Measuring chemicals, sucrose, agar, media, etc.
Weigh boats	Measuring chemicals
pH meter	Measuring and adjusting the pH of media
Hot plate magnetic stirrer	Mixing and heating stock, medium, etc.
Stir bars	Mixing stock solutions and preparing media
Refrigerator/freezer	Storage of hormones, media, stock solutions, etc.
Laminar Flow Transfer Cabinet	Provide a sterile atmosphere for culturing
Autoclave (or pressure cooker)	Sterilizing tools, vessels, media, etc.
Autoclave/hot gloves	Safely removing hot items from autoclave
Syringe and microfilter	Sterilizing non-autoclavable solutions
Bead Sterilizer	Sterilizing the tools during culturing
Thermometer	Measuring the temperature of laboratory and culture room
Aluminum foil	To wrap instruments before sterilization, cover vessels, etc.
Beakers (50, 100, 250, 500, 1000, 2000, 4000, 6000 mL)	Mixing solutions and media
Bottle (25, 50, 100, 500, 1000, 2000 mL)	Storage of stock solutions, sterile distilled water, media
Culture tubes/vessels/boxes	Maintaining plant cultures
Erlenmeyer flask (50, 100, 250, 500, 1000, 2000, 4000, 6000 mL)	Mixing solutions and media
Graduated cylinders	Preparing solutions and media
Forceps	Transferring explants
Scalpel handle and blades	Cutting explants
Scissors	Cutting explants
Pipets (2, 20, 100, 1000 μL; and 1, 5, 10, 25, 50 mL)	Measuring out stock solutions

strategic planning. This designated room facilitates regular meetings with laboratory personnel, fostering effective communication, and promoting team cohesion. Additionally, it provides a dedicated venue for interactions with research and development (R&D) teams, enabling collaborative discussions on innovative techniques, research findings, and project developments. Beyond internal communications, the office also serves as a platform for engaging with external companies. These interactions are vital for understanding market needs, gathering feedback, and establishing partnerships to drive product development and commercialization. Thus, the presence of a meeting room within the laboratory setting not only supports internal coordination and R&D efforts but also strengthens external relationships with industry stakeholders, ultimately contributing to the success and growth of the cannabis tissue culture lab.

6.3.3 Changing Room and Bathroom

One of the key requirements for successful plant tissue culture operations is maintaining a clean space to minimize contamination. It should be recognized that human activity is among the most common source of contamination, often entering facilities on employee shoes and clothing. As such, the integration of specific facilities such as a lab coat room, changing room, and bathroom within a cannabis tissue culture laboratory is crucial for maintaining stringent hygiene standards and minimizing contamination risks. Furthermore, using sticky mats at strategic locations between rooms can effectively capture dirt from footwear. A dedicated lab coat room ensures that laboratory personnel have a designated area to store and change into clean lab coats, thereby reducing the potential transfer of contaminants into the cultivation and experimental areas. Additionally, a changing room provides personnel with a private space to change their clothes and footwear upon entering and exiting the laboratory, further enhancing biosecurity measures. The inclusion of a bathroom equipped with shower facilities is ideal, as it allows personnel to thoroughly cleanse themselves before entering the facility, effectively eliminating any lingering contaminants from their skin and clothing. This is of particular value if they have been working in the plant growth area, increasing the chances of them harboring plant pests. Collectively, these specialized facilities play a pivotal role in preserving the integrity of the cannabis tissue culture environment, safeguarding both experimental outcomes and the overall health and safety of laboratory personnel.

6.3.4 Phytotron Area

In many cases, cannabis tissue culture labs will accept plant material from a variety of sources to be established *in vitro*. It is important to note that success during stage 1 of micropropagation (culture establishment) is highly dependent on the health of the mother plant and the environment in which it was grown. For example, plants coming from outdoor production settings generally harbor more microbes and have much higher contamination rates than plants cultivated in a greenhouse or indoors. Further, from our experience, tissues that have been delayed in shipping often perform poorly *in vitro*, with increased rates of contamination and reduced survival. When receiving plants that are being shipped long distances, are unhealthy, or are coming from outdoor settings, it is often better to establish them in an indoor plant growth facility (aka phytotron) from cutting first. Once they are established and growing well, the new tissue can be more efficiently introduced into culture. For this reason, including a phytotron in the cannabis tissue culture lab is ideal, even if not strictly necessary.

The phytotron can be controlled growth chambers or rooms, but the key requirements are that they provide precise regulation of temperature, humidity, light intensity, and photoperiod, mimicking optimal conditions for plant growth and development. Additionally, the phytotron area should be outfitted with a header house to facilitate activities such as preparing pots, cuttings, pruning, transplanting, and preparing fertilizer solutions. For mother plants, ensuring consistent and ideal environmental conditions is crucial to sustaining their vigor, health, and genetic stability over prolonged periods. The phytotron facilitates the creation of a stable and reproducible environment, minimizing external stressors and fluctuations that could compromise the quality and longevity of mother plants. For further detail on facilities and production practices for controlled environment cultivation of cannabis see (Zheng 2022).

It is ideal to designate a dedicated phytotron for transferring plant materials to the quarantine room. This phytotron should be equipped with a window to allow direct connectivity and transferability between the phytotron and the quarantine room while precluding human traffic directly between the two.

6.3.5 Quarantine Area

As noted above, a paramount concern for any tissue culture lab is maintaining a clean environment. While human activity is a key contributor to introducing pests and microbes, they can also enter the facility on or in the plant material itself. As such, it is highly advised for a cannabis tissue culture laboratory to include a dedicated quarantine area for all materials entering the facility. This serves as the initial line of defense against potential contaminants, preventing the introduction of unwanted pathogens or pests that could compromise the integrity of the lab. By isolating incoming plant material, researchers can meticulously assess and treat them before integration into the controlled environment of the tissue culture laboratory (Reddy 2024). It is essential to have a laminar airflow chamber in the quarantine room.

While the initial surface sterilization procedure is intended to eliminate insects and microbes from the plant material, the process is imperfect, and a quarantine period is required to assess the success. For instance, thrips are a widespread pest in cannabis and pose a significant risk in cannabis tissue culture. These insects lay their eggs within the plant tissue, sometimes enabling them to survive the surface sterilization procedure. After culture initiation, the eggs can hatch and develop into thrips. While the direct damage from the thrips is mild, they are able to escape from culture vessels and enter into others. During this process, they can pick up microbial contaminants and lead to rapid spread among cultures. To address thrips infestations effectively, it is essential to utilize a quarantine area and implement necessary considerations. Incorporating 5 mg/L of orthene, a systemic insecticide, into the growth medium represents a viable strategy for addressing thrips-related issues if regulations permit. Explants should remain in orthene-containing media for a minimum of 5 weeks. Following this period, thrip-related issues are no longer observed, ensuring that the cultures are clean and suitable for subsequent experiments. Overall, any material entering the facility, either from the phytotron or other sources, should first be placed in the quarantine area to monitor for thrips and other potential risks.

6.3.6 Washing Area

The washing zone needs to have spacious sinks equipped with abundant drying space (i.e., drainage platforms, dryers, shelves), storage units, automatic dishwashers, and availability of distilled, double-distilled, and deionized water (Bhatia and Dahiya 2015).

It is always necessary to consider that (i) glassware intended for tissue culture, upon use, must be promptly emptied and submerged in water to prevent the desiccation of media or agar on the glass surface, (ii) glassware used for fixatives and/or corrosive chemicals should be kept separate from other glassware. It is also necessary to autoclave contaminated glassware prior to washing, (iii) upon completion of an experiment, the contents of containers must be promptly disposed of, and (iv) following the dispensation of agar-based media into culture vessels, flasks or beakers should be promptly rinsed to prevent the residual agar from drying before washing.

6.3.7 Media Preparation Area

The media preparation area and layout will vary based on the intended use of the lab. For example, tissue culture research labs will generally require the production of small volumes of various media, while a production lab may require the production of fewer media at much higher volumes. While this will impact some of the specific requirements of the area, there are several key considerations that both have in common.

The media preparation area ought to feature sufficient storage capacity to accommodate glassware, closures, culture vessels, and chemicals essential for media preparation and dispensing. The water utilized in media preparation must meet stringent standards for quality and purity. Specifically, it should adhere to the criteria of type II reagent grade water, ensuring the absence of organic contaminants, gases, and pyrogens, with an electrical conductivity below 1.0 µmho/cm. While tissue culture can be successful using water of lower purity, it should be noted that this can lead to inconsistencies and challenges reproducing protocols. The predominant and favored approach for achieving water purification to type II standards involves deionizing water followed by one or two rounds of glass distillation. Deionization effectively eliminates the majority of ionic impurities, while the distillation process targets the removal of pyrogens, microorganisms, and sizable organic molecules. Alternative techniques for generating type II purity water include membrane filtration, which effectively removes particulate matter and the majority of bacterial contamination through filtration; absorption filtration, where activated carbon is employed to eliminate organic contaminants and free chlorine; and reverse osmosis, a method capable of removing approximately

FIGURE 6.2 *In vitro* growth and development of cannabis plantlets in sub-irrigated dogboxes, depicted in (A) an overview and (B) a closer view within the culture box.

90% of ionized impurities as well as around 9% of particulate, organic, and bacterial matter (Bhatia and Dahiya 2015).

While the choice of basal media is contingent upon various factors such as the *in vitro* culture system (e.g., direct or indirect organogenesis, direct or indirect somatic embryogenesis, and shoot proliferation), as well as genotype and explant type, it is advisable to utilize DKW medium for cannabis shoot proliferation (Page et al. 2021). From our experience, DKW is reliable for cannabis tissue culture and we have successfully used it for over 150 cultivars of cannabis. Undoubtedly, improved media will be developed specifically for cannabis, but until then, DKW is a suitable basal medium.

Upon media preparation, its dispensation into culture vessels ensues. The selection of appropriate culture vessels for cannabis tissue culture is contingent upon the specific objectives of the *in vitro* cultivation (Pepe et al. 2022). For instance, test tubes are suitable for *in vitro* seed germination and establishing nodal explants, whereas petri plates are conducive to callogenesis and somatic embryogenesis (Gautam et al. 2024). Beyond conventional culture vessels, innovative options such as sub-irrigated We-V boxes (Magenta, USA) (Figure 6.2) can be considered robust alternatives for cannabis micropropagation. This system utilizes a liquid medium to sub-irrigate a substrate to feed the plant to facilitate a hedging micropropagation strategy in which plants are repeatedly coppiced to produce multiple flushes of shoot explants. While this can be done to some degree with semi-solid media, it ultimately desiccates while the liquid system allows fresh nutrients to be continually added. From our experience, cannabis is well suited to this approach, producing vigorous and healthy explants. For a detailed discussion of this approach to micropropagation the reader is directed to (Adelberg and Aitken 2023).

6.3.8 Sterilization Room

Sterilization in plant tissue culture labs is generally achieved primarily through heat sterilization using an autoclave. The size and type of autoclave will depend greatly on the scale of the operation and operational decisions such as the use of disposable vs reusable vessels. However, it should be noted that the autoclave often represents a bottleneck in tissue culture lab workflow and should be sized appropriately despite their relatively high cost. In small laboratories, expensive autoclaves can be substituted with basic pressure cookers, supported by selecting disposable vessels to reduce space requirements. Conversely, for large-scale production, it is advisable to install vertical or horizontal autoclaves, and potentially supplement this with dedicated media preparator/dispenser. Dual-door autoclaves, offering direct access to the media storage area, might incur higher costs but effectively mitigate contamination risks.

The sterilization room should be positioned in close proximity to the media preparation room to facilitate seamless workflow. The layout design should prioritize the smooth transition of containers from the washing zone to the media preparation room and finally to the sterilization zone. The sterilization room necessitates walls and flooring capable of enduring steam, heat, and moisture. Installation of an exhaust fan is recommended to expel moist and warm air. Moreover, the exhaust fan should be equipped with an outer casing to prevent infiltration of external air. The cover of the fan must remain closed except during the fan's active operation. An economical option involves transferring the sterilized media to the adjacent room via a hatch window.

6.3.9 Aseptic Transfer Area/Inoculation Room

The primary operational space of utmost importance is the culture transfer room (inoculation room), serving as the central hub for plant culturing. This transfer zone requires meticulous cleanliness and should ideally constitute a distinct room with minimal air turbulence. Smooth surfaces on both walls and floors within the transfer room are essential to facilitate regular cleaning routines. It is advisable to keep doors and windows to a minimum to mitigate the risk of contamination, while ensuring compliance with relevant local safety regulations. It is recommended that a transfer room be equipped with vacuum, compressed air, gas, and electricity sources. No specific lighting specifications are necessary for the transfer room as lights are generally integrated into flow benches.

Although instruments should be autoclaved before the experiments, instruments should be re-sterilized in the glass-bead sterilizers during subcultures to ensure aseptic conditions. It is of particular importance to re-sterilize tools between cultures to ensure potential contaminants or viruses do not spread within the population. Concerning cannabis explant sterilization, while acknowledging the influence of genotype, explant type, and maternal plant maintenance conditions, the predominant sterilization method entails immersing the explants in 70% ethanol for 60 seconds, followed by a 5-minute rinse with sterilized deionized water under laminar airflow conditions. Following this step, explants are subjected to sterilization within the laminar flow hood, involving exposure to 10% (v/v) commercial bleach supplemented with 1 drop of Tween-20 for 12 minutes, followed by three consecutive rinses in deionized water, each lasting 5 minutes. For cannabis shoot proliferation, it is preferable to culture stem segments with a minimum of two nodes in the DKW medium (Page et al. 2021).

Mobile carts with stacked culture containers enabled seamless transportation from the medium storage area to the transfer room, and subsequently to the growth room. As a precautionary safety measure, it is necessary to ensure that fire extinguishers and first aid kits are readily available within the transfer room as flames and sharp tools are often used. In the working area, it is essential to utilize dedicated laboratory footwear and coats. In some instances, UV lights are installed within the transfer room to sanitize the environment. It is imperative to activate these lights solely when the room is unoccupied by individuals and plant materials and change bulbs frequently to maintain efficacy. Safety mechanisms, such as switches, can be implemented to deactivate the UV lights automatically when standard room lighting is engaged. The transfer room serves as the dedicated area for all sterile transfer operations. It is essential to have a laminar airflow chamber, ensuring meticulous precautions are taken to prevent any contamination from entering the culture vessels during inoculation or subculture procedures. Typically, laminar airflow cabinets are sterilized by activating the hood and cleaning the work surface with 70% ethanol for a duration of 30 minutes prior to commencing any tasks under the laminar airflow cabinet. The laminar airflow cabinet ought to feature a positive-pressure ventilation system. This ventilation setup should incorporate a high-efficiency particulate air (HEPA) filter, ideally with a 99.97–99.99% efficiency rating for particles as small as 0.3 micrometers. Additionally, HEPA or standard furnace filters can be installed over laboratory air intakes or furnaces. Generally, air enters the system via a dust filter and subsequently passes through a HEPA filter. Following filtration, the air is directed either outward (in horizontal flow units) or downward (in vertical flow units) across the working surface. This continuous flow of filtered air, free from dust and microbes, effectively prevents the deposition of non-filtered air and particulate matter onto the working area (Bhatia and Dahiya 2015).

6.3.10 Data Collection Area

The establishment of a dedicated data collection room within a plant tissue culture laboratory holds paramount importance for ensuring efficient and accurate data management throughout various stages of experimentation and cultivation. This designated space serves as a centralized hub where researchers can meticulously document and organize essential information related to experimental protocols, growth parameters, and observed outcomes. By systematically recording data in a controlled environment away from the active laboratory workspace, researchers can minimize the risk of errors or discrepancies in data entry and analysis. Additionally, the establishment of standardized data collection protocols through image processing within this dedicated space ensures consistency and reproducibility across experiments, thereby enhancing the reliability and validity of research outcomes in plant tissue culture studies (Reddy 2024).

6.3.11 Molecular Lab

While not critical for the core task of cannabis tissue culture, many labs will also desire a molecular lab. This serves as a critical component for various research and experimentation endeavors. It enables researchers to conduct molecular analyses, including DNA extraction, PCR amplification, flow cytometry analyses, gene expression studies, and genetic transformation experiments. Moreover, the molecular lab plays a pivotal role in monitoring the quality and genetic stability of tissue culture-derived plants, ensuring the reliability and consistency of plant material used for further research or commercial applications. Overall, the molecular lab serves as a cornerstone for advancing plant tissue culture research and applications, bridging the gap between genotype and phenotype to enhance crop productivity and sustainability.

6.3.12 Culture Room

The culture room is a fundamental component of any plant tissue culture laboratory, serving as the primary environment for the maintenance of cultures under controlled conditions. The layout of the culture room should be such that there is no direct access from the exterior of the building, and the media preparation, glassware washing, and storage areas should be situated outside of the culture rooms. Within the culture room, precise control of humidity, temperature, and light is upheld. These environmental parameters are tailored to the specific dimensions of the growth chamber. The culture room holds significant importance as it serves as the environment where plant cultures thrive under carefully controlled conditions to facilitate optimal growth. It is recommended to have multiple growth chambers to cater to diverse culture requirements, considering that various *in vitro* culture procedures (e.g., multiplication, organogenesis, somatic embryogenesis) may exhibit distinct preferences for light and temperature conditions during *in vitro* culture. For germplasm preservation, it is recommended to include a low temperature growth room set to 10°. This will slow plant growth, reducing the labor required to maintain them while also reducing the accumulation of somatic mutations.

If cooling or lighting malfunctions occur in a room, it's advisable to relocate plant cultures promptly to another room to avoid any losses. Additionally, to mitigate contamination risks, the growth room should be designed with a minimal number of doors. All tissue cultures necessitate incubation within meticulously regulated environmental conditions. This entails maintaining a consistent temperature throughout the entirety of the culture room, ensuring absence of hot or cold

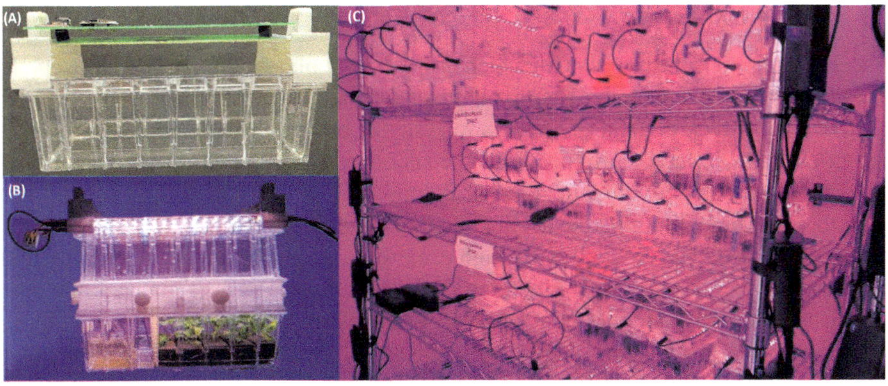

FIGURE 6.3 FinMax programmable LED lighting system. (A) FinMax programmable LED installed at the apex of the WeVitro dogbox, (B) FinMax programmable LED installed at the apex of the sub-irrigated dogboxes, and (C) FinMax programmable LED lighting system within the growth room setting.

spots, with temperatures set between 13 and 30°C. Temperature fluctuations should be minimal, ideally within ±0.3°C, and supplemented with a temperature alert alarm system for monitoring purposes. Humidity levels should be adjustable within a range of 20–98%, with a tight control of ±3%, alongside adequate air circulation. Lighting quality can be achieved via fluorescent or LED sources, with adjustable settings for both intensity and duration. Both lighting and temperature controls should be programmable over a 24-hour cycle (Bhatia and Dahiya 2015).

An additional lighting alternative is the FinMax programmable LED lighting system[1] (Figure 6.3). This modular, Wi-Fi programmable, multi-channel LED light offers precise control over the spectrum provided to explants for organogenesis and regeneration in tissue culture (Pepe et al. 2021). The FinMax programmable LED can be conveniently installed at the apex of the WeV Box (Magenta, USA)[2], enabling independent control of variable fluence rate and spectra LED lighting. This system facilitates more uniform lighting, thereby permitting proper replication and randomization for experimental plant biology, all while enhancing energy efficiency (Shukla et al. 2017). The ideal spectrum for cannabis tissue culture would typically include a balance of blue, red, white, and far-red light. The intensity of light required in tissue culture is typically lower than what mature plants need. Providing around 80–150 μmol/m^2/s of photosynthetically active radiation (PAR) is often sufficient for tissue culture growth. However, it is essential to monitor the plants closely and adjust the light intensity accordingly (Pepe et al. 2021).

Based on space availability and budget considerations, culture rooms can be stored on either stationery or portable shelves. Mobile shelves offer the benefit of accessing cultures from both sides. It's recommended that shelf height does not surpass 2 meters. Tall shelving necessitates the use of step-up stools for placing and retrieving cultures, which can pose risks and consume additional time. In the growth room, illumination primarily comes from lights installed on the shelves. Overhead lighting can be kept to a minimum since it would only be utilized during tasks conducted in the dark cycle. Plant cultures may experience uneven light distribution with traditional downward illumination methods. Directly mounted lights on racks can exacerbate this issue by creating irregular heat distribution, resulting in elevated humidity levels within the culture vessels. Consequently, this may lead to conditions conducive to hyperhydricity. Horizontal lighting presents an alternative option, necessitating fewer lights while offering improved uniformity in illumination (Bhatia and Dahiya 2015). Alternately, recent research has found hyperhydricity in cannabis can be greatly reduced by implementing a bottom cooling system to help reduce culture vessel humidity and leaf wetness.

6.3.13 Acclimatization Area

Culture acclimatization is the final step in which tissue culture plants shift from heterotrophic growth in a high humidity sealed container to being photoautotrophic in a more natural setting. This process can represent a bottleneck in plant tissue culture systems, but from our experience it is not a major challenge in cannabis tissue culture provided the shoots are health and not hyperhydric. For cannabis, if acclimatization is problematic, the best solution is generally to improve the culture system rather than focusing on optimizing the acclimatization conditions.

Regardless, the need for an acclimatization area will depend on the nature of the tissue culture lab. In some cases, plants are shipped out in tissue culture and acclimatized in a customer's facility to ensure the insect and disease-free status of the plants. In these cases, the acclimatization area is not required. However, in many cases plants will be acclimatized before being shipped and a specialized area is needed. The acclimatization area plays a pivotal role in facilitating the successful integration of tissue-cultured plants into natural settings by gradually exposing them to environmental factors such as light intensity, humidity levels, and temperature fluctuations. The controlled conditions within the acclimatization area help mitigate the shock experienced by tissue-cultured plantlets during their transition from *in vitro* to *ex vitro* environments, thereby enhancing their survival rates and overall vigor. Additionally, the acclimatization area provides researchers with an opportunity to monitor the health and growth of plantlets post-culture, allowing for the early detection of any physiological abnormalities or stress-induced responses. Through careful management and optimization of acclimatization protocols within this dedicated space, tissue culture laboratories can ensure the successful establishment of tissue-cultured plants in natural ecosystems, contributing to the conservation and propagation of rare or endangered plant species (Reddy 2024).

6.3.14 Loading and Shipping Rooms

Incorporating specialized loading and shipping rooms within a cannabis tissue culture laboratory is essential for ensuring the efficient and secure transport of micropropagated plants to their designated destinations. A meticulously designed loading room offers a controlled environment that facilitates the safe and organized handling of fragile plant materials, minimizing the risk of damage or stress during the loading process. Complementing this, a dedicated shipping room equipped with state-of-the-art packaging materials and temperature-controlled storage ensures the preservation of plant viability and quality throughout transit. These purpose-built facilities not only streamline logistical operations but also safeguard the vitality of the plants, enhancing the success and efficacy of subsequent cultivation efforts upon arrival. Thus, the thoughtful integration of loading and shipping rooms within the laboratory infrastructure plays a crucial role in facilitating the reliable and effective distribution of high-quality cannabis tissue culture products to a diverse range of stakeholders, including research institutions, commercial growers, and agricultural enterprises (Reddy 2024).

6.4 CONCLUSION

The appropriate configuration of a cannabis tissue culture laboratory not only mitigates contamination risks but also enhances operational efficacy. In fact, a meticulous design of a cannabis tissue culture laboratory is imperative for minimizing operational and energy expenditures. The laboratory must be capable of accommodating equipment utilized across the diverse phases of cannabis tissue culture in an optimally efficient manner. The presence of well-constructed access roads, stable electricity provision, and consistent water access are regarded as paramount factors in site selection. In establishing a commercial cannabis tissue culture laboratory, the consideration extends beyond convenient location to encompass adequate working spaces featuring partitioned walls and sealed ceilings. Interior walls and ceilings necessitate insulation and coverage with water-resistant materials. Windows should be strategically positioned in areas such as washing and media preparation rooms for optimal convenience. The heating infrastructure must have the capacity to sustain a consistent room temperature of 20°C even during the chilliest winter conditions. Water supply pipelines should be linked to both a sanitary sewer and a septic system. Adequate assessment of air conditioning needs is essential. Furthermore, the electrical service capacity must be meticulously calculated to accommodate lighting, equipment, and potential future expansions. A positive-pressure module should be installed to circumvent air intake from outside. The media preparation, transfer, and growth rooms are considered as "clean areas" and should be isolated from other zones, while other areas (e.g., office, meeting, storage, loading, and shipping rooms) can be kept under ordinary conditions. Generally, this comprehensive approach ensures the establishment of a cannabis tissue culture laboratory that meets the highest standards of quality and efficiency.

NOTES

1 https://bigfin.github.io/Prismatic
2 https://wevitro.magentallc.com/vessel

REFERENCES

Adelberg J, Aitken J (2023) Novel approaches to micropropagation, rooting and acclimatization. ActaHortic 1359:1–20. doi:10.17660/ActaHortic.2023.1359.1

Bhatia S, Dahiya R (2015) Laboratory Organization. In: Bhatia S, Sharma K, Dahiya R, Bera T (eds) Modern Applications of Plant Biotechnology in Pharmaceutical Sciences. Academic Press, Boston, pp 109–120. doi:10.1016/B978-0-12-802221-4.00003-0

Gautam N, Faroda P, Agarwal T, Harish, Gupta AK (2024) Effect of microclimatic physical factors on *in vitro* morphogenesis of plants: A systematic review. Discover Agriculture 2 (1):13. doi:10.1007/s44279-024-00022-5

Hesami M, Baiton A, Alizadeh M, Pepe M, Torkamaneh D, Jones AM (2021) Advances and perspectives in tissue culture and genetic engineering of cannabis. International Journal of Molecular Sciences 22 (11):5671. doi:10.3390/ijms22115671

Hesami M, Pepe M, Alizadeh M, Rakei A, Baiton A, Phineas Jones AM (2020) Recent advances in cannabis biotechnology. Industrial Crops and Products 158:113026. doi:10.1016/j.indcrop.2020.113026

Hesami M, Pepe M, Baiton A, Jones AMP (2023) Current status and future prospects in cannabinoid production through *in vitro* culture and synthetic biology. Biotechnology Advances 62:108074. doi:10.1016/j.biotechadv.2022.108074

Monthony AS, Page SR, Hesami M, Jones AMP (2021) The past, present and future of *Cannabis sativa* tissue culture. Plants 10 (1):185. doi:10.3390/plants10010185

Page SRG, Monthony AS, Jones AMP (2021) DKW basal salts improve micropropagation and callogenesis compared with MS basal salts in multiple commercial cultivars of *Cannabis sativa*. Botany 99 (5):269–279. doi:10.1139/cjb-2020-0179

Pepe M, Hesami M, Small F, Jones AMP (2021) Comparative analysis of machine learning and evolutionary optimization algorithms for precision micropropagation of *Cannabis sativa*: Prediction and validation of *in vitro* shoot growth and development based on the optimization of light and carbohydrate sources. Frontiers in Plant Science 12:757869. doi:10.3389/fpls.2021.757869

Pepe M, Leonardos ED, Marie TRJG, Kyne ST, Hesami M, Jones AM, Grodzinski B (2022) A non-invasive gas exchange method to test and model photosynthetic proficiency and growth rates of *in vitro* plant cultures: Preliminary implication for *Cannabis sativa* L. Biology 11 (5):729. doi:10.3390/biology11050729

Reddy J (2024) Plant tissue culture laboratory organisation. In: Reddy J (ed) Plant Tissue Culture. CRC Press, London, pp 51–77. doi:10.1201/9781032712611

Shukla MR, Singh AS, Piunno K, Saxena PK, Jones AMP (2017) Application of 3D printing to prototype and develop novel plant tissue culture systems. Plant Methods 13 (1):6. doi:10.1186/s13007-017-0156-8

Warren JG, Mercado J, Grace D (2019) Occurrence of hop latent viroid causing disease in *Cannabis sativa* in california. Plant Disease 103 (10):2699–2699. doi:10.1094/PDIS-03-19-0530-PDN

Zheng Y (2022) Handbook of cannabis production in controlled environments. CRC Press, Boca Raton. doi:10.1201/9781003150442

7 Cannabis Breeding Strategies

Matt Haddad and Oussama Badad

7.1 CANNABIS BREEDING HISTORY

The origins of *Cannabis sativa* L. breeding, extend back thousands of years to its domestication in the almighty Tibetan plateau. Cannabis history spans over 12,000 years, originating in Central Asia, where it was initially cultivated for its multifaceted utility in food, fiber, and medicine. As ancient civilizations like those in China, India, and Egypt embraced its benefits, cannabis became integral to religious rituals, medicinal practices, and industrial applications such as textile production. Despite facing periods of prohibition and stigmatization in the 20th century, cannabis cultivation persisted clandestinely, leading to the modern era of commercialization and genetic research.

Recent shifts towards legalization worldwide have propelled cannabis cultivation into the mainstream, fostering a thriving industry dedicated to breeding novel strains with optimized attributes. Thanks to advancements in genetic research, scientists now have the ability to manipulate cannabis plant genetics, refining traits such as potency and flavor. As legalization efforts progress and consumer preferences evolve, the future of cannabis breeding hinges on continual innovation, the adoption of sustainable cultivation methods, and the preservation of genetic diversity. These endeavors aim to meet the diverse demands of an expanding market while upholding environmental sustainability.

7.2 CANNABIS BREEDING STRATEGIES

Cannabis breeding strategies encompass a diverse array of approaches aimed at developing cultivars with specific traits tailored to various consumer preferences and market demands. Traditional breeding methods often involve selective breeding, where plants displaying desirable characteristics such as high cannabinoid content, unique terpene profiles, or disease resistance are crossbred to produce offspring with the desired traits. Additionally, techniques like backcrossing, where offspring are crossed with one of the parents to reinforce desired traits while retaining genetic stability, are commonly employed. With advancements in technology, molecular breeding techniques such as marker-assisted selection (MAS) and genomic selection have gained prominence, enabling breeders to identify and select for desired traits more efficiently by analyzing the plant's DNA. Furthermore, innovative approaches like tissue culture propagation and mutagenesis offer opportunities to generate novel genetic variations and accelerate the breeding process. Overall, cannabis breeding strategies continue to evolve, driven by a combination of traditional methods and cutting-edge technologies, to meet the evolving needs of consumers and the industry.

DOI: 10.1201/9781003386438-7

7.2.1 Conventional Breeding

Cannabis conventional breeding refers to the traditional method of breeding cannabis plants through controlled pollination and selection to create desired traits in the offspring. The offspring are grown, and outliers are selected as top performant lines based on potency, flavor, yield etc. the selected plants are propagated vegetatively to maintain the desired trait and have consistent yields over time. This type of breeding led to an extreme bottleneck in cannabis genetic resources due to the restrictive nature of the selection and also contributed to the spread of disease due to clone transfer between cultivators.

Conventional breeding of cannabis begins with the careful selection of parent plants possessing desirable traits such as high cannabinoid content, specific terpene profiles, disease resistance, or favorable growth characteristics. Once chosen, these parent plants undergo controlled pollination, typically through the deliberate transfer of pollen from male to female plants. Techniques like direct pollen application or isolation through "bagging" are employed to ensure controlled pollination and prevent unintended genetic mixing.

Crossbreeding plays a pivotal role in conventional breeding, as breeders intentionally mate different cannabis strains to merge desirable traits from each parent. This deliberate combination can result in a wide spectrum of genetic variations and phenotypic expressions in the offspring. Subsequently, the offspring are grown from the seeds produced through pollination, and meticulous evaluation follows. Breeders carefully observe and test individual plants for specific traits like growth patterns, potency, aroma, taste, and resilience to pests or diseases. Promising offspring exhibiting desired traits are then subjected to a stabilization process, where selective breeding over multiple generations aims to reinforce and maintain the expression of these traits. Throughout this process, both phenotypic and genotypic testing may be conducted to better understand the inheritance patterns of traits, guiding informed decisions in subsequent breeding endeavors. This ongoing cycle of selection, refinement, and improvement characterizes the continuous evolution of cannabis breeding practices.

7.2.2 Backcrossing

Backcrossing is a powerful and very famous tool among cannabis breeders, allowing them to efficiently introduce specific traits into elite cultivars while maintaining the desirable characteristics of the recurrent parent. The process helps balance the need for genetic improvement with the preservation of favorable agronomic traits.

Backcrossing is a fundamental breeding technique employed in seed breeding to incorporate specific desirable traits from one parent, known as the donor parent, into a genetically stable and well-adapted line referred to as the recurrent parent. The primary objective is to preserve the favorable characteristics of the recurrent parent while introducing the desired trait from the donor parent. Initially, parental lines are carefully selected, with the recurrent parent chosen for its advantageous agronomic traits or adaptation to particular environments, while the donor parent possesses the specific trait or gene of interest. The breeding process begins with crossing the recurrent parent with the donor parent to produce the F1 generation, resulting in offspring with a heterozygous genetic makeup, combining traits from both parents.

Subsequent generations involve backcrossing, where an F1 individual is crossed with the recurrent parent to generate backcross (BC) progeny. Through meticulous selection, individuals carrying the desired trait from the donor parent are identified and further backcrossed to the recurrent parent in successive generations. Each backcross iteration aims to retain the desired trait while progressively recovering the genetic background of the recurrent parent, except for the genomic region harboring the desired trait. Marker-assisted selection may be employed to expedite the identification of individuals carrying the target gene associated with the desired trait. The process continues for multiple generations until near isogenic lines (NILs) are developed, homozygous for the desired trait and genetically almost identical to the recurrent parent. Thorough verification and testing ensure the stability and consistent expression of the desired trait across diverse environmental conditions. Once stability is confirmed, the near isogenic lines are released for further breeding programs, research, or commercial utilization, serving as valuable genetic resources for enhancing the performance of adapted cultivars with specific traits.

7.2.3 Near Isogenic Lines (NILs)

One of the most used breeding techniques used in cannabis breeding today and what caused a big genetic diversity bottleneck is near isogenic lines **(NILs)** are developed through a meticulous breeding approach aimed at creating lines that are nearly genetically identical, with the exception of targeted genomic regions. The process begins with the selection of parental lines: a genetically diverse population or cultivar serving as the donor line, carrying the desired trait or gene, and a recurrent parent, typically an elite line lacking the trait. Through crossing and selection in the S1 generation, plants with the desired trait are identified. Subsequent backcrossing with the recurrent parent gradually replaces the genomic background of the recurrent parent with that of the donor parent, while retaining the desired trait. Marker-assisted selection may be employed to expedite the identification of individuals carrying the target gene. This iterative backcrossing and selection process continues until the genomic background of the recurrent parent is nearly isogenic with the donor parent, except for the specific genomic region containing the trait of interest. Once stability is verified, the developed NILs are released for research or commercial use, offering valuable tools for studying the effects of individual genes or traits under different conditions.

Case Study

Cultivar A has elite genetics with susceptibility to PM, on the other hand, the cultivar B (landrace) doesn't manifest any commercial traits but has a nonfunctioning MLO locus. See Figure 7.1.

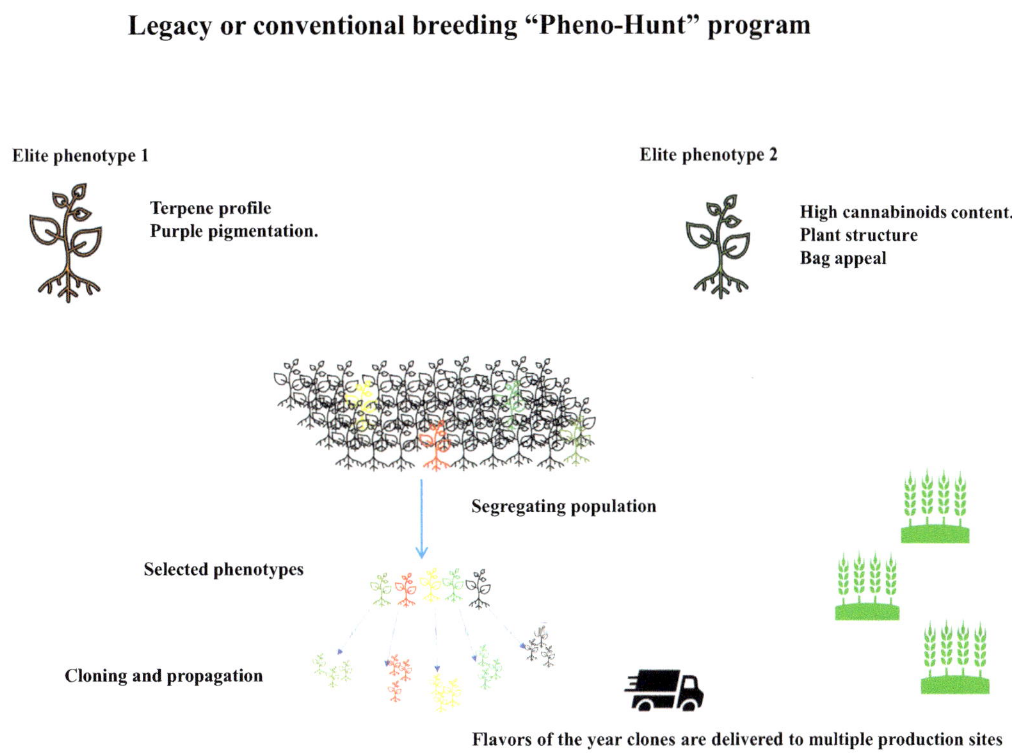

FIGURE 7.1 Schematic representation of a pheno-hunt strategy.

7.2.4 F1 Hybrid Breeding Program

The approach to breeding F1 hybrid seeds entails carefully crossbreeding two distinct, purebred parental lines to generate offspring with desirable traits. This process involves several key steps. Initially, suitable parental lines (P1 and P2) are selected based on their complementary advantageous characteristics, such as disease resistance or high yield potential. Subsequently, these chosen parental lines undergo extensive inbreeding over multiple generations to establish genetically stable and uniform lines, ensuring the fixation of desired traits within each line. Following this, the selected parental lines are crossed to produce the F1 generation, characterized by a combination of traits inherited from both parental lines. F1 hybrids often demonstrate hybrid vigor, displaying superior traits compared to either parent, such as increased yield or disease resistance. Finally, the F1 hybrid seeds are mass-produced for commercial distribution, although it's important to note that saving and replanting these hybrid seeds won't yield offspring with consistent characteristics due to their hybrid nature.

Creating inbred lines involves a meticulous breeding strategy aimed at establishing genetically stable and uniform individuals through successive rounds of self-pollination or sib-pollination. Initially, parental lines are selected from a genetically diverse population, prioritizing individuals with robust traits. Isolation measures are implemented to prevent unintended cross-pollination, ensuring controlled mating. Through controlled pollination techniques, the F1 generation is produced, representing a genetic amalgamation of the selected parents. Subsequent generations undergo self-pollination or sib-pollination, facilitating the selection and propagation of individuals with desired traits. With each generation, the proportion of homozygous alleles increases, enhancing genetic stability. Evaluation methods such as genetic testing or phenotypic observations verify the degree of homozygosity, confirming the development of highly uniform inbred lines. Continuous monitoring and maintenance efforts are essential to safeguard genetic stability and prevent unintended cross-pollination, paving the way for the release of stable inbred lines for commercial or research purposes. These lines serve as valuable genetic resources, commonly utilized as parents in hybrid breeding programs or as stable and predictable assets in scientific studies. See Figure 7.2.

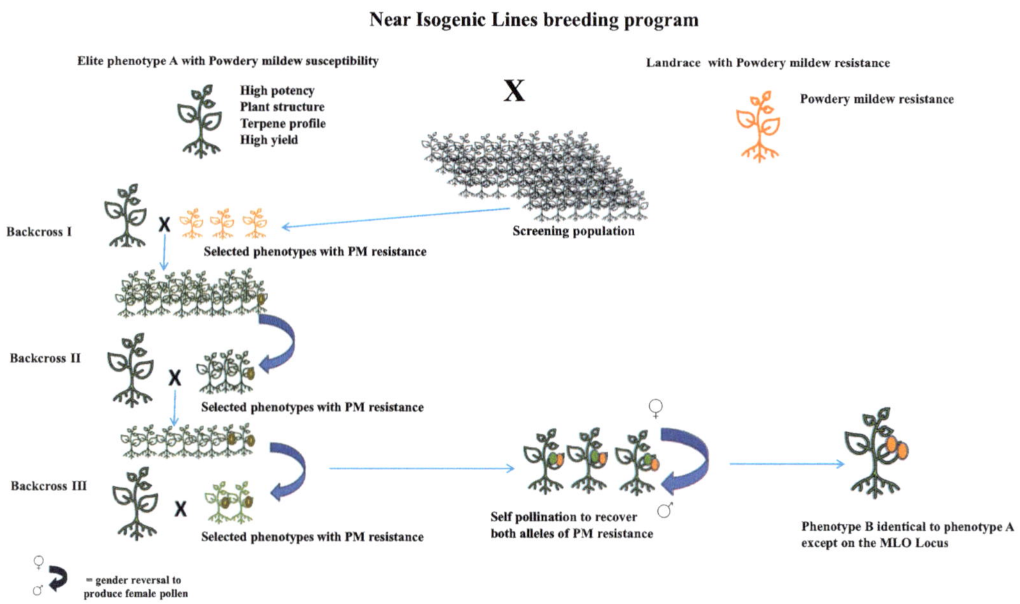

FIGURE 7.2 Schematic representation of a real case study.

7.2.5 Variation Breeding

Variation breeding encompasses several breeding strategies aimed at enhancing genetic diversity within a population. Genetic variation is crucial for adaptation, resilience, and the development of improved crop varieties. Here's an example of strategies related to enhancing genetic variation in breeding programs:

Mutation breeding is a powerful technique utilized in plant breeding to deliberately induce genetic mutations in plants in the form of deletion, insertion, and single nucleotide polymorphism aiming to develop new and enhanced varieties. This approach relies on the deliberate introduction of genetic variation by exposing seeds or plant tissues to mutagenic agents such as chemicals or radiation. These agents induce random changes in the DNA sequence, leading to mutations that can impact various traits crucial for crop improvement, including yield, disease resistance, nutritional content, and other agronomically significant characteristics.

Following mutagenesis, mutant populations are generated, comprising seeds or plant tissues harboring induced mutations. Screening methods are then employed to identify individuals with desirable mutations, which may involve observing phenotypic changes or utilizing molecular markers. Selected mutants are further characterized and evaluated for the expression of the desired traits, such as improved yield, disease resistance, or changes in plant morphology. Optionally, backcrossing with elite cultivars may be conducted to combine beneficial mutations with other desirable traits while retaining the original genetic background.

Promising mutants undergo extensive field trials to assess their performance across different environments, collecting data on agronomic performance, disease resistance, and other relevant characteristics. Once a mutant with desirable traits is thoroughly evaluated and deemed stable, it can be released as a new variety, subject to regulatory approvals and certification processes. Furthermore, mutation breeding serves as a valuable research tool for studying gene function, regulation, and plant responses to environmental stresses. Continuous improvement is achieved through successive rounds of mutagenesis to generate additional genetic diversity and further enhance traits. Overall, mutation breeding has proven successful in creating novel varieties with improved characteristics across various crops, playing a vital role in advancing agriculture through the introduction of genetic diversity and the exploitation of naturally occurring mutations. See Figure 7.3.

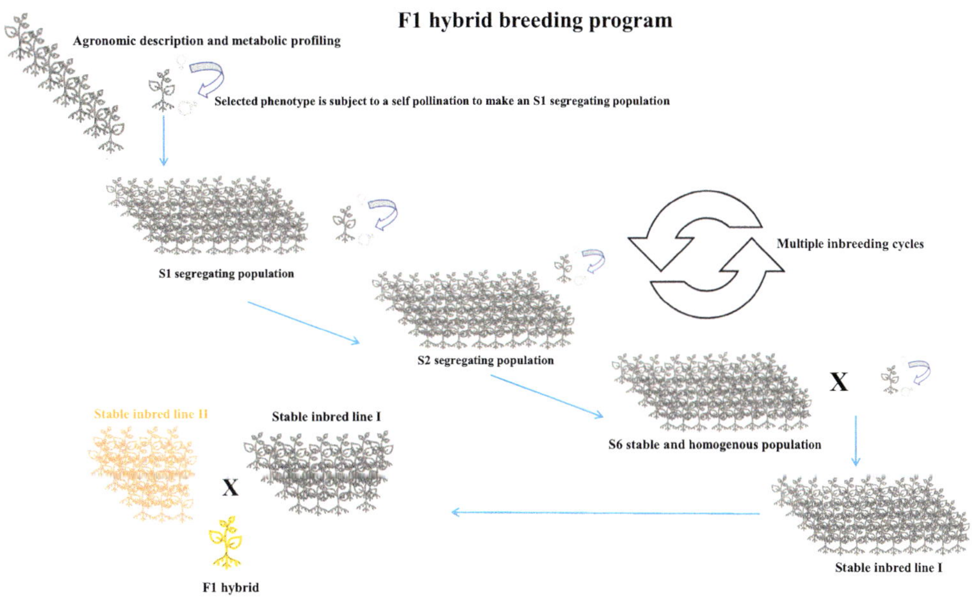

FIGURE 7.3 Schematic representation of an F1 hybrid breeding chart.

7.2.6 Ploidy Manipulation and Pollen-Resistant Triploid Development

Creating triploid seeds involves manipulating the chromosome count of plants to produce offspring with three sets of chromosomes, a condition known as triploid. This breeding strategy typically entails crossing a diploid parent (2n =20 chromosomes) with a tetraploid parent (2n=40), resulting in triploid offspring (3n=30). These triploid plants are often sterile due to difficulties in chromosome pairing during meiosis, leading to the formation of seedless fruits through parthenocarpy. While triploid plants cannot produce viable gametes, they can still be propagated through vegetative means, such as cuttings or tissue culture. This technique is particularly valuable in horticulture, where seedless fruits are preferred, as seen in crops like watermelon, banana, and certain grape varieties.

The first step to pollen-resistant triploids is making a stable tetraploid plant with four sets of chromosomes per cell. A tetraploid could be achieved through the application of anti-mitotic agents. The process begins by selecting a stable inbred line diploid parent with desirable traits, treating its seeds or tissues with colchicine to induce polyploidy, and establishing polyploid plants. To verify the ploidy level of the treated tissue or seeds, flow cytometry and chromosome counting needs to be performed. After validation, these tetraploid plants are crossed with the original diploid parent, resulting in triploid offspring.

In some cases, chromosome doubling leads to an induced variation due to splicing sites rearrangement like the case in watermelon. Thus, further selection and stabilization efforts are necessary to maintain and enhance desirable traits in tetraploid progeny over generations, facilitating their utilization in breeding programs to develop new cultivars with improved qualities. Despite its potential, the success of creating stable tetraploids depends on various factors, including the plant species, timing of colchicine application, its concentration, the lab protocols, and the ability to maintain the fragile early weeks of plant development. Tetraploids have a slow growth habit and a fascinating mushroom shaped thick and short stock and ancient feel cotyledon in the first 4 to 6 weeks after seed treatment.

With industrial hemp for fiber and grain gaining momentum, it is vital to develop triploid genetic to withstand the clouds of pollen generated by neighbor farms growing cannabis for food, feed, textile, and construction material fields including greenhouse and indoor facilities in the regular season. there are potential advantages associated with triploids in plant breeding.

Sterility: Triploid plants are often sterile or exhibit a statistically significant reduced fertility due to their odd number of chromosomes, which can be advantageous in certain agricultural settings. Sterility can prevent unwanted cross-pollination with nearby plants, which is particularly useful in seedless fruit production or preventing accidental pollination of female cannabis plants by male plants, ensuring that female plants focus their energy on producing flowers rather than seeds.

Vigor and Growth: Triploid plants may exhibit increased vigor and growth compared to diploid plants. This phenomenon, known as triploid vigor or heterosis, can result in larger plant size, enhanced biomass production, and improved overall performance. In agricultural contexts, triploid crops often show increased yields and resilience to environmental stresses, which can be beneficial for commercial cultivation.

Seedlessness: Triploid cannabis plants are often seedless or produce significantly fewer seeds compared to diploid plants. This characteristic is desirable for certain cannabis products, particularly in the production of high-quality flower buds for consumption. Seedless buds are preferred by consumers due to their smoother smoke, better flavor, and higher cannabinoid content, making triploid cannabis potentially valuable for premium cannabis markets.

Uniformity: Triploid plants may exhibit greater uniformity in terms of traits such as growth rate, flowering time, and cannabinoid content. This uniformity can simplify cultivation practices and management, leading to more consistent crop yields and product quality.

See Figure 7.4.

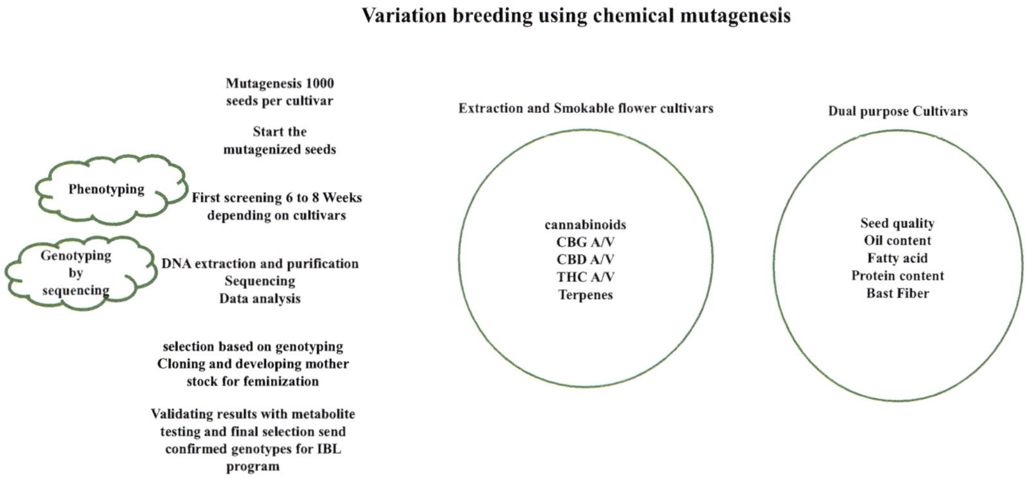

FIGURE 7.4 Schematic representation of variation breeding using EMS mutagenesis.

7.2.7 Breeding Auto Flowering

Breeding auto flowering cannabis cultivars requires a precise fusion of the auto flowering trait from *Cannabis ruderalis* with photoperiodic lines, a process meticulously executed by breeders.

Cannabis ruderalis, emerges as an intriguing yet less-explored member of the Cannabis lineage. Originating amidst the harsh climates of Central Asia, notably within the terrains of Russia, Siberia, and Central Europe, this robust plant has evolved to withstand the unforgiving conditions characteristic of northern latitudes. Its designation, "*ruderalis*," stemming from the Latin term "*rudera*," which denotes "debris" or "neglected places," underscores its remarkable adaptability to thrive in disturbed environments such as roadside margins and abandoned lots.

Distinguished by its genetic blueprint, *Cannabis ruderalis* possesses unique traits that set it apart within the cannabis family. Most notably, its auto flowering mechanism, prompting flowering based on chronological age rather than the traditional reliance on light cycles, marks a significant divergence. This adaptive feature facilitates the rapid completion of *Cannabis ruderalis* life cycle, rendering it particularly suited to short growing seasons and environments with limited daylight exposure.

Characterized by its diminutive stature, sparse foliage, and delicate leaves, *Cannabis ruderalis* presents a notable departure from the morphology of *Cannabis sativa* and *indica* varieties. Its comparatively lower levels of cannabinoids, particularly THC, differentiate it from its counterparts renowned for their psychoactive and medicinal attributes. Nevertheless, the rich genetic diversity inherent in *Cannabis ruderalis* offers promise to breeders seeking to incorporate its distinctive traits, notably auto flowering, into hybrid cultivars tailored for both recreational and therapeutic use. Despite its historical overshadowing, *Cannabis ruderalis* is now garnering recognition within the cannabis breeding community. As the industry progresses, a deeper understanding of the origins and genetic makeup of *Cannabis ruderalis* promises to catalyze innovative cultivar development, enable adaptation to diverse climatic conditions, and spur advancements in cultivation methodologies.

Cannabis ruderalis, renowned for its auto flowering characteristic, flowers based on age rather than light cycles. To incorporate this advantageous trait into photoperiodic lines, breeders first select *Cannabis ruderalis* landraces/lines with desirable auto flowering properties, typically characterized by compact stature and rapid life cycle. Concurrently, they identify photoperiodic strains prized for traits like high THC content or unique flavors. These strains undergo strategic crossbreeding,

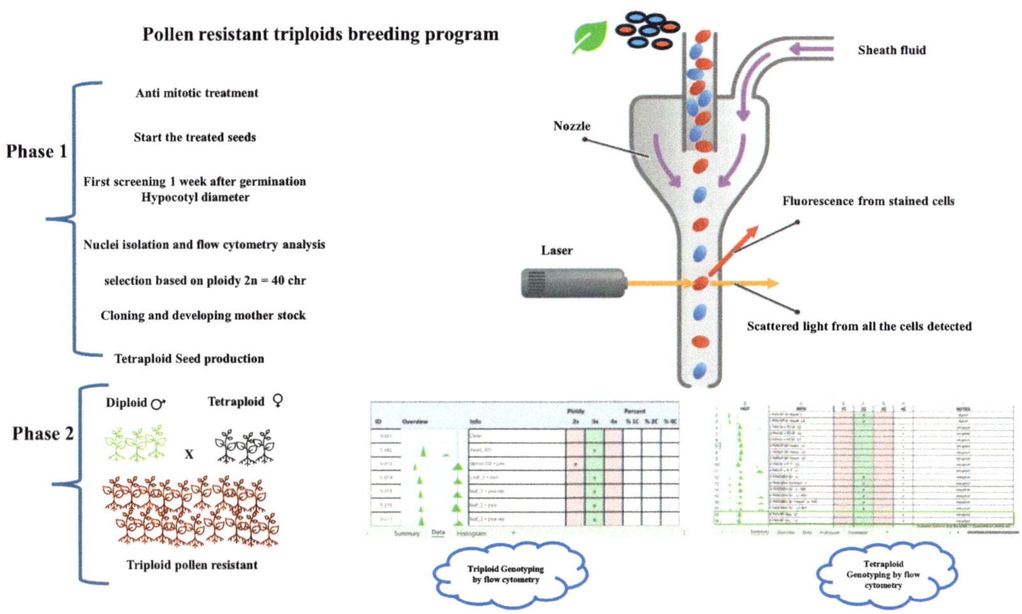

FIGURE 7.5 Schema of the development of pollen-resistant triploids.

resulting in a first-generation hybrid (F1) inheriting the coveted auto flowering trait from the ruderalis lineage.

In the F1 generation, breeders meticulously isolate individuals exhibiting auto flowering characteristics independent of light cycles. Through successive backcrosses with the original photoperiodic lines, breeders aim to consolidate the auto flowering trait while preserving desirable attributes from both parental lines. Extensive selection and stabilization procedures ensure consistent auto flowering traits across subsequent generations. These stabilized lines are subjected to thorough testing for potency, yield, flavor, and resistance to pests or diseases. Breeders may further incorporate crossbreeding to introduce genetic diversity, enhancing hybrid vigor and producing novel varieties tailored to thrive independently of light cycles, thus catering to diverse cultivation needs and environments. See Figure 7.5.

7.2.8 Marker-Assisted Selection (MAS)

Marker-Assisted Selection (MAS) stands at the forefront of modern breeding techniques, leveraging molecular markers to streamline the identification and selection of plants harboring specific desirable traits. This innovative method offers breeders a more efficient pathway to pinpoint and cultivate plants with sought-after genetic characteristics compared to conventional breeding approaches. MAS unfolds through a systematic series of steps: initially, breeders the traits of interest, ranging from disease resistance to yield, establishing the genetic foundation upon which selection will be based. Subsequently, molecular markers associated with these traits are identified, including DNA sequences like QTLs, SNPs or copy number facilitating precise genetic tracking.

Following marker identification, a genetic map is constructed, illuminating the genomic locations of these markers and elucidating their linkage to the target traits. With this framework established, the breeding population undergoes genotyping, wherein individual plants are analyzed to ascertain the presence or absence of specific marker alleles associated with the desired traits. Statistical analyses then assess the correlation between marker alleles and trait expression,

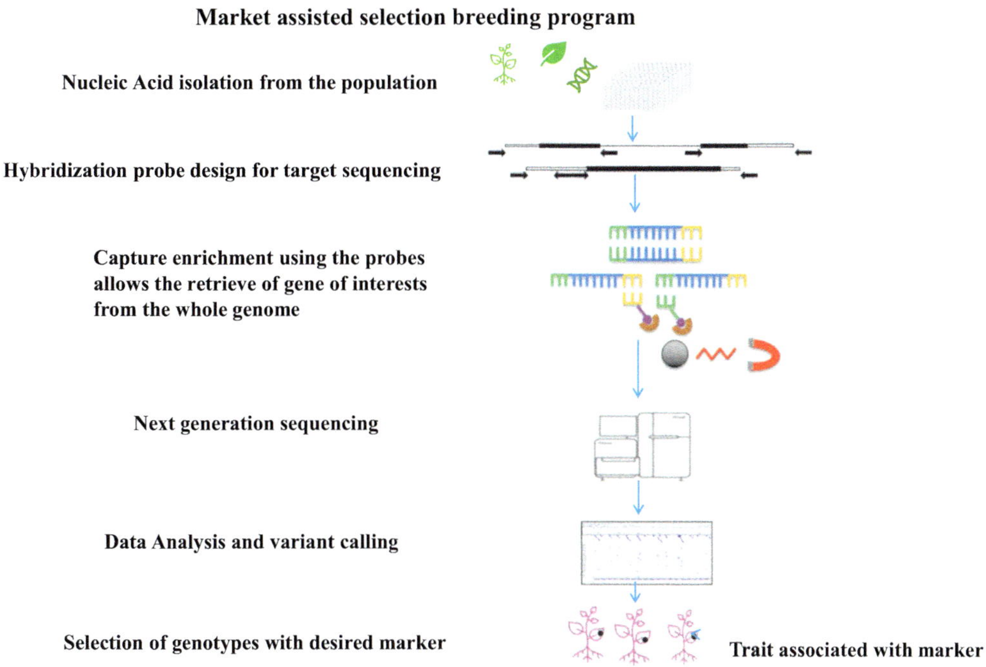

FIGURE 7.6 Schema of application of molecular markers in cannabis breeding.

guiding the selection of plants carrying marker alleles closely tied to the target traits. Through iterative cycles of crossbreeding and progeny genotyping, MAS not only accelerates the breeding process but also ensures the selection of superior individuals adept at expressing desired traits under real-world conditions, marking a pivotal advancement in plant breeding methodologies. See Figure 7.6.

8 Integrated Pest Management Practices in Cannabis

Matthew Gates

8.1 INTRODUCTION: PLANT STRESS DYNAMICS

Land plants have evolved a mainly sessile lifestyle which generally precludes them from mobile escape of environmental and biotic hazards, at least post-germination. Despite this circumstance they have become one of the most successful clades of life, and a prolific source of carbohydrates through which other organisms acquire. Several fundamental physiological traits contribute to this achievement, primarily those biological processes represented in the Growth-Defense Tradeoff model whereby plants respond to multiple stressors, often simultaneously and requiring diametrically different solutions, in a proportional manner based on a sophisticated signaling system that commandeers and reprioritizes metabolic resources normally used for development[1]. Solar radiation, temperature, hydration, salinity, substrate compaction, and composition, herbivory, disease agents, as well as primary and secondary metabolism are all common stress factors that must be managed as parsimoniously as possible while remaining capable of procreation under natural contexts. Cultivated premises are different, imposing selection pressures related to maximizing various end-product traits like the aroma, flavor, and edibility of typically delicate and vulnerable biomass yield that diverges from the ancestral defensive context and capabilities. While human tolerance for cosmetic damage, atypically formed produce, and other performance goals under guided conditions are usually much more extreme than in nature, it must be acknowledged that desirable domestication traits are still generally manipulated; wrought from a collection of genes developed through vertiginously fierce selection against total annihilation across several majorly catastrophic extinction events and all the smaller interactions that persist today in an unbroken line spanning half a billion years or so[2]. Across millions of years and thousands of kilometers what would become cannabis spread and adapted throughout Eurasia[3] enduring polar deserts, xeric shrublands, and both temperate and tropical forests during which regional ice ages ebbed and flowed, evolving in tandem with its hardy relatives like *Humulus* (Hop), *Celtis*, *Trema*, *Parasponia*, and more basally, *Aphananthe*[4]. Within a brief evanescent span of the last several millenia cannabis-human symbiosis coalesced in various regions as cultivation for many functions, honing the proportions of mass and volume for floral, seed, stem, and other tissues[5]. Most saliently in the past few decades modern cultivated cannabis has become globally established, turgescent with colorful odiferous inflorescences that are densely bracted and tomentose trichomes that swell and sometimes ooze with metabolites. Along with this expansion in range and mass, so too has exposure to environmental and biological threats increased, particularly the concentration of compatible pests and intensification of their adaptation to cultivation contexts requiring a systematic, multidomain approach to management for considerations molecular to ecosytemic[6].

 DOI: 10.1201/9781003386438-8

The main processes by which plants resist biotic stressors generally begin at the site of physical contact with epidermal tissues and the various biota and substances residing therein that may influence pest success. Upon detection, which may be influenced by pest and environmental factors, intra- and intercellular signaling triggers the reprogramming of cellular events[7], carried out by genes from the endogenous genome[8], augmented by interaction with symbiotic, specifically mutualistic, microbes—from the collective microbiome—and primarily mutualistic multicellular life and its associated microbial symbionts[9]. Reactive measures degrade, disable, and may eventually defeat a pest through an array of effects like directly damaging the pest with a group of complementary chemical agents, or indirectly by recruiting predators or parasites. At the molecular level, the plant genome and its associated microbiome form a "hologenome" composed of all the genes and subsequently cellular interactions influenced by them that occur together collectively[10]. In turn, at the cellular level and above, it is this enmeshed networked ecosystem—the holobiont—that mitigates biotic and abiotic threats of the plant directly by sustaining a primed status for certain stress responses: increased or altered production, transport, and sequestration of pigments or structural polymers and specific antibiotic factors, disrupting herbivores or microbial pathogens, as well as more indirect influences such as the exchange of resources like water, nutrients, as well as microbial symbionts[11]. When present, these beneficial factors mitigate to various degrees the full gamut of stressors plants experience daily.

8.2 MAJOR AGENTS OF BIOTIC STRESS

Biotic stressors exist in many forms, and myriad organisms of diverse capability typically inflict stressful effects such as through the production of molecular agents or those of microbial symbionts. Molecular agents like various proteins and other chemical compounds are called effectors and comprise a significant arsenal of not only weapons but means of signaling, for plants as well as their mutualists and parasites[12]. Various microbes in the plant microbiome can influence each other discretely and directly[13]: some plant parasites may form mutualistic interactions with plant mutualists, emphasizing the importance of specific biodiversity composition of the cultivation context.

Microbial agents of biotic stress are often plant parasites that as a normal course of development use plant metabolic resources for sustenance, replication, or shelter, such as a virus that induces copies of proteins, DNA, or RNA made from host-derived molecules or an obligately biotrophic fungus like powdery mildews that rely on living host cells to manipulate into providing nutrients and suppressing the local tissue's immune system[14]. Some mutualist interactions can become parasitic or have parasitic effects contextually[15]. A mycorrhizal fungus can develop a relationship with one or more host plants, vectoring soilborne nutrients, hydration, as well as metabolites from other plants and itself, which might give some hosts a greater competitive advantage than others due to individual differences in colonization dynamics. The presence of some mychorrizae can elicit immune signaling changes in a host that decrease or increase the virulence of plant viruses[16], while the dynamics of bacteriophages[17], viroids[18], mycoviruses[19], and other viral entities[20], as well as virovores[21] encountered in the phyto- and rhizomicrobiome influence not only microbiota but mutualist and parasitic insects, mites, and other organisms.

Arthropods constitute the majority of herbivorous plant parasites, with around half a million herbivorous insects comprising a quarter of currently known metazoan species[22]. Co-evolution with the earliest land plants as the first multicellular animals to colonize land fundamentally affected the course of plant evolution as mutualists and parasites. Since the first arthropods are estimated to have been detrivores and detrivory exposes an organism to microbes in the decaying organic matter which can process plant material, plants have had to develop responses to a combined animal-microbe threat very early[23]. Herbivorous arthropods like mites and insects deploy these agents and their effectors, which directly disrupt plant processes, especially immune response, through feeding, secretion, and elimination primarily. Folivorous beetles, larval moths, and butterflies, and hemiptera like aphids represent significant herbivorous insect lineages and common crop pests which closely co-evolved with land plants and radiated with the now dominant flowering plants[24].

8.3 ECOLOGICAL THEORY OF IPM—CANNABIS CONSIDERATIONS

Although many subjects and processes concerning it are relatively ancient, integrated pest Management (IPM) as a named concept emerged in the mid-1900s during the advent of a more sophisticated ecological understanding of Earth, particularly where agriculture intersected with the environment and societal considerations, and an acknowledgment of environmental harm associated with noxious pesticidal chemistries and other ecologically unsustainable practices. Essentially, it combines empirical knowledge from natural sciences with agricultural and societal contexts to develop and implement salubrious pest mitigation strategies that are as minimally disruptive to the environment and people as possible[25]. The consideration of natural processes in everything from the location and physical site's layout to technology and cultivation goals is integral. Knowledge codified from research informs decisions and plausibility of goals, including the sustainability of a pest management strategy. Refined information collection and processing systems have made traditional low-energy, low-tech practices substantially more precise and in the case of automation, stridently less physically laborious in many contexts while enhancing what an individual can achieve in the same period of time such as with crop scouting[26].

By exploiting natural weaknesses of pests, they can be disrupted more efficiently, and in ways that are less likely to be ecologically disruptive, such as with the application of targeted biocontrol organisms—though they must be investigated for efficacy and non-target effects that may make them ecologically disruptive as well. Understanding the strength and capabilities of pests informs and facilitates preventative and reactive measures. From an ecosystemic perspective, the community of commonly encountered pests in cannabis represents an ever-present threat mosaic across which individual pest species may require certain contingencies but the simultaneous prevention of and potential reaction to many different pest species complicates integrated pest management procedure.

The social aspect of IPM specifically and agriculture more broadly, is often overlooked or misunderstood, and represents an important obstacle to adaptive or well-intentioned reformative prospects of pest management because the gestalt complexity of thousands or millions of individuals and such collective desires and actions is necessarily cumbersome, guided by laws and similar ordinances that can at times be circumspect or trammelling. Cannaculture has been especially restricted in this regard across many societies, particularly in the last century or so, as prohibitionist fervor has been stoked despite historical utility in the production of textiles, food, and medical or spiritual products. However, it is at this interface between agricultural and societal mores that improvements to pest management strategies can be made more seamless at local, provisional, national, and international levels. Some pests cannabis shares with other crops like some Budworm Moths (*Helicoverpa zea, H. armigera*)[27] or various bud rot pathogens (*Botrytis* sp., *Fusarium* sp.)[28] that are so noxious and insidious that international cooperation and several billions of dollars of both preventative and curative management procedures annually still results in large crop losses with the potential for greater cost in new locations[29]. They are often invasive, and in exotic locations benefit from a dearth of stressors like predators and parasites that allow them to swell in number, sometimes displacing historical pests or native plant life which disrupts the ecological landscape over short or long periods.

As cannabis cultivation becomes more widely adopted both privately and commercially, its cultivation and physical presence will influence the local environment to include other commercial crops and vice versa. Diverging with a common ancestor of hop plants (*Humulus*), cannabis may be a choice vector for pests of this crop specifically, and certain pests like Hop Latent Viroid (HLVd) and Hop Powdery Mildew (*Podosphaera macularis*)[30] have been shown to use cannabis as a host obstreperously and modestly, respectively. Often, this is because close relatives share similar physiological processes, which may render some adaptations in one host population more compatible with the other. Hop Powdery Mildew is a designated quarantine pest in the pacific northwest of the United States of America where hop production is concentrated, with tight controls levied towards the movement of potential material containing it while cannacultural efforts have flourished in recent years. Recognition and acceptance as an agricultural commodity may give cannabis cultivation several protections and infrastructural support, but it may also impose limitations in cases where its commercial presence legitimately threatens other established crops. In a similar but more imperiled vein, Hop Latent Viroid already summarily truncated the hop cultivars suitable for cultivation as there are no conventional solutions available besides preventative measures like selecting tolerant or resistant plants. The infamously salient and seemingly capricious viroid has spread throughout commercial and private cannabis populations and may represent a contamination risk for hop cultivation or fledgling industries of cannaculture. Integrated Pest Management strategy would have to accommodate the reality of these societal pressures that exist outside the purely cultivation-oriented sphere. Influence of safeguarding measures on commercial or private cannabis cultivation could have a substantial effect on the richly ebullient cultural community that propelled cannabis from prohibition.

8.4 CANNABIS PEST THREAT MOSAIC—PREVENTING POTENTIAL PROBLEMS

Dozens of organisms colonize, consume, or otherwise make use of cannabis as a resource, from viral particles of several nanometers in length to large mammal several meters in height. The most diverse and common group of organisms that can be observed unaided are arthropods such as insects or mites. Pestiferous fungi, especially necrotrophic pathogens that fester on wounds or floral tissue such as *Botrytis*, *Diaporthe*, and *Sclerotinia* are highly destructive, while viruses and viroids maintain a swelling threat to cannaculture, chiefly in the form of Hop Latent Viroid, Beet Curly Top Virus[31], and Cucumber Mosaic Virus[32] among others though the Cannabis virome is poorly studied. Fortunately, modern techniques for the detection and prevention of viral and other microbial pathogens both at the level of cultivation and agronomic research allow economically efficient and strategically informed decision making that has never been available in agricultural history. Similarly, molecular technologies aimed at genetic sequencing elucidate the physiological development of cannabis populations cultivated currently and those that are not, allowing the assessment of various levels of tolerance to biotic and abiotic factors. Such assessments aid in breeding projects by succinctly confirming the presence and degree by which genes confer temperature, salinity, and specific pest tolerances, desirable growth, and metabolic profiles, as well as efficient ways to breed them into cultivars resulting in fortified populations that can be empirically tested. Until recently, purported proclamations of resistance to pests and environmental hazards have not been rigorously tested in cannabis as in other crops[33], relying primarily on the confidant assertions of cultivators, often leading to failures in crop performance that are increasingly untenable.

The arthropoda, for example, encompass organisms with an armored exoskeleton and often multiple legs; the most infamous in agriculture being the exceptionally diverse mites and insects which themselves branch from lineages that last shared a common ancestor more than 500 million years ago[34]. Predating land plants these groups, and in the case of insects, their direct ancestors, adapted relatively quickly to plant hosts[35], co-evolving with primitive ferns and conifers and strongly radiating with the success of flowering plants which make up most food crops and includes cannabis therein.

The primary herbivorous mites that concern cannabis crops are contained in three families: the broad mite of Tarsonemidae, the Hemp Russet Mite (*Aculops cannabicola*) of Eriophyidae, and several spider mites of Tetranychidae such as the Two-spotted Spider Mite (*Tetranychus urticae*) and Clover Mite (*Bryobia praetiosa*). Several other omnivorous and predatory mite groups like Phytoseiidae and Laelapidae contribute a spectrum of mutualistic and commensalic interactions in the form of regulating herbivore population and nutrient bioturbation. As research continues to compile more observations it is likely that additional taxa will be added.

Russet Mites (Eriophyidae) have some of the smallest genomes of any animal (e.g. *Aculops lycopersici*, 32.5 megabase pairs), with physiological traits differing greatly from related mites and other arachnids, specifically their worm-like body and four total legs at maturity, perhaps facilitated by this extreme genetic reduction. Many species have been documented in air traps and field extractions, with a tendency for high physical similarity and narrow host range sometimes necessitating the pairing of individual and host for accurate identification. Across their diverse hosts, russet mites are typically inoffensive, and colonization causes very little physiological trauma while a select fraction of species, mainly recorded on crop or pest plants, severely injure and warp their host tissue through phyotoxic saliva, the constituents and mode of action for which is being investigated but in some cases gnarls, puckers, or galls by manipulating hormonal signaling system.

Such is the case for the Hemp Russet Mite *Aculops cannabicola* which induces a range of crinkled, twisted, and stunted developments in leaf, stem, and flower tissue that remains permanently and may give the mites shelter from the elements and concealment from some potential predators like phytoseiid mites such as *Amblyseius swirskii* and *Neoseiulus cucumeris* commonly available as agents of biocontrol for various russet mite species. Individuals can reach maturity within several days in warm conditions and reproduce quickly, spreading short distances across the foliar surface and longer distances on fomites such as clothes and equipment, as well as in feral cannabis population reservoirs, cuttings, and with very similar bodies to other russet mites, long distance dispersal on wind currents is also common. Hemp Russet Mite colonies develop and radiate rapidly, with dense populations capable of severely damaging individual plants and can result in significant plant loss.

The advent of viroid detection in plants during the late 1970s radically changed perceptions about viral capabilities and origins[36], influencing theories about the prebiotic world in origin of life research[37], and while plants were the first documented reservoir for these RNA-based replicants plant pathogenic fungi appear to readily retain them[38]. Other viroid-like elements have been found associated with bacteria found in the human oral microbiome across earth[39], reinforcing their purported ubiquity, which could have implications for agricultural biosecurity such as with cannabis.

The Cocadviroid Hop Latent Viroid (HLVd) is highly pernicious and expansive, commonly associated with both hop and cannabis populations, but has also been detected in Tomato, Cucumber, Chrysanthemum, Tobacco, and Thale Cress[40]. Unique dilemmas associated with HLVd include its capricious onset and expression of symptoms, high prevalence and spread across cannabis cultivation zones[41], and recalcitrant solutions owing to its simple RNA structure that makes it virtually identical to plant RNA for the purposes of cellular recognition as well as targeting, making it difficult to render options that would affect viroid material without harming native plant RNA. With limited curative options like the proposition of thermotherapy, HLVd represents an existential threat to cannabis populations both rare and historical, and may truncate commercial and private cultivation towards a limited selection that are tolerant or resistant, such as with hop.

Cannaculture must reconcile with the diverse matrix of stressors—eustressors that have positive outcomes and distressors that have negative outcomes nominally[42], informed by ecological dynamics between cannabis and these biotic factors at a fine degree of molecular interaction across varied environmental contexts. For the most efficient mitigation of biotic stress, a holistic multifaceted approach that integrates robust deployment of synergetic preventative and reactive actions at various scales and disintegrates the disparate capabilities of pests and other stressors maximizes the effect of management strategy.

REFERENCES

(1) Monson, R. K.; Trowbridge, A. M.; Lindroth, R. L.; Lerdau, M. T. Coordinated Resource Allocation to Plant Growth–Defense Tradeoffs. *New Phytologist* **2022**, *233* (3), 1051–1066. https://doi.org/https://doi.org/10.1111/nph.17773.

(2) Clark, J. W.; Hetherington, A. J.; Morris, J. L.; Pressel, S.; Duckett, J. G.; Puttick, M. N.; Schneider, H.; Kenrick, P.; Wellman, C. H.; Donoghue, P. C. J. Evolution of Phenotypic Disparity in the Plant Kingdom. *Nat Plants* **2023**, *9* (10), 1618–1626. https://doi.org/10.1038/s41477-023-01513-x.

(3) Zhang, Q.; Chen, X.; Guo, H.; Trindade, L. M.; Salentijn, E. M. J.; Guo, R.; Guo, M.; Xu, Y.; Yang, M. Latitudinal Adaptation and Genetic Insights Into the Origins of Cannabis Sativa L. *Front Plant Sci* **2018**, *9*.

(4) Osterberger, E.; Lohwasser, U.; Jovanovic, D.; Ruzicka, J.; Novak, J. The Origin of the Genus Cannabis. *Genet Resour Crop Evol* **2022**, *69* (4), 1439–1449. https://doi.org/10.1007/s10722-021-01309-y.

(5) Hourfane, S.; Mechqoq, H.; Bekkali, A. Y.; Rocha, J. M.; El Aouad, N. A Comprehensive Review on Cannabis Sativa Ethnobotany, Phytochemistry, Molecular Docking and Biological Activities. *Plants* **2023**, *12* (6). https://doi.org/10.3390/plants12061245.

(6) Cranshaw, W.; Schreiner, M.; Britt, K.; Kuhar, T. P.; McPartland, J.; Grant, J. Developing Insect Pest Management Systems for Hemp in the United States: A Work in Progress. *J Integr Pest Manag* **2019**, *10* (1), 26. https://doi.org/10.1093/jipm/pmz023.

(7) Li, P.; Lu, Y.-J.; Chen, H.; Day, B. The Lifecycle of the Plant Immune System. *CRC Crit Rev Plant Sci* **2020**, *39* (1), 72–100. https://doi.org/10.1080/07352689.2020.1757829.

(8) Han, G.-Z. Origin and Evolution of the Plant Immune System. *New Phytologist* **2019**, *222* (1), 70–83. https://doi.org/https://doi.org/10.1111/nph.15596.

(9) Kemen, E. Microbe–Microbe Interactions Determine Oomycete and Fungal Host Colonization. *Curr Opin Plant Biol* **2014**, *20*, 75–81. https://doi.org/https://doi.org/10.1016/j.pbi.2014.04.005.

(10) Carrier, T. J.; Reitzel, A. M. The Hologenome Across Environments and the Implications of a Host-Associated Microbial Repertoire. *Front Microbiol* **2017**, *8*. https://doi.org/10.3389/fmicb.2017.00802.

(11) Batstone, R. T. Genomes within Genomes: Nested Symbiosis and Its Implications for Plant Evolution. *New Phytologist* **2022**, *234* (1), 28–34. https://doi.org/https://doi.org/10.1111/nph.17847.

(12) Leach, J.; Triplett, L.; Argueso, C.; Trivedi, P. Communication in the Phytobiome. *Cell* **2017**, *169*, 587–596. https://doi.org/10.1016/j.cell.2017.04.025.

(13) Mesny, F.; Hacquard, S.; Thomma, B. P. H. J. Co-evolution within the Plant Holobiont Drives Host Performance. *EMBO Rep* **2023**, *24* (9), e57455. https://doi.org/https://doi.org/10.15252/embr.202357455.

(14) Kuhn, H.; Kwaaitaal, M.; Kusch, S.; Acevedo-Garcia, J.; Wu, H.; Panstruga, R. Biotrophy at Its Best: Novel Findings and Unsolved Mysteries of the Arabidopsis-Powdery Mildew Pathosystem. *Arabidopsis Book* **2016**, *2016* (14). https://doi.org/10.1199/tab.0184.

(15) Rovenich, H.; Thomma, B. P. H. J. Battle of the Bugs: How an Oomycete Pathogen Shapes the Microbiota of Its Host. *New Phytologist* **2023**, *239* (6), 2064–2066. https://doi.org/https://doi.org/10.1111/nph.19133.

(16) Miozzi, L.; Vaira, A. M.; Catoni, M.; Fiorilli, V.; Accotto, G. P.; Lanfranco, L. Arbuscular Mycorrhizal Symbiosis: Plant Friend or Foe in the Fight Against Viruses? *Front Microbiol* **2019**, *10*. https://doi.org/10.3389/fmicb.2019.01238.

(17) Q, T. P.; Karthik, A. Biogeochemistry Goes Viral: Towards a Multifaceted Approach To Study Viruses and Biogeochemical Cycling. *mSystems* **2021**, *6* (5), 10.1128/msystems.01138-21. https://doi.org/10.1128/msystems.01138-21.

(18) Adkar-Purushothama, C. R.; Perreault, J.-P. Current Overview on Viroid–Host Interactions. *WIREs RNA* **2020**, *11* (2), e1570. https://doi.org/https://doi.org/10.1002/wrna.1570.

(19) Hao, F.; Ding, T.; Wu, M.; Zhang, J.; Yang, L.; Chen, W.; Li, G. Two Novel Hypovirulence-Associated Mycoviruses in the Phytopathogenic Fungus Botrytis Cinerea: Molecular Characterization and Suppression of Infection Cushion Formation. *Viruses* **2018**, *10* (5). https://doi.org/10.3390/v10050254.

(20) Jansson, J. K. Soil Viruses: Understudied Agents of Soil Ecology. *Environ Microbiol* **2023**, *25* (1), 143–146. https://doi.org/10.1111/1462-2920.16258.

(21) DeLong, J. P.; Van Etten, J. L.; Al-Ameeli, Z.; Agarkova, I. V; Dunigan, D. D. The Consumption of Viruses Returns Energy to Food Chains. *Proc Natl Acad Sci* **2023**, *120* (1), e2215000120. https://doi.org/10.1073/pnas.2215000120.
(22) Simon, J.-C.; d'Alençon, E.; Guy, E.; Jacquin-Joly, E.; Jaquiéry, J.; Nouhaud, P.; Peccoud, J.; Sugio, A.; Streiff, R. Genomics of Adaptation to Host-Plants in Herbivorous Insects. *Brief Funct Genomics* **2015**, *14* (6), 413–423. https://doi.org/10.1093/bfgp/elv015.
(23) Terra, W. R.; Ferreira, C. Evolutionary Trends of Digestion and Absorption in the Major Insect Orders. *Arthropod Struct Dev* **2020**, *56*, 100931. https://doi.org/https://doi.org/10.1016/j.asd.2020.100931.
(24) Shelomi, M.; Wipfler, B.; Zhou, X.; Pauchet, Y. Multifunctional Cellulase Enzymes Are Ancestral in Polyneoptera. *Insect Mol Biol* **2020**, *29* (1), 124–135. https://doi.org/https://doi.org/10.1111/imb.12614.
(25) Dara, S. K. The New Integrated Pest Management Paradigm for the Modern Age. *J Integr Pest Manag* **2019**, *10* (1), 12. https://doi.org/10.1093/jipm/pmz010.
(26) Fuentes, S.; Tongson, E.; Unnithan, R.; Gonzalez Viejo, C. Early Detection of Aphid Infestation and Insect-Plant Interaction Assessment in Wheat Using a Low-Cost Electronic Nose (E-Nose), Near-Infrared Spectroscopy and Machine Learning Modeling. *Sensors* **2021**. https://doi.org/10.3390/s21175948.
(27) Cosner, J. B.; Grant, J. F. Influence of Varieties of Hemp, Cannabis Sativa (Rosales: Cannabaceae), and Fertilization Rates on Damage Caused by Corn Earworm, Helicoverpa Zea (Lepidoptera: Noctuidae). *Environ Entomol* **2024**, *53* (1), 26–33. https://doi.org/10.1093/ee/nvad063.
(28) Punja, Z. K.; Ni, L. The Bud Rot Pathogens Infecting Cannabis (Cannabis Sativa L., Marijuana) Inflorescences: Symptomology, Species Identification, Pathogenicity and Biological Control. *Canadian J Plant Pathol* **2021**, *43* (6), 827–854. https://doi.org/10.1080/07060661.2021.1936650.
(29) Kriticos, D. J.; Ota, N.; Hutchison, W. D.; Beddow, J.; Walsh, T.; Tay, W. T.; Borchert, D. M.; Paula-Moreas, S. V; Czepak, C.; Zalucki, M. P. The Potential Distribution of Invading Helicoverpa Armigera in North America: Is It Just a Matter of Time? *PLoS One* **2015**, *10* (3), e0119618.
(30) Weldon, W.; Ullrich, M.; Smart, L.; Smart, C.; Gadoury, D. Cross-Infectivity of Powdery Mildew Isolates Originating from Hemp (Cannabis Sativa) and Japanese Hop (Humulus Japonicus) in New York. *Plant Health Prog* **2020**, *21*, 47–53. https://doi.org/10.1094/PHP-09-19-0067-RS.
(31) Creamer, R.; Simpson, A.; Rheay, H. T.; Brewer, C. E. Interactions of Beet Leafhopper (Hemiptera: Cicadellidae), Vector of Beet Curly Top Virus, and Hemp in New Mexico. *Environ Entomol* **2024**, *53* (1), 11–17. https://doi.org/10.1093/ee/nvad069.
(32) Miotti, N.; Passera, A.; Ratti, C.; Dall'Ara, M.; Casati, P. A Guide to Cannabis Virology: From the Virome Investigation to the Development of Viral Biotechnological Tools. *Viruses* **2023**, *15*. https://doi.org/10.3390/v15071532.
(33) Stack, G. M.; Cala, A. R.; Quade, M. A.; Toth, J. A.; Monserrate, L. A.; Wilkerson, D. G.; Carlson, C. H.; Mamerto, A.; Michael, T. P.; Crawford, S.; Smart, C. D.; Smart, L. B. Genetic Mapping, Identification, and Characterization of a Candidate Susceptibility Gene for Powdery Mildew in Cannabis Sativa. *Mol Plant-Microbe Interactions* **2023**, *37* (1), 51–61. https://doi.org/10.1094/MPMI-04-23-0043-R.
(34) Condamine, F. L.; Clapham, M. E.; Kergoat, G. J. Global Patterns of Insect Diversification: Towards a Reconciliation of Fossil and Molecular Evidence? *Sci Rep* **2016**, *6* (1), 19208. https://doi.org/10.1038/srep19208.
(35) Von Reumont, B.; Edgecombe, G. Crustaceans and Insect Origins. In: *Evolution and Biogeography of the Crustacea, The Natural History of Crustacea*, Volume 8; **2020**. Oxford University Press. https://doi.org/10.1093/oso/9780190637842.003.0005.
(36) Durzyńska, J.; Goździcka-Józefiak, A. Viruses and Cells Intertwined since the Dawn of Evolution. *Virol J* **2015**, *12* (1), 169. https://doi.org/10.1186/s12985-015-0400-7.
(37) Forgia, M.; Navarro, B.; Daghino, S.; Cervera, A.; Gisel, A.; Perotto, S.; Aghayeva, D. N.; Akinyuwa, M. F.; Gobbi, E.; Zheludev, I. N.; Edgar, R. C.; Chikhi, R.; Turina, M.; Babaian, A.; Di Serio, F.; de la Peña, M. Hybrids of RNA Viruses and Viroid-like Elements Replicate in Fungi. *Nat Commun* **2023**, *14* (1), 2591. https://doi.org/10.1038/s41467-023-38301-2.
(38) Wei, S.; Bian, R.; Andika, I. B.; Niu, E.; Liu, Q.; Kondo, H.; Yang, L.; Zhou, H.; Pang, T.; Lian, Z.; Liu, X.; Wu, Y.; Sun, L. Symptomatic Plant Viroid Infections in Phytopathogenic Fungi. *Proc Natl Acad Sci* **2019**, *116* (26), 13042–13050. https://doi.org/10.1073/pnas.1900762116.

(39) Zheludev, I. N.; Edgar, R. C.; Lopez-Galiano, M. J.; Peña, M. de la; Babaian, A.; Bhatt, A. S.; Fire, A. Z. Viroid-like Colonists of Human Microbiomes. *bioRxiv* **2024**, 2024.01.20.576352. https://doi.org/10.1101/2024.01.20.576352.

(40) Atallah, O. O.; Yassin, S. M.; Verchot, J. New Insights into Hop Latent Viroid Detection, Infectivity, Host Range, and Transmission. *Viruses* **2024**, *16* (1). https://doi.org/10.3390/v16010030.

(41) Punja, Z. K.; Wang, K.; Lung, S.; Buirs, L. Symptomology, Prevalence, and Impact of Hop Latent Viroid on Greenhouse-Grown Cannabis (Cannabis Sativa L.) Plants in Canada. *Canadian J Plant Pathol* **2024**, *46* (2), 174–197. https://doi.org/10.1080/07060661.2023.2279184.

(42) Berni, R.; Thiry, M.; Hausman, J.-F.; Lutts, S.; Guerriero, G. Eustress and Plants: A Synthesis with Prospects for Cannabis Sativa Cultivation. *Horticulturae* **2024**, *10* (2). https://doi.org/10.3390/horticulturae10020127.

9 Comprehensive Genomic Analysis and Genetic Diversity Assessment of Cannabis Cultivars by Whole Genome Sequencing

Athar Hussain, Riaz Ahmad, Ajith Anand and Angel Fernandez Marti

9.1 INTRODUCTION

Cannabis is an erect annual herb that belongs to the family Cannabaceae (Malabadi, Kolkar, & Chalannavar, 2023). Its cultivation, initially documented in China as a source of fiber, has spread throughout the world. Cannabis is one of the oldest plants, used medicinally for over 10,000 years in China, as documented by Emperor Shen Neng. Recently, the plant has regained interest and popularity for its potential therapeutic effects (Zhang et al., 2020). There are many ways to classify cannabis species. From a practical perspective, cannabis is classified into four types: wild, fiber, oilseed, and psychoactive types. In this chapter, when referring to the plant, we will refer to it using its scientific genus name, *Cannabis*, and species-specific terminologies such as *indica*, *sativa*, *hemp,* and *drug type (marijuana)*. From a legal and regulatory standpoint, most countries where cannabis is produced choose to regulate it according to the amount of psychoactive cannabinoid delta-9-tetrahydrocannabinol (THC) produced plants with less than 0.3% of THC are classified as *hemp*, while plants that produce 0.3% or more THC are categorized as *drug type* (M. Hesami et al., 2021; Monthony, Page, Hesami, & Jones, 2021). Taxonomically, cannabis is recognized as either hemp or marijuana (Small, 2017). Based on the degree of domestication, cannabis is classified as wild, domesticated, and intermediate types (Small, 2017). Studies show significant genetic diversity between marijuana and hemp lines, with *Cannabis sativa* less variable and relatively more homogeneous than *Cannabis indica* (Lynch et al., 2016). For several decades, scientists have been investigating tools to explore and analyze the cannabis genome. Molecular markers have been used to identify sexual phenotypes, determine chemotype-defining factors, and understand the genetic diversity of *Cannabis* species (Zhang et al., 2020).

Genetic diversity in plants is usually estimated using DNA sequences (polymorphisms among varieties) and cytological and morphological markers (El-Mansy et al., 2021). However, morphological characteristics are often influenced by the environment, making molecular markers more stable and popular (Nadeem et al., 2018). These markers provide a reliable method to assess the evolution and conservation of varieties. Events such as inbreeding and evolution can alter allele frequency and reduce genetic diversity, so it is crucial to accurately estimate the correlation among different germplasm resources (Govindaraj, Vetriventhan, & Srinivasan, 2015). This ensures high-efficiency utilization and management and maintains adequate genetic variability for breeding various plant varieties.

DOI: 10.1201/9781003386438-9

Genetic diversity and population structure analyses are commonly used to examine various plant species. For instance, an analysis of 1,151 ramie germplasms using SSR and phenotypic markers revealed that the genetic diversity of wild germplasms is higher than that of domesticated ones, clustering the population into two subpopulations (Zhang et al., 2020). Similarly, the genetic diversity of 50 populations of confectionery sunflowers in Iran showed they could be subdivided into two subpopulations (Jannatdoust et al., 2016). Additionally, a population structure analysis of 109 Brazilian common bean accessions indicated two distinct subpopulations (Valentini, Gonçalves-Vidigal, Elias, Moiana, & Mindo, 2018).

The genetic diversity of different cannabis germplasm resources has also been estimated using molecular markers. For example, a single-hexanucleotide short tandem repeat (STR) named NMI101 was used to analyze the distribution of 93 processed seeds (Zhang et al., 2020). An analysis of 81 marijuana and 43 hemp samples using single nucleotide polymorphisms (SNPs) revealed significant genetic differences at the genome-wide level, showing that hemp is genetically more similar to the *Cannabis indica* type than the *Cannabis sativa* type (Zhang et al., 2020). Expressed sequence tag SSRs (EST-SSRs) were used to assess the genetic diversity of 115 hemp germplasm resources, which were divided into four groups (Gao et al., 2014). SSR molecular markers are proven to be highly polymorphic, with genomic SSRs being more polymorphic and stable than EST-SSRs (Ahmad, Wang, Pan, Sharif, & Gao, 2018; Gupta et al., 2003).

Next-generation sequencing (NGS) plays a critical role in the whole genome comparison and comparative genomics of cannabis cultivars and strains. NGS technologies allow for the comprehensive analysis of the *Cannabis* genome at a high resolution, enabling researchers to identify genetic variations and understand the genetic relationships between different strains and cultivars. This technology facilitates the discovery of novel genes, regulatory elements, and genetic markers that can be used to improve cannabis breeding programs. NGS also aids in studying the evolutionary history and genetic diversity of cannabis, providing insights into the adaptation and domestication processes of the plant. While many studies focus on the classification of cannabis, including population structure and genealogical classification, the relationship between Chinese and foreign strains remains unclear. To identify the relatedness between strains accurately, more precise molecular markers are needed to analyze their population diversity. NGS enhances this process by offering a detailed and comprehensive view of the genetic makeup of cannabis, thereby supporting more effective and informed breeding and conservation strategies (M. Hesami et al., 2020; Hurgobin et al., 2021).

9.2 MATERIALS AND METHODOLOGY

Total DNA extraction from 10 THC cannabis cultivars and one hemp cultivar was extracted (Candy, GMO, Gelato, Garlic Bud, Silver, Crescendo, Chili Verd, Blue Dream, Banana, and Hemp). Initially, the plant leaves were ground into a fine powder using liquid nitrogen to facilitate cell disruption (Doyle & Doyle, 1987; Sika et al., 2015). The total genomics DNA was extracted using the CTAB method (Li, Wang, Yu, Wang, & Zhou, 2013). The quality and quantity of the extracted DNA were assessed using spectrophotometric methods and agarose gel electrophoresis (Lucena-Aguilar et al., 2016). For resequencing, entire genomes of plants and animals using Illumina platforms through the sequencing by synthesis (SBS) mechanism, the methodology adopted a systematic approach encompassing DNA quantification, library preparation, clustering, and sequencing (Lucena-Aguilar et al., 2016). Initially, DNA integrity was examined via agarose gel electrophoresis (Sambrook & Russell, 2001), and concentration was determined using the Qubit® DNA Assay Kit (Scientific, 2010). Genomic DNA was then fragmented to uniform sizes, processed, and ligated with Illumina sequencing adapters (Teer et al., 2010). PCR amplification enriched DNA fragments, yielding final library constructs that were purified using the AMPure XP system (Henn et al., 2010). Quality control measures, including size distribution analysis and real-time PCR for molarity measurements, ensured uniformity and integrity (Agilent Technologies, 2016). Libraries meeting predefined criteria

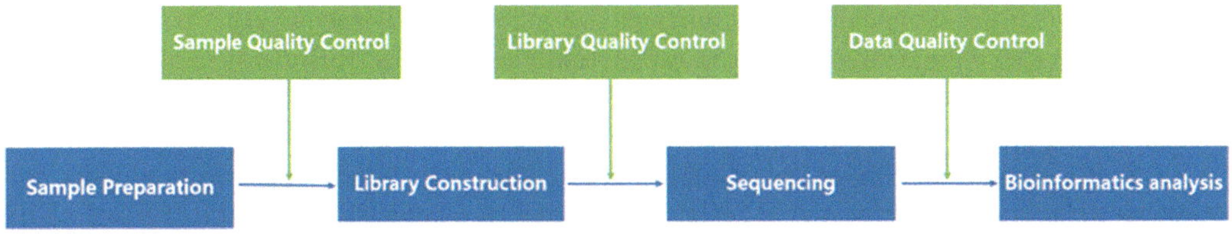

FIGURE 9.1 Graphical representation of sample preparation to resequencing pipeline.

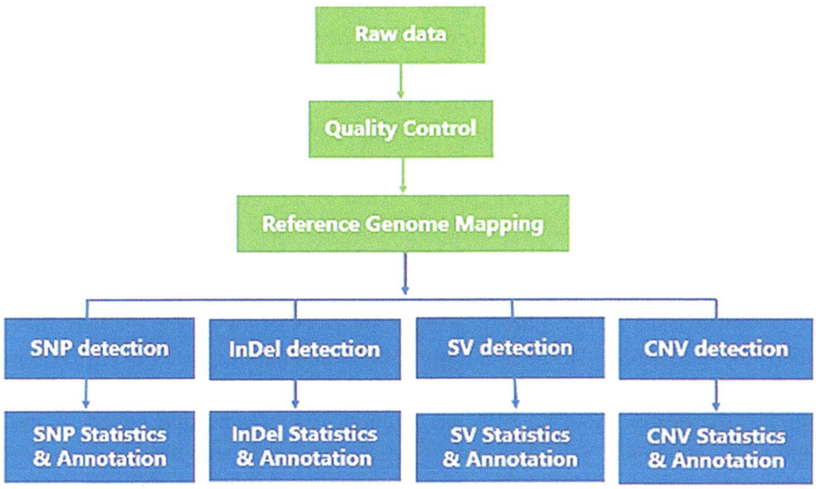

FIGURE 9.2 Bioinformatics analysis pipeline flow chart.

were sequenced on Illumina platforms, ensuring the generation of high-quality sequencing data (Bentley et al., 2008) (**Figure 9.1**).

The bioinformatics analysis pipeline employed various software tools for comprehensive processing of sequencing data, starting with quality control using Fastp (de Sá et al., 2018). Clean reads were mapped to the reference genome assembly *Cannabis sativa* (GCF_900626175.2) using BWA, producing SAM files (Woods, 2023), which were then converted to BAM format, sorted, and duplicates removed using SAMtools (Lo Giudice, Tangaro, Pesole, & Picardi, 2020). Picard was used to merge multiple BAM files into a single BAM file per sample (Lo Giudice et al., 2020). Detection and filtration of SNPs and InDels were conducted using SAMtools, structural variants identified using Breakdancer (Chen et al., 2009), and CNVs detected with CNVnator (Kim, Park, Jo, Koo, & Park, 2017). Finally, variations were annotated using ANNOVAR (Yang & Wang, 2015) (**Figure 9.2**).

9.3 RESULTS OF GENOMICS STUDIES OF CANNABIS CULTIVARS

9.3.1 Genetic Makeup and Reference Genome Mapping of Cannabis Strains

In our study of cannabis genetic diversity, we have analyzed 11 cultivars: Candy, GMO, Gelato, Garlic Bud, Silver, Crescendo, Chili Verd, Blue Dream, Banana, and Hemp. In these cultivars, hemp had the highest GC content at 36.025%, followed closely by Gelato at 35.855% and Garlic Bud at 35.735% (**Figure 9.3**). On the other hand, Chili Verd had the lowest GC content at 35.285%, with Blue Dream and GMO following closely behind. The mapping rate percentage refers to how well the DNA sequences align with a reference genome, *Cannabis sativa* (GCF_900626175.2). Banana

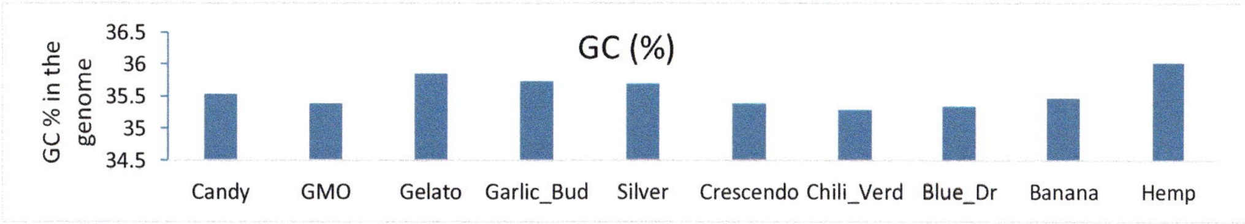

FIGURE 9.3 GC % in the genome of Cannabis cultivars.

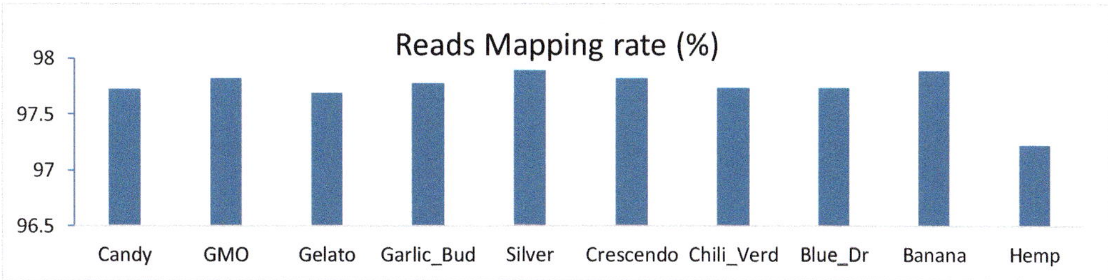

FIGURE 9.4 Reads mapping rate of cannabis cultivars with reference genomes.

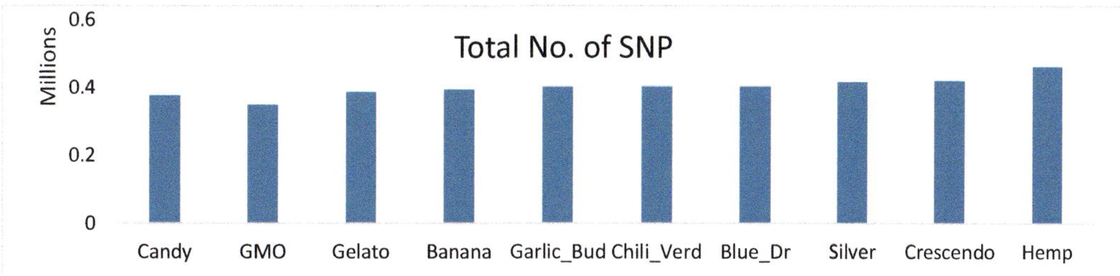

FIGURE 9.5 Total SNPs found in each cultivar w.r.t. reference genome.

had the highest mapping rate at 97.89%, followed by Silver and Crescendo at 97.89% and 97.82%, respectively. Hemp had the lowest at 97.23% (**Figure 9.4**).

9.3.1.1 Single Nucleotide Polymorphism (SNP) Diversity Among Cultivars

In the SNP detection and annotation, we found variations among different cannabis cultivars. Hemp showed the highest total mutations, with 3,135,943, followed by Silver with 2,622,860, and Blue Dream with 2,580,473. On the contrary, the lowest total mutations were observed in GMO, with 2,091,373, followed by Candy with 2,313,868, and Gelato with 2,091,373 (**Figure 9.5**).

Regarding mutations present upstream of genes, hemp exhibited the most with 132,221, followed by Silver with 112,601, and Crescendo with 109,162. Conversely, GMO had the fewest upstream mutations, with 91,015, followed by Gelato with 98,074, and Candy with 99,671 (**Figure 9.6**).

In terms of downstream mutations, hemp had the highest count, with 126,956, followed by Silver with 107,320, and Crescendo with 103,660. GMO again had the lowest count, with 86,147, followed by Gelato with 93,505, and Candy with 95,281 (**Figure 9.6**).

Exonic mutations, which occur within coding regions, were most abundant in hemp, with 211,676, followed by Silver with 172,257, and Crescendo with 167,338. Conversely, GMO had the fewest exonic mutations, with 142,024, followed by Gelato with 152,731, and Chili Verde with 155,932. Similarly, non-synonymous mutations, which can alter protein function, were highest in hemp, with

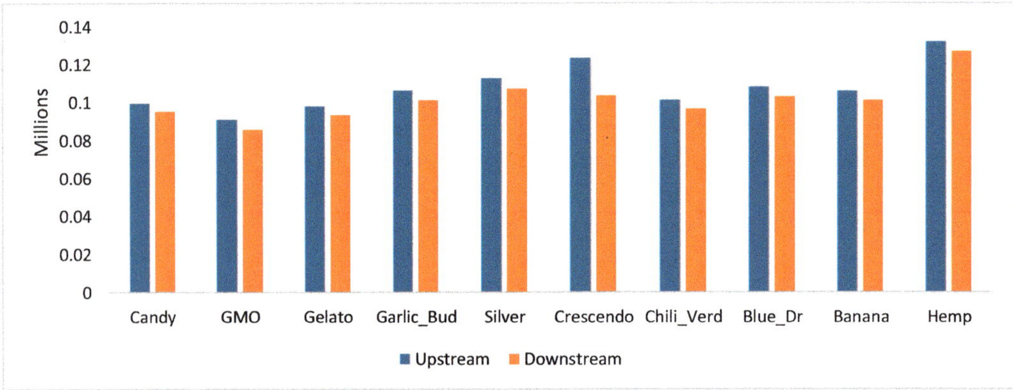

FIGURE 9.6 SNPs found in the upstream and downstream region of genes in cannabis cultivars w.r.t. reference genome.

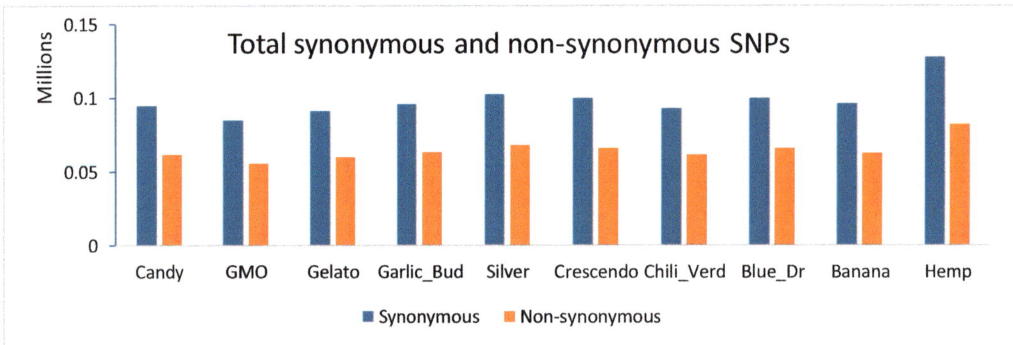

FIGURE 9.7 Synonymous and non-synonymous SNPs found in the exonic region of genes in cannabis cultivars w.r.t. reference genome.

82,307, followed by Silver with 68,284, and Blue Dream with 66,247. GMO had the fewest non-synonymous mutations, with 55,842, followed by Gelato with 60,228, and Candy with 61,595. This data sheds light on the genetic diversity present within these cannabis cultivars (**Figure 9.7**).

Single nucleotide polymorphisms (SNPs) can result in two types of base substitutions: transversions (purine to pyrimidine or vice versa) and transitions (purine to purine or pyrimidine to pyrimidine changes) each having distinct impacts on genomic stability and function. In SNP mutation frequency, which includes mutations such as T:A>G:C, T:A>A:T, C:G>T:A, C:G>G:C, and C:G>A:T, we observed that hemp exhibited the highest frequency, followed by Silver and Blue Dream genomes. Conversely, the lowest frequencies were observed in GMO, followed by Gelato and Candy genomes. Specifically, the C:G>A:T mutation was more prevalent in Candy, followed by Gelato, among the genomes analyzed (**Figure 9.8**).

Furthermore, the heatmap illustrates the genetic relatedness among various cannabis cultivars based on their distance matrix. Each value in the heatmap represents the genetic distance between pairs of cultivars, with lower values indicating higher relatedness and vice versa. Starting with Banana, it is most closely related to Gelato (0.31), Chili Verd (0.33), and Garlic Bud (0.34), with the least genetic similarity to Hemp (0.37). Blue Dream shows its closest genetic relationship with Gelato (0.31) and Crescendo (0.32), while also being least related to Hemp (0.37). Candy shares the greatest genetic similarity with Crescendo (0.29) and the least with Hemp (0.38). Chili

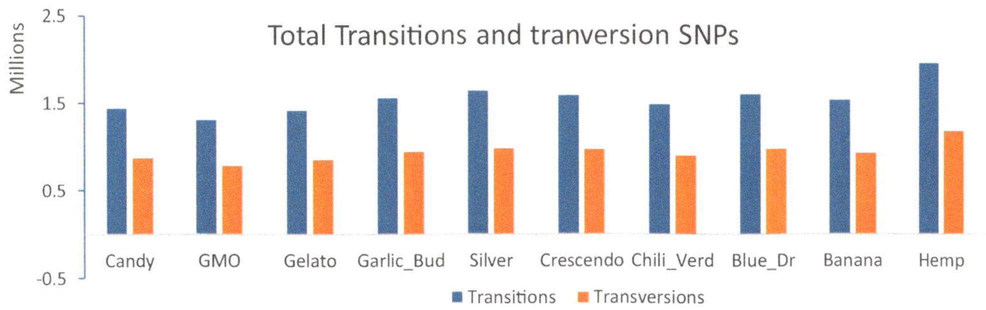

FIGURE 9.8 Transversions and Transitions SNPs found in the exonic region of genes in cannabis cultivars w.r.t. reference genome.

Verd, on the other hand, is also most similar to Gelato (0.28) and Garlic Bud (0.30), and least to Hemp (0.34). Crescendo has its closest genetic relationship with Gelato (0.25) and Garlic Bud (0.29), while being most distinct from Hemp (0.34). Garlic Bud is notably similar to Gelato (0.27) and Crescendo (0.29), with the least similarity to Hemp (0.36). Gelato shows a close genetic relationship with Crescendo (0.25), and Garlic Bud (0.27) and the least with Hemp (0.32). Hemp stands out as the most genetically distinct cultivar, showing the highest genetic distance values with all other cultivars, including 0.37 with Banana, 0.37 with Blue Dream, 0.38 with Candy, 0.34 with Chili Verd, 0.34 with Crescendo, 0.36 with Garlic Bud, and 0.32 with Gelato. Overall, the genetic distances range between 0.25 and 0.38, indicating a moderate level of genetic similarity among the cultivars, with hemp being an outlier (**Figure 9.9**).

Similarly, the phylogenetic tree visualizes the evolutionary relationships among various cannabis cultivars, shedding light on their genetic divergence and relatedness. The tree indicates that Banana is relatively distinct but shares a common ancestor with Blue Dream, Candy, and Crescendo, suggesting a closer genetic relationship with these cultivars. Blue Dream, positioned closely to Banana, also shows a high degree of relatedness to Candy and Crescendo, indicating a shared evolutionary path and significant genetic material in common. Candy, closely related to Crescendo, and Gelato shares a branching point with these cultivars, suggesting a recent common ancestor. Chili Verd branching out earlier, indicating some genetic divergence from the other cultivars.

However, hemp stands out as the most distinct branch, diverging early from all other cultivars, indicating a significantly different genetic profile. This aligns with hemp's known genetic and phenotypic differences. Moreover, the tree highlights two main clusters: one containing Banana, Blue Dream, Candy, Crescendo, Gelato and Garlic Bud which are closely related, and another containing Chili Verd, and Hemp, which show more genetic divergence but still share a common ancestor. The phylogenetic information is crucial for breeding programs, genetic conservation, and studies on cannabis diversity. The close genetic relationships within clusters suggest potential candidates for crossbreeding to enhance specific traits, while the distinct genetic profile of hemp could be valuable for breeding programs aimed at introducing unique traits or studying genetic diversity (**Figure 9.10**).

9.3.1.2 Insertion and Deletions (InDels) Diversity Among Cultivars

Our examination of mutations, insertions, and deletions, both upstream and downstream, revealed interesting findings. Hemp exhibited the highest counts across these categories, followed by Crescendo and Silver, while Candy, GMO, and Gelato showed the lowest counts. Specifically, total mutations were most prevalent in Hemp (463,313), followed by Crescendo (421,980) and Silver (415,517), with GMO (348,763), Candy (377,002), and Gelato (386,066) exhibiting lower counts.

When considering total insertions, Hemp again showed the highest number (236,077), followed by Crescendo (219,597) and Silver (217,679), whereas GMO (183,163), Candy (197,662), and

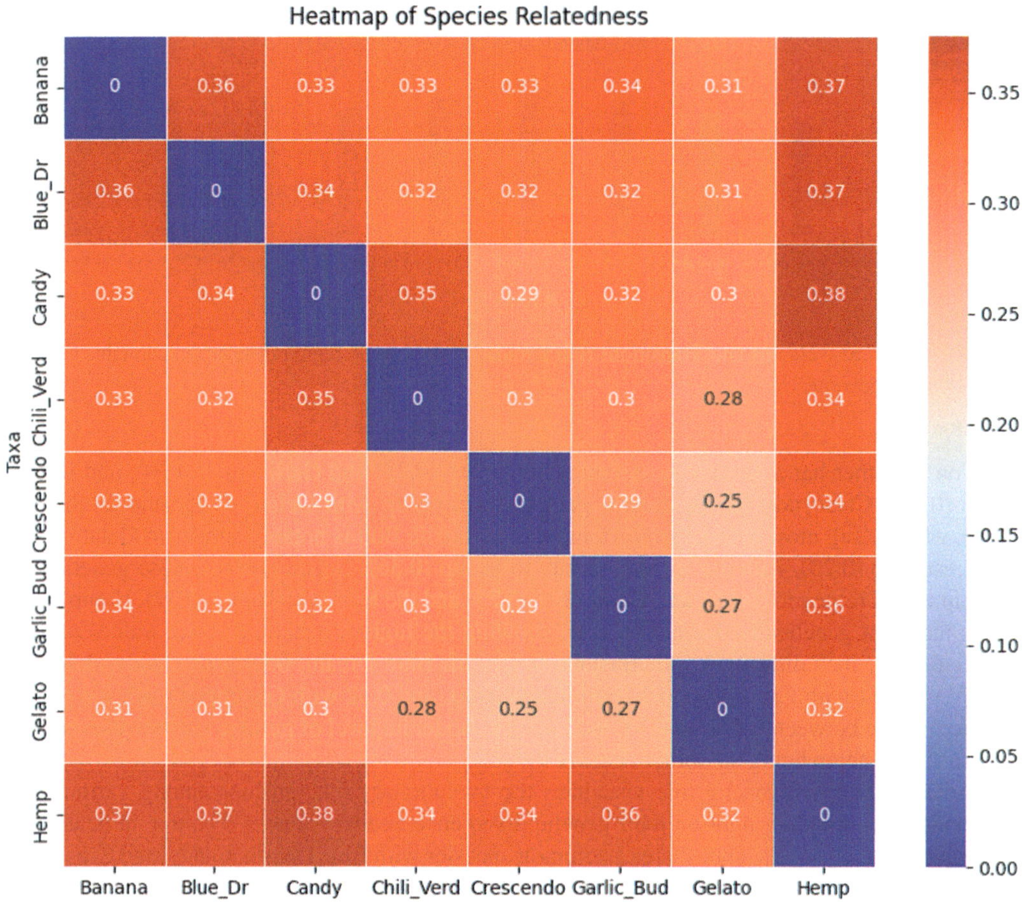

FIGURE 9.9 SNP heatmap distance matrix showing genetic relatedness among various cannabis cultivars.

Gelato (201,340) had fewer insertions. Similarly, total deletions were most frequent in Hemp (226,527), followed by Crescendo (201,841) and Silver (197,221), while GMO (165,214), Candy (178,858), and Gelato (184,373) had lower counts (**Figure 9.11**).

Moreover, upstream mutations were most abundant in Hemp (34,047), followed by Crescendo (30,982) and Silver (30,752), with lower counts in GMO (26,261), Candy (28,470), and Gelato (28,665). Likewise, downstream mutations were highest in Hemp (29,913), followed by Silver (26,609) and Crescendo (26,565), while GMO (22,297), Candy (24,383), and Gelato (24,383) exhibited fewer downstream mutations (**Figure 9.12**).

Additionally, an InDels (Insertions/Deletions) distance heatmap reveals the genetic relatedness between cannabis cultivars. The values in the matrix represent the InDels distance between each pair of cultivars. Lower values indicate a smaller number of insertions or deletions and thus more genetic similarity, and higher values indicate more insertions/deletions and less genetic similarity. Moreover, the heatmap uses a color scale where red indicates higher InDels distance (less related) and blue indicates lower InDels distance (more related).

Visualizing the genetic relatedness of cannabis cultivars, this heatmap depicts InDels distances. Banana shares the closest genetic ties with Gelato (0.26), Garlic Bud, Crescendo, and Chili Verd (0.27), while Hemp stands out as the most distinct (0.33). Similarly, Blue Dream finds its closest relative in Gelato and Garlic Bud (0.25) but farthest from Hemp (0.33). The pattern continues with Candy (closest to Gelato at 0.23 and Crescendo at 0.24). Gelato itself shares the closest bond with

FIGURE 9.10 Phylogenetic tree constructed from SNP data showing relatedness among eight genomes of cannabis cultivars.

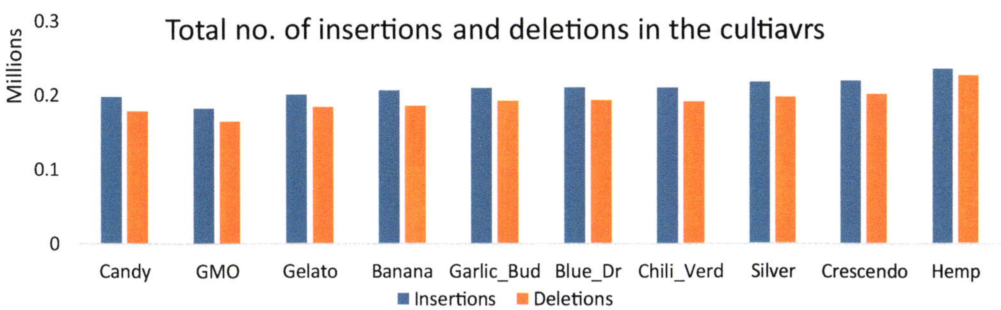

FIGURE 9.11 Total no. of insertions and deletions found in the cannabis cultivars w.r.t. reference genome.

Crescendo and Garlic Bud (0.20). Interestingly, InDels distances here are larger than those typically seen in SNP matrices, suggesting significant insertions or deletions between cultivars. This genetic diversity within the cultivars (0.20 to 0.33) offers exciting potential for breeding programs. Hemp's distinct genetic makeup (0.29 to 0.33) makes it valuable for studying cannabis diversity, while the close relationships between other cultivars (0.20 to 0.27) open doors for targeted crossbreeding to cultivate desired traits (**Figure 9.13**).

Similarly, the phylogenetic analysis of InDels sequences revealed the genetic relationships among eight cannabis cultivars. The resulting phylogenetic tree demonstrated a moderate degree of kinship between all cultivars. Interestingly, the branching pattern identified two distinct clusters. The first cluster grouped Banana, Hemp, and Chili Verde, suggesting a close evolutionary relationship and

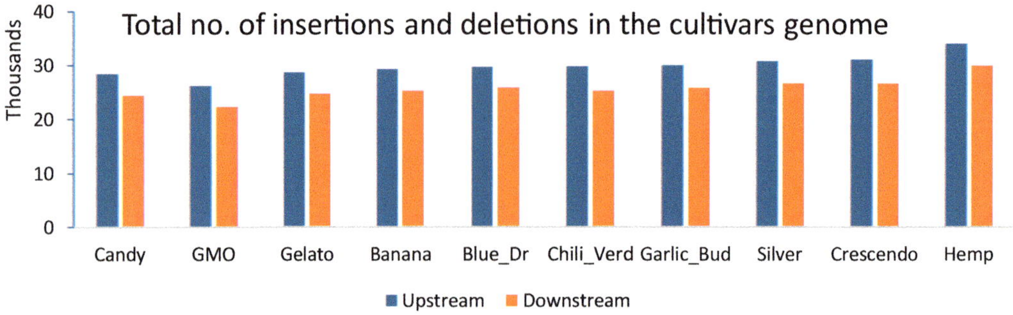

FIGURE 9.12 Total no. of insertions and deletions found in the upstream and downstream region of genes in cannabis cultivars w.r.t. reference genome.

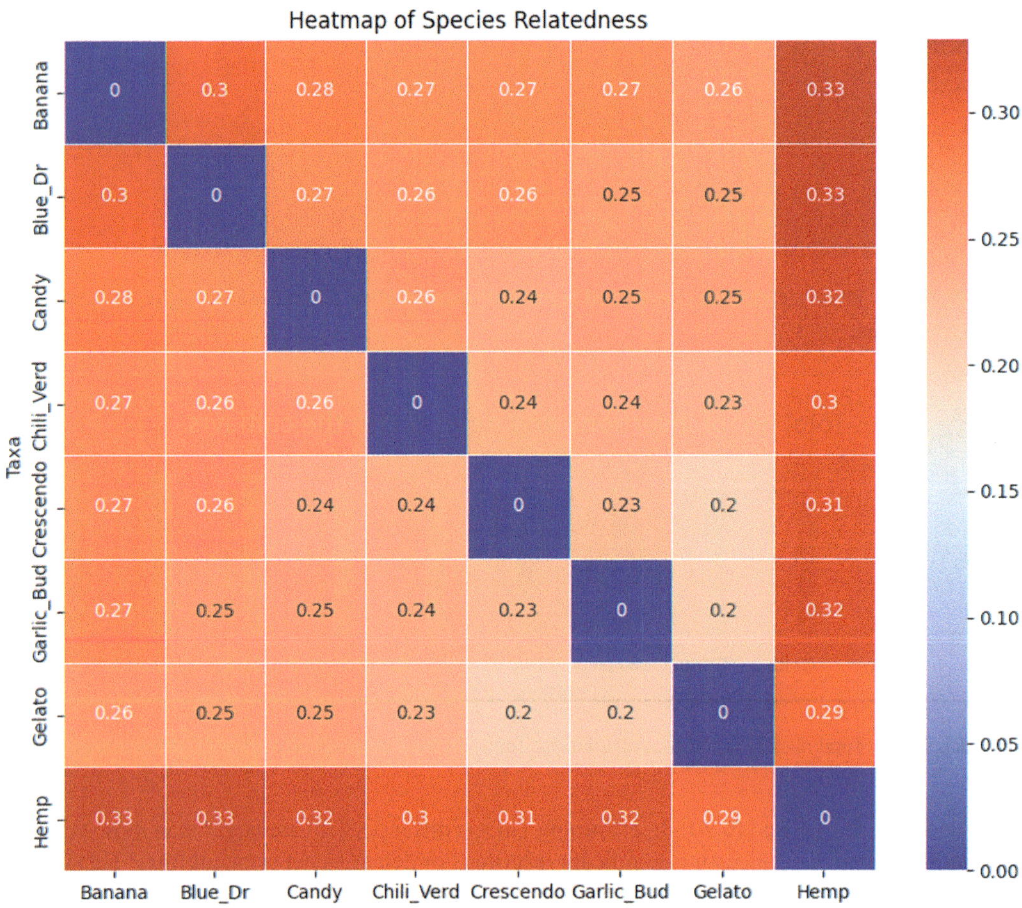

FIGURE 9.13 Heatmap of InDels distance among cannabis cultivars.

a likely shared recent ancestor. The second cluster, comprised of Crescendo, Gelato, and Garlic Bud, indicated a more recent divergence from the common ancestor compared to the first cluster. Notably, Blue Dream and Candy diverged from the ancestral lineage earlier than the other cultivars, suggesting a more distant relationship. This phylogenetic tree provides valuable insights into the

FIGURE 9.14 Phylogenetic tree constructed from InDels data showing relatedness among eight genomes of cannabis cultivars.

genetic diversity of cannabis cultivars, highlighting both shared ancestry and unique evolutionary trajectories within the studied varieties (**Figure 9.14**).

9.3.1.3 Structural Variants (SV) Diversity Among Cultivars

The analysis of structural variants (SVs) shows that the highest total SVs were observed in Crescendo (30,449), followed by Chilli Verd (30,271) and Hemp (29,935), while the lowest were in GMO (23,554), Garlic Bud (24,105), and Gelato (24,991). Upstream SVs were most frequent in Hemp cultivars (1,045), followed by Chilli Verd (915) and Candy (862), with the lowest counts in Garlic Bud (630), Silver (658), and Banana (676). Downstream SVs were highest in Hemp (842), followed by Candy (729) and Chilli Verd (724), and lowest in Garlic Bud (522), Banana (564), and Silver (565) (Figure 9.15).

SV insertions were most common in Silver (56), Banana (52), and Candy (27), and least common in GMO (1), Hemp (1), and Chilli Verd (3). Deletions in SVs were most frequent in Hemp (12,694), Chilli Verd (11,782), and Candy (11,052), while the lowest were in Garlic Bud (8,252), Banana (8,854), and Gelato (9,022). SV inversions were highest in Chilli Verd (2,956), Candy (2,801), and Hemp (2,664), and lowest in Garlic Bud (2,258), Gelato (2,345), and Banana (2,499). SV intra-chromosomal translocations (ITX) were most present in Chilli Verd (3,129), followed by Silver (3,075) and Candy (3,040), while the lowest were in Garlic Bud (2,612), GMO (2,707), and Gelato (2,813). SV inter-chromosomal translocations (CTX) were most frequent in Hemp (11,698), Banana (13,427), and Blue Dr (11,026), with the fewest in Candy (8,157), GMO (8,950), and Gelato (10,805) (Figure 9.16).

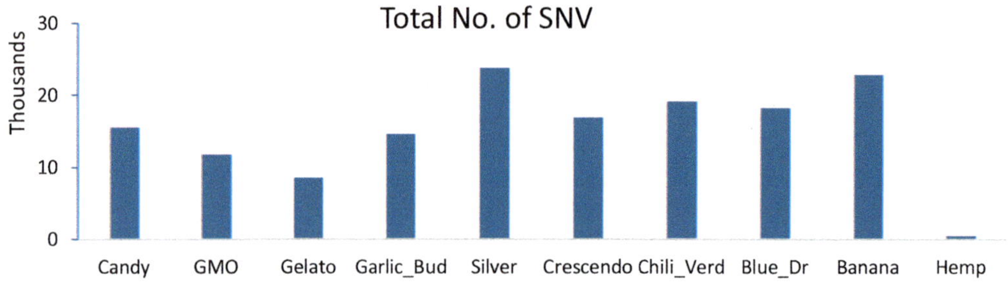

FIGURE 9.15 No. of structural variants found in different cultivars of cannabis w.r.t. reference genome.

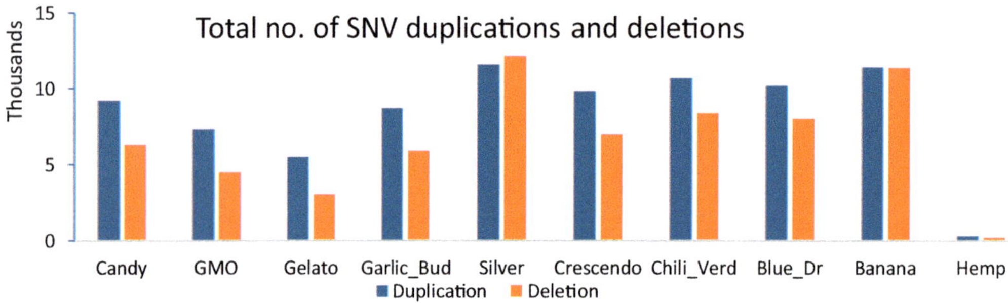

FIGURE 9.16 Total no. of structural variants (SNV) duplications and deletions found in the genome of cannabis cultivars w.r.t. reference genome.

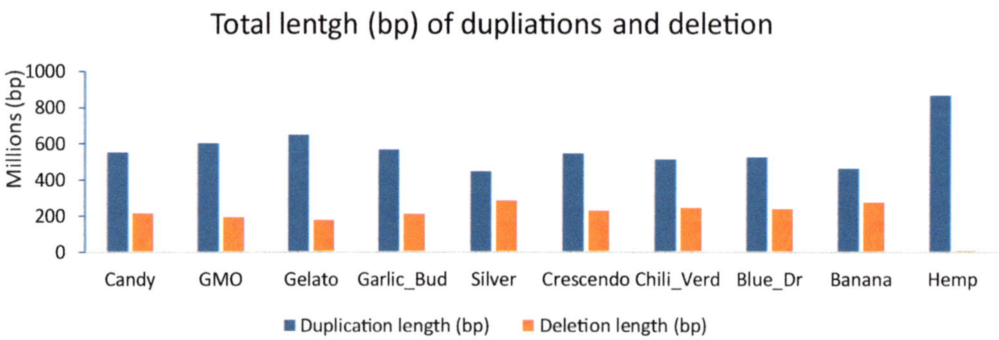

FIGURE 9.17 Total length of duplications and deletion in base pair (bp) found in genome of cannabis cultivars w.r.t. reference genome.

Moreover, four random cultivars; Banana, Blue Dream, Candy, and Chili Verd, we observed distinct length distribution patterns across various ranges. For sequences between 1–100 base pairs (bps), all four cultivars exhibited almost identical lengths. In the 100–200 bps range, Banana showed the highest length, while Candy had the lowest. For lengths between 200–700 bps, Candy had the highest length distribution. However, in the 800–1200 bps range, Candy's length started to align more closely with the other samples. For sequences longer than 1200 bps, Candy's length was lower compared to the other samples, with Chili Verd showing the highest length in this range (Figure 9.17).

9.3.1.4 Copy Number Variants (CNV) Diversity Among Cultivars

The analysis of copy number variations (CNV) mutations revealed that the highest total CNVs were found in Silver (23,748), followed by Chilli Verd (19,046) and Banana (2,282), while the lowest were in GMO (180), Hemp (523), and Gelato (85,520). For upstream CNVs, Silver had the highest count (471), with Banana (414) and Chilli Verd (348) following, and the lowest counts were observed in Hemp (0), Gelato (143), and GMO (173). Regarding downstream SNVs, Silver again had the most (355), followed by Banana (343) and Chilli Verd (282), with the fewest in Hemp (3), Gelato (100), and GMO (159) (Figure 9.18).

Duplication events were most frequent in Silver (11,604), Banana (11,430), and Chilli Verd (10,670), while the fewest were in Hemp (319), Gelato (5,481), and GMO (7,318). Deletions were highest in Silver (12,144), followed by Banana (11,390) and Chilli Verd (8,376), with the lowest in Hemp (204), Gelato (3,041), and GMO (4,486). The length of duplications was highest in Hemp (866,007,600), followed by Gelato (646,015,700) and GMO (605,164,700), while the shortest were in Silver (446,731,200), Banana (462,492,600), and Chilli Verd (510,815,700). Lastly, the deletion lengths in CNVs were greatest in Silver (282,962,100), Banana (274,406,700), and Chilli Verd (242,449,200), and smallest in Hemp (6,095,900), Gelato (172,881,000), and GMO (191,512,500) (Figure 9.19).

These findings shed light on the genetic diversity and variation present within different cannabis strains, providing valuable insights for further research and breeding efforts.

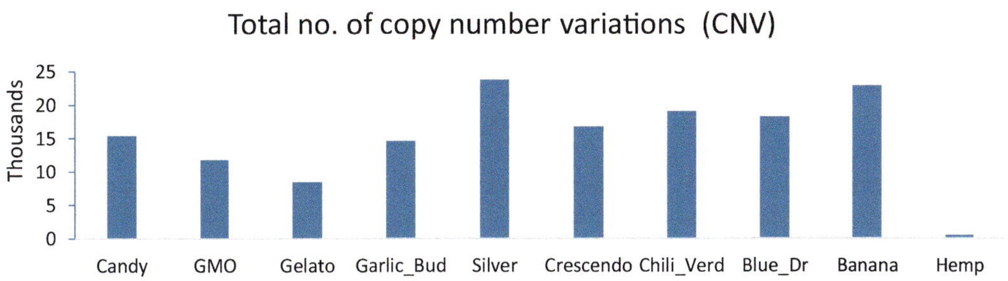

FIGURE 9.18 Total copy number variants found in the genome of cannabis cultivars w.r.t. reference cannabis genome.

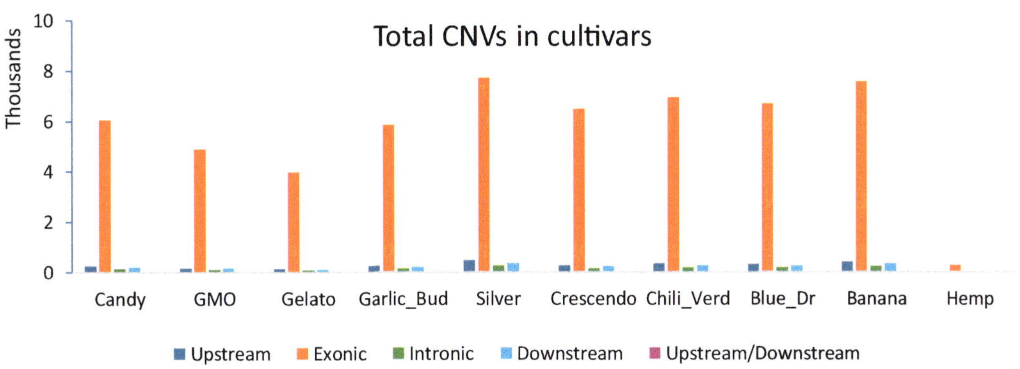

FIGURE 9.19 Total CNVs found in upstream, downstream, intronic and exonic regions of cannabis cultivars w.r.t. reference genome.

9.4 DISCUSSION

Cannabis, an ancient plant with documented medicinal use dating back over 10,000 years, has garnered renewed interest for its therapeutic potential in recent years (Pisanti & Bifulco, 2019). Our study, alongside existing literature, contributes to a deeper understanding of cannabis genetics, highlighting its genetic diversity, taxonomic classification, and therapeutic implications. Taxonomically, cannabis is classified into various types, including wild, fiber, oilseed, and psychoactive types (Small, 2017). Although it's a significant plant species, but the legal restrictions severely hindered its breeding and development, limiting its market competitiveness to less than 200,000 hectares globally. The recent relaxation of drug laws increased interest in its agricultural potential, though progress remained slow due to limited genomic understanding. A comprehensive cannabis pangenome study, involving 181 new and 12 previously released genomes, revealed high genetic diversity and structural variation, identifying key gene variants linked to cannabinoid production and suggesting untapped potential and the likely existence of wild relatives in Asia (Lynch et al., 2024).

Molecular markers, such as single nucleotide polymorphisms (SNPs) and short tandem repeats (STRs), have been instrumental in deciphering the genetic diversity and population structure of cannabis germplasm resources (Zhang et al., 2020). Our study corroborates these findings, revealing significant genetic variations among ten cannabis cultivars through comprehensive analysis of SNPs, insertions and deletions (InDels), copy number variations (CNVs), and structural variants (SVs).

The identification of SNPs and InDels in our study provides insights into specific genetic variations within each cultivar, reflecting their evolutionary relationships and potential phenotypic differences. Additionally, analysis of CNVs and SVs elucidates structural alterations in the cultivar genomes, contributing to their distinct traits and properties. These findings underscore the importance of advanced sequencing technologies, such as next-generation sequencing (NGS), in unraveling the complex genomic landscape of cannabis (M. Hesami et al., 2021). Furthermore, the literature highlights the therapeutic potential of cannabinoids, particularly THC and CBD, in treating various conditions, including chronic pain, inflammation, and neurological disorders (Giacoppo, Mandolino, Galuppo, Bramanti, & Mazzon, 2014; Pagano et al., 2022). Cannabinoids interact with the endocannabinoid system (ECS), regulating various physiological functions and offering promising avenues for medical research (Sallaberry & Astern, 2018). Our study, in conjunction with literature data, underscores the need for further research to explore both the positive and negative effects of cannabinoids on human health. Moreover, the integration of our empirical data with existing literature enhances our understanding of cannabis genetics and its implications for research, breeding, and therapeutic applications. By leveraging molecular markers and advanced sequencing technologies, researchers can develop informed breeding strategies and conservation efforts to harness the full potential of cannabis for medical and industrial purposes. In conclusion, the synergy between our study findings and literature data underscores the multifaceted nature of cannabis genetics and its significance in medicine and agriculture. Continued advancements in genomic analysis and sequencing technologies hold promise for unlocking the therapeutic and commercial potential of cannabis, paving the way for future research and innovation in this field.

9.5 CONCLUSION

In summary, our comprehensive analysis of ten cannabis cultivars based on whole genome sequencing has revealed significant insights into their genetic diversity and variation. Exonic mutations, particularly those within coding regions, were found to be most abundant in Hemp, followed by Silver and Crescendo, while GMO exhibited the fewest exonic mutations. Similarly, non-synonymous mutations, which can impact protein function, were highest in Hemp and lowest in GMO. This data underscores the rich genetic diversity present within these cannabis cultivars and provides valuable information for further research and breeding programs. Furthermore, our investigation into single nucleotide polymorphisms (SNPs) highlighted distinct mutation frequencies among the cultivars, with Hemp exhibiting the highest frequency followed by Silver and Blue Dream, while GMO, Gelato, and Candy showed lower frequencies. Additionally, structural variants (SVs) analysis revealed varying counts and types of SVs across the cultivars, contributing to their genomic diversity. Overall, the examination of SNPs, InDels, CNVs, and SVs, along with the phylogenetic relationships depicted through SNP data, demonstrates the complex genetic landscape of cannabis cultivars. These findings provide valuable insights for understanding genetic diversity, evolution, and potential applications in cannabis breeding and research.

REFERENCES

Agilent Technologies (2016). PCR Catalog.

Ahmad, A., Wang, J.-D., Pan, Y.-B., Sharif, R., & Gao, S.-J. (2018). Development and use of simple sequence repeats (SSRs) markers for sugarcane breeding and genetic studies. *Agronomy, 8*(11), 260.

Bentley, D. R., Balasubramanian, S., Swerdlow, H. P., Smith, G. P., Milton, J., Brown, C. G., … Smith, A. J. (2008). Accurate whole human genome sequencing using reversible terminator chemistry. *Nature, 456*(7218), 53–59. doi: 10.1038/nature07517

Chen, K., Wallis, J. W., McLellan, M. D., Larson, D. E., Kalicki, J. M., Pohl, C. S., ... & Mardis, E. R. (2009). BreakDancer: An algorithm for high-resolution mapping of genomic structural variation. *Nature Methods, 6*(9), 677–681.

de Sá, P. H., Guimarães, L. C., das Graças, D. A., de Oliveira Veras, A. A., Barh, D., Azevedo, V., … Ramos, R. T. (2018). Next-generation sequencing and data analysis: strategies, tools, pipelines and protocols. In *Omics Technologies and Bio-Engineering* (pp. 191–207). Elsevier.

Doyle, J. J., & Doyle, J. L. (1987). A rapid DNA isolation procedure for small quantities of fresh leaf tissue. *Phytochemical Bulletin, 19*, 11–15.

El-Mansy, A. B., Abd El-Moneim, D., ALshamrani, S. M., Safhi, F. A., Abdein, M. A., & Ibrahim, A. A. (2021). Genetic diversity analysis of tomato (*Solanum lycopersicum* L.) with morphological, cytological, and molecular markers under heat stress. *Horticulturae, 7*(4), 65.

Gao, C., Xin, P., Cheng, C., Tang, Q., Chen, P., Wang, C., … Zhao, L. (2014). Diversity analysis in *Cannabis sativa* based on large-scale development of expressed sequence tag-derived simple sequence repeat markers. *PloS One, 9*(10), e110638.

Giacoppo, S., Mandolino, G., Galuppo, M., Bramanti, P., & Mazzon, E. J. M. (2014). Cannabinoids: new promising agents in the treatment of neurological diseases. *19*(11), 18781–18816.

Govindaraj, M., Vetriventhan, M., & Srinivasan, M. (2015). Importance of genetic diversity assessment in crop plants and its recent advances: an overview of its analytical perspectives. *Genetics Research International, 2015*(1), 431487.

Gupta, P. K., Rustgi, S., Sharma, S., Singh, R., Kumar, N., & Balyan, H. (2003). Transferable EST-SSR markers for the study of polymorphism and genetic diversity in bread wheat. *Molecular Genetics and Genomics, 270*, 315–323.

Henn, M. R., Sullivan, M. B., Stange-Thomann, N., Osburne, M. S., Berlin, A. M., Kelly, L., … Weiand, M. (2010). Analysis of high-throughput sequencing and annotation strategies for phage genomes. *PloS One, 5*(2), e9083.

Hesami, M., Baiton, A., Alizadeh, M., Pepe, M., Torkamaneh, D., & Jones, A. M. P. (2021). Advances and perspectives in tissue culture and genetic engineering of cannabis. *International Journal of Molecular Sciences, 22*(11). doi: 10.3390/ijms22115671

Hesami, M., Pepe, M., Alizadeh, M., Rakei, A., Baiton, A., & Jones, A. M. P. (2020). Recent advances in cannabis biotechnology. *Industrial Crops and Products, 158*, 113026.

Hurgobin, B., Tamiru-Oli, M., Welling, M. T., Doblin, M. S., Bacic, A., Whelan, J., & Lewsey, M. G. (2021). Recent advances in *Cannabis sativa* genomics research. *New Phytologist, 230*(1), 73–89.

Jannatdoust, M., Darvishzadeh, R., Ziaeifard, R., Ebrahimi, M. A., Maleki, H. H., Gholinezhad, E., & Hatamnia, A. A. (2016). Analysis of genetic diversity and population structure of confectionery sunflower (Helianthus annuus L.) native to Iran. *Journal of Crop Science and Biotechnology, 19*, 37–44.

Kim, B.-Y., Park, J. H., Jo, H.-Y., Koo, S. K., & Park, M.-H. J. P. O. (2017). Optimized detection of insertions/deletions (INDELs) in whole-exome sequencing data. *PLoS One, 12*(8), e0182272.

Li, J., Wang, S., Yu, J., Wang, L., & Zhou, S. (2013). A modified CTAB protocol for plant DNA extraction. *Chinese Bulletin of Botany, 48*(1), 72.

Lo Giudice, C., Tangaro, M. A., Pesole, G., & Picardi, E. (2020). Investigating RNA editing in deep transcriptome datasets with REDItools and REDIportal. *Nature Protocols, 15*(3), 1098–1131.

Lucena-Aguilar, G., Sánchez-López, A. M., Barberán-Aceituno, C., Carrillo-Avila, J. A., López-Guerrero, J. A., & Aguilar-Quesada, R. (2016). DNA source selection for downstream applications based on DNA quality indicators analysis. *Biopreservation and Biobanking, 14*(4), 264–270.

Lynch, R. C., Padgitt-Cobb, L. K., Garfinkel, A. R., Knaus, B. J., Hartwick, N. T., Allsing, N., ... Michael, T. P. (2024). Domesticated cannabinoid synthases amid a wild mosaic cannabis pangenome. *bioRxiv*, 2024.2005.2021.595196. doi: 10.1101/2024.05.21.595196

Lynch, R. C., Vergara, D., Tittes, S., White, K., Schwartz, C., Gibbs, M. J., ... Kane, N. C. (2016). Genomic and chemical diversity in Cannabis. *Critical Reviews in Plant Sciences, 35*(5-6), 349–363.

Malabadi, R. B., Kolkar, K., & Chalannavar, R. (2023). *Cannabis sativa*: Ethnobotany and phytochemistry. *International Journal of Innovation Scientific Research and Review, 5*(2), 3990–3998.

Monthony, A. S., Page, S. R., Hesami, M., & Jones, A. M. P. (2021). The Past, Present and Future of *Cannabis sativa* Tissue Culture. *Plants (Basel), 10*(1). doi: 10.3390/plants10010185

Nadeem, M. A., Nawaz, M. A., Shahid, M. Q., Doğan, Y., Comertpay, G., Yıldız, M., ... Labhane, N. (2018). DNA molecular markers in plant breeding: current status and recent advancements in genomic selection and genome editing. *Biotechnology & Biotechnological Equipment, 32*(2), 261–285.

Pagano, C., Navarra, G., Coppola, L., Avilia, G., Bifulco, M., & Laezza, C. (2022). Cannabinoids: therapeutic use in clinical practice. *International Journal of Molecular Sciences, 23*(6), 3344.

Pisanti, S., & Bifulco, M. (2019). Medical Cannabis: A plurimillennial history of an evergreen. *Journal of Cellular Physiology, 234*(6), 8342–8351.

Sallaberry, C. A., & Astern, L. (2018). The endocannabinoid system, our universal regulator. *Journal of Young Investigators, 34*(6).

Sambrook, J., & Russell, D. W. J. M. C. (2001). *A Laboratory Manual* (Detection of DNA in agarose gels, pp. 5–14). Cold Spring Harbor Laboratory Press.

Scientific, T. F. (2010). *Thermo Scientific Pierce Electrophoreis Technical Handbook: Featuring Thermo Scientific Gelcode Staining Kits*. Thermo Scientific.

Sika, K. C., Kefela, T., Adoukonou-Sagbadja, H., Ahoton, L., Saidou, A., Baba-Moussa, L., ... Gachomo, E. W. (2015). A simple and efficient genomic DNA extraction protocol for large scale genetic analyses of plant biological systems. *Plant Gene, 1*, 43–45.

Small, E. (2017). Classification of *Cannabis sativa* L. in relation to agricultural, biotechnological, medical and recreational utilization. In *Cannabis sativa L.-Botany and Biotechnology* (pp. 1–62). Springer.

Teer, J. K., Bonnycastle, L. L., Chines, P. S., Hansen, N. F., Aoyama, N., Swift, A. J., ... Green, E. D. (2010). Systematic comparison of three genomic enrichment methods for massively parallel DNA sequencing. *Genome Research, 20*(10), 1420–1431.

Valentini, G., Gonçalves-Vidigal, M. C., Elias, J. C. F., Moiana, L. D., & Mindo, N. N. A. (2018). Population structure and genetic diversity of common bean accessions from Brazil. *Plant Molecular Biology Reporter, 36*, 897–906.

Woods, P. O. N. D. (2023). *Phenotype to Genotype and Back in Emerging and Established Crop Species.* Colorado State University.

Yang, H., & Wang, K. (2015). Genomic variant annotation and prioritization with ANNOVAR and wANNOVAR. *Nature Protocols, 10*(10), 1556–1566.

Zhang, J., Yan, J., Huang, S., Pan, G., Chang, L., Li, J., ... Peng, D. (2020). Genetic diversity and population structure of cannabis based on the genome-wide development of simple sequence repeat markers. *Frontiers in Genetics, 11*, 958.

10 Exploring Genomic and Transcriptomic Data and Available Resources for Cannabis Research

Mireia Uranga, Chen Zhang, Pablo Romero, Miriam Caballero Cerveró, Antonio Santiago Pajuelo, David Navarro-Payá, Gastón A. Pizzio and José Tomás Matus

10.1 INTRODUCTION: BACKGROUND AND HISTORICAL CONTEXT

Cannabis is one of the first domesticated plant crops, being cultivated for more than 6,000 years. Initially considered as a plant mostly used for religious rituals and alternative medicine practices, its cultivation later became industrialized for fiber production. Now, its organs, i.e., stems, seeds, flowers, and leaves, and their derived products, are involved in clinical research studies and in several industrial applications, including oil and medicinal compound extractions (House et al., 2010; Abot, 2012). Cannabis has an important recreational use, which, depending on the country, may have from mild restrictions to strict bans. This is due to the accumulation of tetrahydrocannabinolic acid (THCA), a psychotropic cannabinoid that accumulates in trichomes of flowers and leaves. In addition to THCA, more than 110 additional cannabinoids are known to be produced (e.g., cannabidiolic acid, or the first committed compound of the cannabinoid pathway, cannabigerolic acid), some of which have been studied for their pharmacological and therapeutic properties (Di Marzo, 2006). Cannabinoids are specialized metabolites that share precursor intermediates with terpenes, a group of volatile compounds responsible for the characteristic *bouquet* of cannabis flowers. The cannabis aroma industry has, in fact, expanded recently and conquered the flavor and scent industries. Because of all of the above-stated uses of cannabis, understanding the molecular mechanisms underpinning essential biological processes, such as plant growth, seed development, flowering or specialized metabolite production, becomes of great importance. In the era of big analytics and systems biology, the acquisition of data that can describe and predict cannabis biology and physiology results of much interest for the research- and industry-related communities (reviewed by Romero et al., 2021). We can now use large amounts of data and explore (epi)genomes and transcriptomes to identify genes related to important traits. These genes could be potentially used as markers for breeding purposes. The omics revolution of cannabis has allowed several achievements in this regard, some of which will be described here.

10.2 CANNABIS IN THE F.A.I.R. GENOMICS ERA

Several public and private initiatives have generated and examined next and third generation sequencing data (mostly genomics and transcriptomics), aimed at understanding cannabis physiological

DOI: 10.1201/9781003386438-10

and metabolic diversity and evolution. A good portion of this public and private data follows findable, accessible, interoperable and reusable (F.A.I.R.) principles. Public repositories such as the European Nucleotide Archive (ENA) or the Sequence Read Archive (SRA) and Gene Expression Omnibus (GEO), both located within the National Center for Biotechnology Information (NCBI) platform, have been the preferred databases. However, more recent species-specific databases are currently being generated (*see Section 9.5*). Cannabis genomics data has accumulated at increasing rates in the last 12 years, after the release of the first genome assemblies (Figure 10.1A). The highest rates are observed starting in 2015, coinciding with the legalization of medical cannabis in the majority of the states in the United States of America, and following until 2021. Most of the public data has been generated with short-read technologies (e.g., Illumina) and more than 80% of all the data corresponds to genome sequences (Figure 10.1B).

10.3 COMPARATIVE STUDIES OF CANNABIS DIVERSITY AT THE GENOME LEVEL

The *Cannabis* genus is a member of the Cannabaceae family, situated within the Rosales clade. The diverse uses of cannabis in human history have prompted a large diversification of the species, with numerous hybridization episodes between landraces (domesticated and locally adapted), ferals (domesticated naturalized) and wild cannabis plants. This has led to a complex taxonomic and functional classification due to the enormous variability in growth habits, responses to the environment, and in the production of specialized metabolites (e.g., cannabinoids and terpenoid volatile compounds). As a result, and taking advantage of the speed-up advances in genomics, there has been a big interest in sequencing large collection of genotypes to establish whether different domestication episodes have occurred, or whether *Cannabis sativa* is a single-species genus, or is subdivided in *C. sativa, C. indica* and *C. rudelaris* species, as suggested by Hillig (2005). The whole-genome resequencing of more than 100 worldwide accessions allowed Ren et al. (2021) to demonstrate that cannabis was initially domesticated in East Asia around the Neolithic period and that modern marijuana (drug) and hemp cultivars diverged from an ancestral gene pool presently embodied by landraces and ferals.

To date, more than 20 different cultivars have been genome-sequenced and annotated in order to release high-quality and gold-standard genome resources to the scientific community. Some of these have taken advantage of long-read sequencing technologies (e.g., PacBio or GridION) to obtain better scaffolded and more contiguous chromosome-scaled assemblies (Table 10.1). Although extensive whole genome sequencing data exist for *Cannabis*, most of it comes from modern hemp or drug strains with limited genetic diversity and unknown geographic origins. Balant et al. (2024) conducted a detailed and novel sampling of wild-growing and landrace *Cannabis*, using Hyb-Seq data to identify three main genetic groups with clear phylogeographic patterns despite significant admixture. Their findings support those of Ren et al. (2021) regarding the treatment of *Cannabis* as a monotypic genus, but, in contrast, show that hemp and drug-type cultivars consistently align with their geographic origins.

10.3.1 Genomes to Resolve the Genetic Origin of Cannabis Chemotypes

Drug-type cannabis possesses a high THCA amount while in fiber-type hemp the THCA content is less than 0.2%. Beyond this simple classification, cannabis plants can be grouped according to their chemotype, i.e., based on the predominant cannabinoids being produced in their flowers. This represents a more "practical" classification, firstly adopted for cannabinoid-focused breeding (Aizpurua-Olaizola et al., 2016). The genetics laying behind this classification was explained only after different chemotypes were genome-sequenced. Type I, II and III correspond to THCA-rich, cannabidiolic acid (CBDA)/THCA-balanced, and cannabidiol (CDBA)-rich chemotypes,

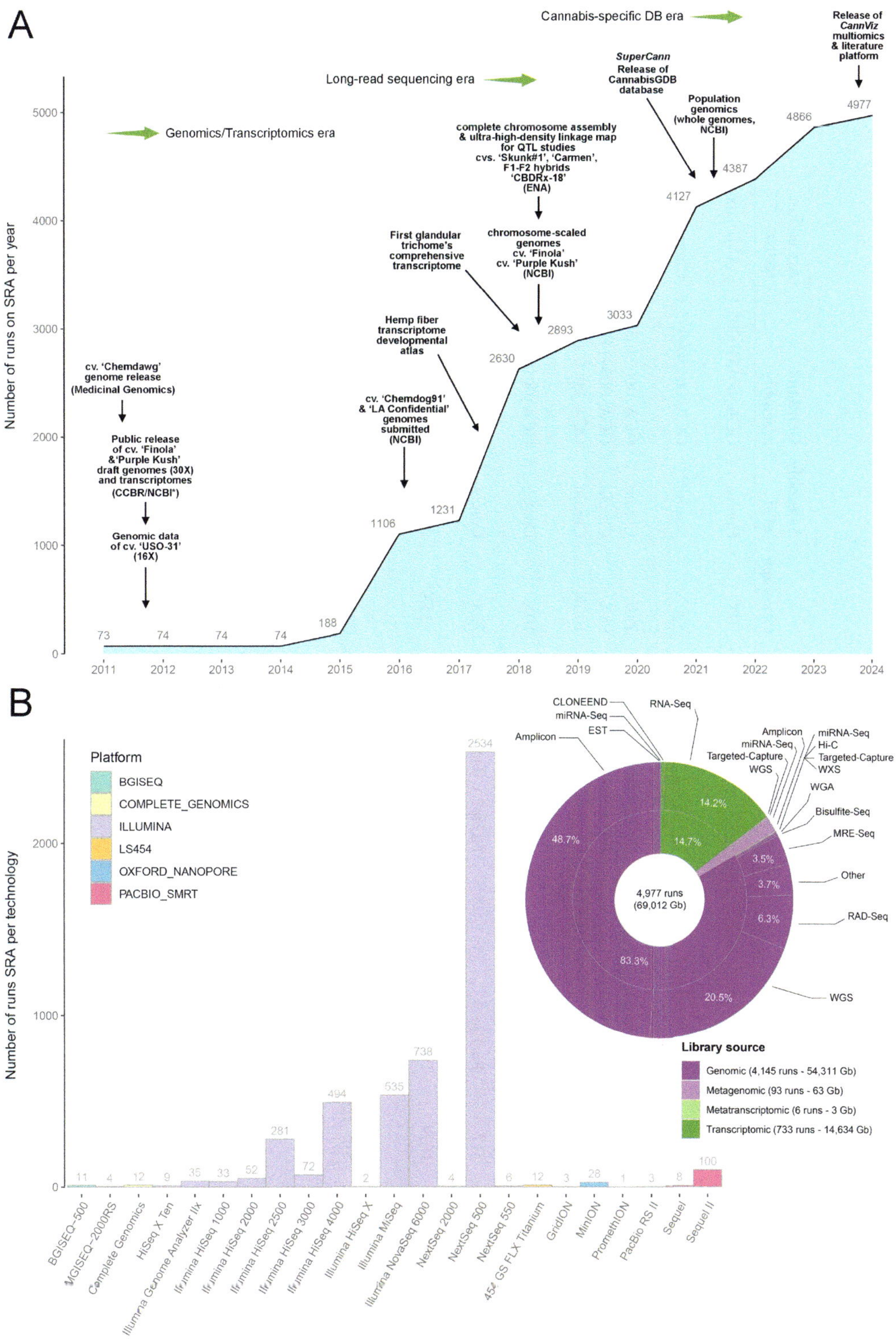

FIGURE 10.1 Cannabis genomics data publicly available in the Sequence Read Archive (SRA). (A) Accumulative trend of SRA runs per year. (B) Public data according to platform (left panel) and library source (right panel). Dating up to 1st May 2024.

TABLE 10.1
Available Genome References of *Cannabis sativa*

Cultivar (Chemotype); Release	Size (Mb)	Seq method	Assembly method	Coverage	Assembly level	GenBank assembly	Accession NCBI	SRA Bioproject
LA Confidential (THC-rich); 2016	595	Roche 454	Newbler v.May 2011	50X	Contig	GCA_001510005.1	ASM151000v1	PRJNA297710
Chemdog 91 (THC-rich); 2016	286	Illumina GA II	CLCBio v.5	300X	Scaffold	GCA_001509995.1	Chemdog91_175268	PRJNA297710
PurpleKush (THC-rich); 2018	892	PacBio	F ALCON v. 0.4.0	79X	Chrom.	GCA_000230575.5	ASM23057v5	PRJNA73819
Finola (CBD-rich); 2018	1010	PacBio	FALCON v. 1.8.6	98X	Chrom.	GCA_003417725.2	ASM34772v2	PRJNA73819
CBDRx-18* (CBD-rich); 2019	876	Illumina Hiseq / PacBio Sequel	Customized pipeline**	100X	Chrom.	GCA_900626175.1	cs10	PRJEB29284
Cannatonic (THC/CBD-balanced); 2016	586	PacBio	FALCON v. February-2016	10X	Contig	GCA_001865755.1	ASM186575v1	PRJNA350523
Pineapple Banana Bubba Kush (THC rich, male); 2017	512	PacBio	HGap2 v. January 2015	72X	Contig	GCA_002090435.1	ASM209043v1	PRJNA378470
Jamaican Lion DASH (THC/CBD balanced, high β-Caryophyllene content); 2019	1074	PacBio Sequel	FALCON Unzip v. Oct-2018	125X	Contig	GCA_003660325.2	Oct15$_3$.7Mb_N50_Jamaican_Lion_Assembly	PRJNA486541
Jamaican Lion Type II (father), β-caryphyllene dom Type II (mother) and Type III (off spring); 2020	876- 1000	PACBIO_SMRT (Sequel II) / Illumina NovaSeq 6000	FALCON Unzip v.0.3.0	30X- 98X	Chrom.	-	-	PRJNA575581
Tibetan Wild ecotype, isolate "JL"; 2020	808	PACBIO_SMRT (Sequel)/ Illumina HiSeqxTen	Canu v. 1.5; SmartDenovo v. 1.5; WTDBG v. 2.2; & others	153X	Chrom.	GCA_013030365.1	ASM1303036v1	PRJNA562042
Cannbio-2; 2020	914	PacBio Sequel	RaGOO v. 1.1	86X	Chrom.	GCA_016165845.1	Cannbio-2	PRJNA667278
Abacus; 2022	796	PacBio Sequel	FALCON Unzip v. FEB-2019	83X	Chrom.	GCA_025232715.1	Csat_AbacusV2	PRJNA870966
Pink pepper; 2023	770	Oxford Nanopore GridION; PacBio RSII; Illumina NovaSeq	NextDenovo v. 2.3.1; SMARTdenovo v. SEP-2022	299X	Chrom.	GCA_029168945.1	ASM2916894v1	PRJNA890313

respectively (Small and Beckstead, 1973), while type IV is CBGA-rich (Fournier et al., 1987) and type V is cannabinoid-free (de Meijer et al., 2009). The first genotypes to be genome-sequenced were chemotype I cv. "Purple Kush" ("PK") and "Chemdawg" and chemotype III cv. "Finola" (Figure 10.1, Table 10.1). By using these genomes and comparing the aligned and reconstructed transcriptomes of "PK" and "Finola", van Bakel et al. (2011) suggested possible mechanisms of the inheritance of cannabinoid profiles, corroborating the two-loci model for the genetic control of THCA:CBDA ratios. Here, it was demonstrated that genes related to cannabinoid and precursor (e.g., hexanoate and MEP/GPP) pathways were less expressed in hemp. In addition, a functional Δ9-tetrahydrocannabinolic acid synthase (*THCAS*) gene was exclusively identified in the drug type transcriptome, while a cannabidiolic acid synthase (*CBDAS*) replaced the *THCAS* in the "Finola" transcriptome (several pseudogenes were found though).

Aiming to redefine the causal association between each landrace/cultivar and their cannabinoid chemotype, several genome sequencing projects have emerged in the past years (reviewed in Romero et al., 2021). Grassa et al. (2018) and Laverty et al. (2019) identified a high number of transposable elements close to *THCAS/CBDAS* gene clusters. This suggested transposon-mediated duplications followed by neofunctionalization events that affected THCAS activity. These modified enzymes would later evolve in CBDAS and CBCAS activities still holding vestigial THCA synthesis ability, explaining the presence of low THC levels in hemp. Using both long and short-read sequencing, McKernan et al. (2020) generated a highly contiguous assembly of female and male genomes of cv. "Jamaican Lion", including the male Y chromosome. Sequencing and annotating a total of 42 cannabis genomes allowed the authors to identify extensive copy number variation in cannabinoid synthesis, which were proposed as responsible for chemotypes I, II and III.

Obtaining well-resolved genome assemblies allowed first to propose that high-cannabidiol cultivars arise from the integration of hemp-type *CBDAS* gene clusters in marijuana-type genome backgrounds. Despite the *CBDAS* and *THCAS* gene clusters have been well-described, the evolution and inheritance of drug potential requires additional population genomics studies, for example, by resequencing a large set of drug and hemp types. The work of Ren et al. (2021) provides here a great contribution after mapping the reads of 104 genomes in the chemotype II cv. "Jamaican Lion DASH" that presents full-length coding sequences of *THCAS* and *CBDAS* genes. The study confirms the chemotype classification, it validates the work of Grassa et al. (2018) and the idea of the loss of function of the two main cannabinoid synthase genes during domestication. It also supports the hypothesis of Wenger et al. (2020), describing a genetic model of inheritance based on *CBDAS* genotyping.

10.4 TRANSCRIPTOMICS DATA OPENS NEW TOOLS FOR CANNABIS IMPROVEMENT

Transcriptomics is a common toolbox for gene discovery. Since the first gene expression atlas was released (cv. "PK" and "Finola"; van Bakel et al. 2011), which included roots, stems, shoots, and different stages of flower development, several transcriptome comprehensive datasets have been produced. Braich et al. (2019) gathered an extensive transcriptome atlas of cv. "Cannbio-2" that included 71 vegetative and reproductive tissues but most importantly added isolated trichomes (by vortexing in liquid nitrogen followed by centrifugation) and both female/male flowers. The work of Livingston et al. (2020) is also considered here a milestone due to the technical innovation in isolating different types of trichomes and at different developmental stages. Since then, sophisticated methods have arisen, including the use of deep learning methods together with RNA profiling to identify trichome development regulators. The following studies have been selected as they offer comprehensive insights into cannabis' biology, revealing essential gene expression dynamics and novel regulatory pathways.

10.4.1 Gene Regulatory Networks of Trichome Development in Cannabis

Trichomes are typically described as structures originating from epidermal cells, existing in both unicellular and multicellular forms. In cannabis, they are distributed across various aerial parts of the plant, with the highest concentration found on the floral buds of female plants. Here, we find both glandular and non-glandular types based on their secretion capacity and morphology (Braich et al., 2019).

Glandular trichomes (GTs) are epidermal outgrowths characterized by the presence of a secretory head or gland, a supporting stalk, and a base that joins to neighboring epidermal cells (Huchelmann et al., 2017; Schuurink & Tissier, 2020). They are responsible for synthesizing and releasing large quantities of specialized metabolites, many known for their pharmaceutical properties. Around 30% of all vascular plants have GTs, including members of the Solanaceae, Labiatae, Asteraceae, and Cannabaceae families. In cannabis, secretory GTs are found in bracts and flowers of female plants, i.e., the primary site for cannabinoid production. Based on morphology, they are classified into three types (Hammond & Mahlberg, 1973; Livingston et al., 2020); (i) bulbous GTs, rather small and low, with limited storage of specialized metabolites; (ii) sessile GTs, with a very short base and a globular head with eight secretory cells; and (iii) stalked GTs, which possess a multicellular stalk and a larger storage head comprising 12–16 secretory cells. Sessile GTs store high proportions of cannabinoids and sesquiterpenes, whereas stalked GTs produce secretions rich in cannabinoids and monoterpenes (Mahlberg & Eun, 2004; Andre et al., 2016; Livingston et al., 2020).

Some studies have suggested that the capacity of trichomes to accumulate large quantities of specialized metabolites is related to peculiarities in their cell wall architecture. It has been reported that stalked GTs present loosely bound polysaccharides in their cell walls that interact with metabolite droplets, creating a hydrophobic-hydrophilic interface that promotes the catalytic activity of cannabinoid synthases (Livingston et al., 2021). Further studies have revealed the existence of a metabolically active cell syncytium, termed "supercell", where cytoplasmic bridges are widespread across cell walls and the polarization of organelles towards the apical surface favors metabolite secretion towards the storage cavity (Livingston et al., 2022, 2023). In addition, the absence of cytoplasmic connections between the supercell and the surrounding tissue prevents metabolite backflow from the trichome, boosting even more their accumulation.

In cannabis, several pieces of evidence indicate that sessile GTs are an earlier developmental stage of stalked GTs (Livingston et al., 2020; Punja et al., 2023) (Figure 10.2). Once bract growth is initiated, epidermal cells differentiate into sessile GTs, and as female inflorescences mature, the latter are converted into bulbous and stalked GTs. In the last stages of flowering, the density of stalked GTs increases dramatically until they become the predominant trichome type. The enlargement of stalks only begins when the head has acquired its final size. However, not all sessile GTs are converted into stalked GTs. This asynchronous cell division causes variations in stalk length from just a few to 1000 μm, frequently leading to the misidentification of both trichome types due to their similar appearance.

Traditional methods for observing differences in trichome morphology and density are based on manual inspection and visual counting, which are time-consuming and often linked to human errors (Schellmann et al., 2002; Maes et al., 2008). Deep learning has been claimed as a powerful alternative for studying plant development and phenotypic traits (Nevavuori et al., 2019; Yamaguchi et al., 2020; Tassis et al., 2021). Two newly released reports have used deep learning algorithms to specifically detect and quantify trichomes in cotton (Luo et al., 2023) and cannabis (Sutton et al., 2023). However, the coexistence of multiple GT types in a single sample of cannabis flowers hinders an accurate characterization of trichome diversity. This limitation was recently overcome by the setup of a Cannabis Glandular Trichome Detection Model (CGTDM), a deep learning model that combines object detection and pixel-level segmentation to reliably detect the divergences among cannabis GTs (Huang et al., 2024). First, researchers assembled an extensive dataset of scanning

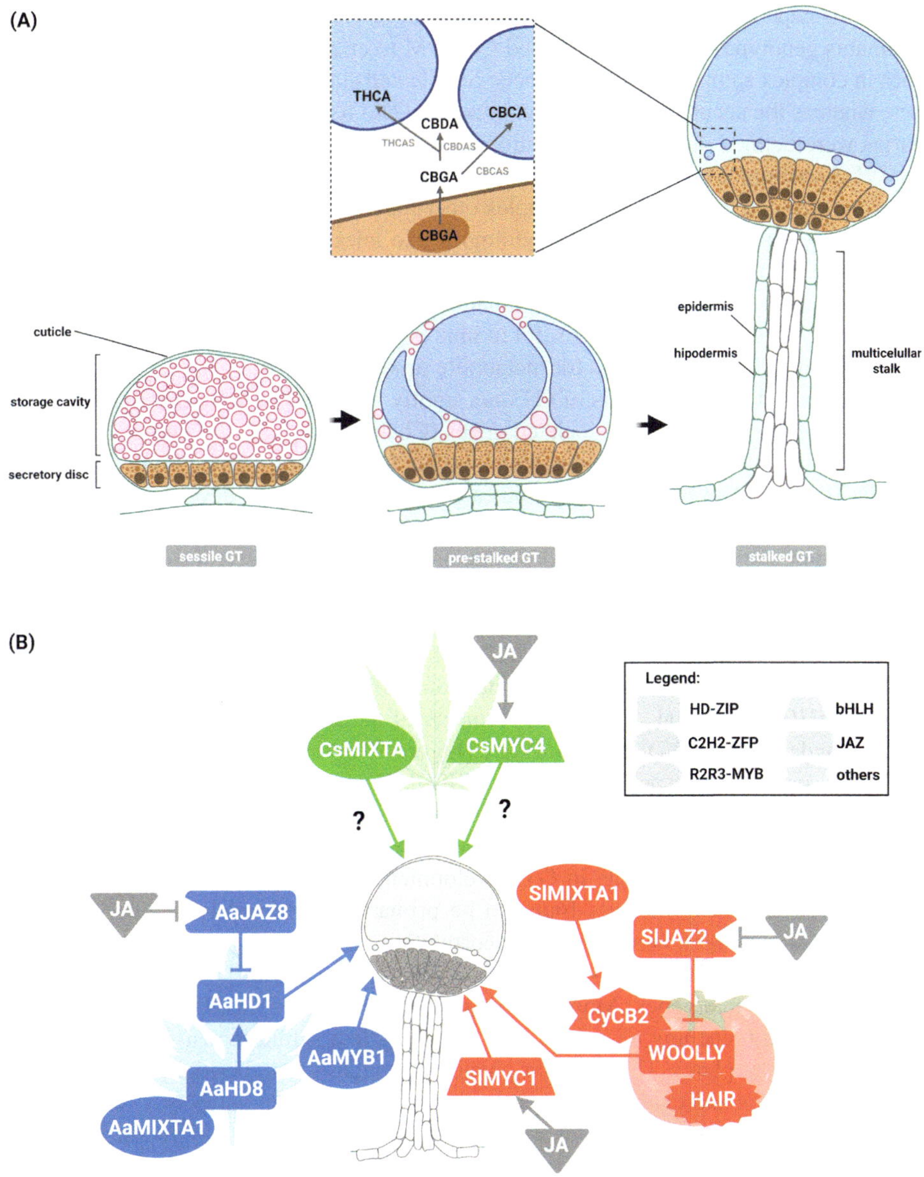

FIGURE 10.2 Glandular trichome (GT) initiation and development in cannabis. (A) Illustrative representation of sessile to stalked GT transition. Sessile GTs (left) are dispersed in aerial parts of the plant, and they have a very short base with a globular head comprising eight secretory cells. They produce numerous small secretions of red-shifted autofluorescence that are enriched in sesquiterpenes with lower proportions of cannabinoids. As female inflorescences develop, sessile GTs are converted into pre-stalked (center) and stalked GTs (right). Both trichome types contain an enlarged disk formed by 12–16 cells that produce secretions with a greater proportion of monoterpenes than sesquiterpenes. As the multicellular stalk lifts the globular head, the storage cavity gets filled with a giant droplet of blue-shifted autofluorescence and high cannabinoid content. Legend in the amplified region within stalked GTs: the brown oval refers to the plastid, and the blue circles represent the secretory vesicles. (B) Main transcriptional factors (TFs) regulating GT initiation in tomato (right, red), *A. annua* (left, blue) and cannabis (top, green). Arrows and blunt ends indicate up- and down-regulation, respectively. Different TF types are detailed in the legend. CBGA, cannabigerolic acid; CBCA, cannabichromenic acid; CBDA, cannabidiolic acid; THCA, tetrahydrocannabinolic acid. Figure created with ©Biorender.

electron microscope images capturing the development of GTs during an 8-week flowering period in 13 cannabis genotypes. Then, they applied CGTDM to effectively distinguish between the three GT types in complex samples of female bracts. Sessile and stalked GTs were correctly identified in all cases, whereas the accuracy was slightly reduced (92%) for bulbous GTs. Because the trichome boundaries were defined during the annotation phase, the model showed an outstanding performance for GT discrimination with a significant reduction in time and effort. However, it was challenging to categorize overlapping GTs in samples other than female flowers that were not part of the model's training dataset. Additional methodologies like infrared and hyperspectral imaging could further improve the discrimination potential of CGTDM in a wide range of cannabis tissues (Mishra et al., 2021; Wang et al., 2021). Such a combination of artificial intelligence and existing computational technologies seems particularly useful to improve the assessment of plant maturity and identify the optimal developmental stages for metabolite production.

The initiation and morphogenesis of GTs are tightly regulated by transcriptional factors (TFs) that control the expression of GT-associated genes (Chalvin et al., 2020; Feng et al., 2021). Gene regulatory networks identified in various plant species have shown that the homeodomain leucine zipper (HD-ZIP) and R2R3-MYB TF families are fundamental in GT development (Figure 10.2). Tomato possesses four distinct secretory GTs, each with unique characteristics regarding anatomy, tissue localization, metabolic pattern, and TF-mediated regulation. Several evidences indicate that the initiation of type I capitate GTs in tomato involves the formation of a complex comprising the HD-ZIP protein *Woolly* and the C2H2 zinc-finger protein *Hair*, which in turn regulates the expression of cyclin B2 (*SlCycB2*) (Figure 10.2, right panel; Yang et al., 2011; Chang et al., 2018). Conversely, GT initiation and artemisinin production in *Artemisia annua* is controlled by the HD-ZIP protein *AaHD1* (Yan et al., 2017) and members of the R2R3-MYB family such as *AaMYB1* (Matías-Hernández et al., 2017) and *AaMIXTA1* (Figure 10.2, left panel; Shi et al., 2018).

MIXTA-like proteins are especially relevant R2R3-MYB members since they have been reported to coordinate trichome initiation in various plant species (Jaffé et al., 2007; Oshima et al., 2013; Ying et al., 2020). Recently, the MIXTA1 homolog from cannabis (*CsMIXTA*) was characterized as the first TF with a potential role in GT development in this species (Figure 10.2, top panel; Haiden et al., 2022). *CsMIXTA* was found to be primarily expressed in female inflorescences, where its expression pattern showed a strong correlation with that of geranyl pyrophosphate synthase (GPPS) and olivetolic acid cyclase (OAC), two enzymes that generate the precursors of cannabinoids and monoterpenes. Moreover, ectopic expression of *CsMIXTA* in tobacco favored the enlargement and branching of stalked GTs, resulting in a 3-fold increase in trichome density compared to wild-type plants. This peculiar phenotype suggests that, besides coordinating the formation of stalked GTs and cannabinoid accumulation, *CsMIXTA* might play a role in the development of other trichome types in cannabis. Altogether, these findings highlight *CsMIXTA* as a candidate gene to modulate GT formation and cannabinoid production via genetic engineering or CRISPR-Cas editing approaches.

Aiming to expand the limited knowledge of gene regulatory networks governing cannabis GT morphogenesis, Huang et al. (2024) carried out a transcriptome profiling on female cannabis flowers to detect temporal variations in gene expression patterns. 647 TFs from 49 distinct families were differentially expressed during the 8-week flowering period, including relevant families such as WRKY, R2R3-MYB, bHLH and ERF. Most of the TFs showed high expression levels between weeks 4 and 7, followed by a significant decrease in week 8, thus indicating that female flower development is finely orchestrated at the transcriptional level. Intriguingly, this pattern concurs with critical changes in GT density during week 4. In the same study, authors identified a co-expression module encompassing various MeJA-responsive genes associated with stalked GT density and cannabinoid accumulation. Interactions between specific TFs and jasmonic acid (JA) or its derivative methyl-jasmonate (MeJA) have been reported to modulate GT initiation and development in diverse plant species (Hua et al., 2021; Kobayashi et al., 2010; Yan et al., 2017; Zhang et al., 2019).

For instance, *SlMYC1* is a JA-responsive TF belonging to the basic helix-loop-helix (bHLH) that positively regulates type VI GT development in tomato (Xu et al., 2018). By studying the temporal and tissue expression of the *bHLH* genes in cannabis, Huang et al. (2024) found that only *CsMYC4* showed GT-specific expression (Figure 10.2, top panel). Interestingly, *CsMYC4* appeared to be phylogenetically related to *SlMYC1*, and its overexpression triggered GT formation in tomato. Furthermore, exogenous MeJA treatment in cannabis significantly up-regulated *CsMYC4* transcript levels, leading to an increase in trichome density. These findings postulate *CsMYC4* as an essential regulator of MeJA-mediated GT formation in cannabis.

10.4.2 Studies for Determining the Genetic Basis of Aroma

One of the limitations of chemotype classification relies on the lack of knowledge in the genomic background other than cannabinoid synthase activity. While the cannabis plant is predominantly recognized for its production of cannabinoids, notably THC and CBD, its resin also contains a diverse array of terpenoids, phenylpropanoids/polyketides, acyl sugars, and fatty acid derivatives. In addition to giving aroma and distinctive flavors to cannabis plants, terpenes possess antibacterial, anti-inflammatory, analgesic, anti-anxiety, anxiolytic, and sedative properties.

Monoterpenes and sesquiterpenes are accountable for much of the odor of cannabis flowers and naturally contribute to the distinctive aroma qualities of cannabis-derived products. Metabolomic approaches have been used for understanding terpene variation and abundance in hundreds of cannabis cultivars and strains (Fischedick et al. 2010; Fischedick 2017; Richins et al. 2018; Allen et al. 2019; Booth et al. 2020). While cannabis plants are known to produce over 100 terpenes, the typical number observed in flower samples falls within the range of 30 to 35, with the majority being identifiable using established standards. In addition, terpene composition of cannabis resin exhibits considerable variability attributable to genetic, environmental, and developmental influences.

Cannabis offers a wide range of genotypes mostly being generated in the last three decades by human selection. Many recent domestication events have focused on scent as a very important trait for selection (Booth et al. 2020). Therefore, the genotypic variation in cannabis is often accompanied by changes in terpene profiles among cultivars. Mono and sesquiterpenes are synthesized by the activity of terpene synthases (TPS) which in cannabis are encoded in a gene family of more than 40 genes (Allen et al. 2019; Xu et al. 2024). It is generally known that intraspecific diversity in the several plant *TPS* gene families, generated by gene duplication and neofunctionalization, can explain metabolic variation leading to aroma differences between cultivars.

With the advent of massively parallel sequencing and significant reductions in cost per base, transcriptomics approaches have been harnessed to elucidate the relationship between *TPS* expression and terpene profiles, aiming to predict and comprehend terpene biosynthesis from a genetic perspective. Zager et al. (2019) constructed gene co-expression networks from isolated glandular trichome cell type-specific transcriptomes and characterized several *TPSs*. Livingston et al. (2020) also identified *TPS* genes while studying transcriptomes of isolated floral trichomes and functionally characterized two of them as important activities related to metabolic specialization of stalked trichomes. Yeo et al. (2022) also identified and characterized a few more *TPS* based on comparative transcriptome analyses.

Despite the advances in TPS-related research in cannabis, most transcriptomic studies were conducted in the absence of easily accessible genome assemblies and annotations. Corresponding accession numbers were typically provided in NCBI-related formats, which may not offer the most straightforward traceability for staying updated on TPS research and their functions. Furthermore, cannabis TPS nomenclature varied across studies. Taking advantage of the current transcriptomic studies performed in *C. sativa*, we have compiled available TPS information (Table 10.2), including identifiers or names from different genome annotations and studies, along with *TPS* expression

TABLE 10.2
Comprehensive Overview of the *C. sativa* Terpene Synthase (TPS) Family

	Xu et al., 2024				Allen et al., 2019 (cv. Purple Kush)	
New Name	cv. CBDRx	Gene Name	Expression pattern explored by RNA-seq	Expression validation in qPCR	gene name	Expression pattern explored by RNA-seq
TPS1	XM_030630973.1	TPS1	highly expressed at seed			
TPS2	XM_030631298.1	TPS2				
TPS3	XM_030635217.1	TPS3	highly expressed at flower		TPS49	highly expressed in root
TPS4	XM_030635393.1	TPS4	highly expressed at flower		TPS63	expressed at shoot, stem, and root
TPS5	XM_030635402.1	TPS5	highly expressed at bract		TPS62	expressed at flower and shoot
TPS6	XM_030635401.1	TPS6	highly expressed at leaf		TPS61	expressed at flower, shoot, and stem
TPS7	XM_030641433.1	TPS7	highly expressed at stem		TPS24	highly expressed at stem, and also expressed at flower and root
TPS8	XM_030644589.1	TPS8	highly expressed at leaf		TPS13-like2	highly expressed at shoot
TPS9	XM_030644482.1	TPS9			TPS30	expressed at flower and shoot
TPS10	XM_030645709.1	TPS10				
TPS11	XM_030645710.1	TPS11	highly expressed at leaf and stem		TPS52	expressed at flower, shoot, and root
TPS12	XM_030645438.1	TPS12	highly expressed at root	highly expressed at root, and also expressed at bract	TPS6, TPS11-like	highly expressed at root
TPS13	XM_030645439.1	TPS13	expressed at flower, bract, and stem	highly expressed at bract, and also expressed at flower, leaf, seed, and root	TPS6-like, TPS33	expressed at flower, shoot and stem
TPS14	XM_030644559.1	TPS14	highly expressed at root		TPS12, TPS13	TPS12 is highly expressed at root

This Study, Pfam Search				
Data source: NCBI (GCF_900626175)	Data source: SuperCan			
cv. CBDRx	cv. CBDRx	cv.Purple Kush	Genbank Accession Number	Main Product
XP_030486833.1		CsPK_10G0014280	MT295506 (cv.Purple Kush, Booth et al., 2020)	
XP_030487158.1	CsCBD_02G0016660			
XP_030491077.1	CsCBD_02G0022400	CsPK_06G0035410		
XP_030491253.1	CsCBD_02G0040580	CsPK_06G0013530	MN967475 (cv. Lemon Skunk, Booth et al., 2020)	Linalool/ Nerolidol
XP_030491262.1	CsCBD_02G0040600	CsPK_06G0013550	MK801763 (cv. Black Lime, Zager et al., 2019)	Nerolidol/ Linalool
			MK801764 (cv. Valley Fire, Zager et al., 2019)	
			MN967473 (cv. Chocolope, Booth et al., 2020)	Linalool
XP_030491261.1	CsCBD_02G0040620	CsPK_06G0013570, CsPK_07G0010860		
XP_030497293.1, XP_030497294.1	CsCBD_04G0008840	CsPK_00G0007610	MT295505 (cv.Purple Kush)	
XP_030500449.1, XP_030500450.1	CsCBD_05G0001240	CsPK_00G0068360		
XP_030500342.1	CsCBD_05G0001250		KY014558 (cv. Purple Kush, Booth et al., 2017)	(Z)-β-ocimene
XP_030501569.1	CsCBD_05G0001260	CsPK_00G0103840		
XP_030501570.1, XP_030501571.1, XP_030501572.1, XP_030501573.1		CsPK_00G0031350, CsPK_01G0001780		
XP_030501298.1	CsCBD_05G0001810	CsPK_00G0062020, CsPK_01G0002180		
XP_030501299.1		CsPK_01G0002190	KY014563 (cv. Finola, Booth et al., 2017)	(E)-β-ocimene
			KY014562 (cv. Finola, Booth et al., 2017)	
XP_030500419.1, XP_030500420.1	CsCBD_05G0002120	CsPK_01G0002620	KY014559 (cv. Purple Kush, Booth et al., 2017)	

(continued)

TABLE 10.2 (Continued)
Comprehensive Overview of the *C. sativa* Terpene Synthase (TPS) Family

	Xu et al., 2024				Allen et al., 2019 (cv. Purple Kush)	
New Name	cv. CBDRx	Gene Name	Expression pattern explored by RNA-seq	Expression validation in qPCR	gene name	Expression pattern explored by RNA-seq
TPS15	XM_ 030644770.1	TPS15			TPS42	
TPS16	XM_ 030644768.1	TPS16	highly expressed at flower and bract	highly expressed at bract, and also expressed at flower	TPS1	highly expressed at flower
TPS17	XM_ 030644766.1	TPS17	highly expressed at flower and bract	highly expressed at bract, and also expressed at flower and leaf	TPS17	expressed at flower
TPS18	XM_ 030645191.1	TPS18	highly expressed at flower and bract	highly expressed at bract, and also expressed at flower and leaf		
TPS19	XM_ 030652072.1	TPS19	highly expressed at flower		TPS5	expressed at flower and shoot
TPS20	XM_ 030652515.1	TPS20	highly expressed at root	highly expressed at root, and also expressed at leaf	TPS37	highly expressed at root
TPS21	XM_ 030652514.1	TPS21	highly expressed at root	highly expressed at root	TPS11	highly expressed at root
TPS22	XM_ 030652513.1	TPS22	highly expressed at root	highly expressed at leaf, and also expressed at bract, root, seed, and flower	TPS36, TPS38	highly expressed at root
TPS23	XM_ 030655054.1	TPS23	expressed at flower and bract		TPS4, TPS9	expressed at flower and shoot
TPS24	XM_ 030621835.1	TPS24			TPS44	highly expressed at flower, and also expressed at shoot and stem
TPS25	XM_ 030622484.1	TPS25	highly expressed at root		TPS50	
TPS26	XM_ 030622955.1	TPS26	highly expressed at flower and bract	highly expressed at flower and bract, and also expressed at stem, leaf, seed, and root	TPS20	highly expressed at flower

This Study, Pfam Search				
Data source: NCBI (GCF_900626175)	Data source: SuperCan			
cv. CBDRx	cv. CBDRx	cv.Purple Kush	Genbank Accession Number	Main Product
XP_030500630.1				
XP_030500628.1, XP_ 030500629.1	CsCBD_ 05G0002130	CsPK_01G0002600	ABI21837.1 (cv. Skunk, Gunnewich et al., 2007) MK801766 (cv. Canna Tsu, Zager et al., 2019)	(-)-limonene
XP_030500626.1, XP_ 030500627.1				
XP_030501051.1	CsCBD_ 05G0002170			
XP_030507932.1	CsCBD_ 07G0011980	CsPK_08G0016550, CsPK_ 00G0070520	MN967481 (cv. Purple Kush, Booth et al., 2020) MN967484 (cv. Purple Kush, Booth et al., 2020) KY014560 (cv. Finola, Booth et al., 2017)	Myrcene/ Bisabolol Geraniol/ Himachalane Myrcene
XP_030508375.1	CsCBD_ 07G0012000	CsPK_05G0014100, CsPK_ 05G0014090		
XP_030508374.1	CsCBD_ 07G0012010	CsPK_05G0014110	MN967474 (cv. Purple Kush, Booth et al., 2020)	Terpinolene
XP_030508373.1	CsCBD_ 07G0012020	CsPK_00G0019420, CsPK_ 05G0014120, CsPK_ 05G0014130, CsPK_ 00G0019430, CsPK_ 05G0014080		
XP_030510914.1	CsCBD_ 08G0010020	CsPK_02G0023200	KY014555 (cv. Finola, Booth et al., 2017)	β-caryophyllene, α-humulene
XP_030477695.1	CsCBD_ 08G0022270	CsPK_02G0006870		
XP_030478344.1	CsCBD_ 08G0022280	CsPK_02G0006690, CsPK_ 00G0115000		
XP_030478815.1	CsCBD_ 08G0026790	CsPK_02G0001780, CsPK_ 09G0016390	MN967483 (cv. Afghan Kush, Booth et al., 2020) MN967482 (cv. Purple Kush, Booth et al., 2020) KY014556 (cv. Finola, Booth et al., 2017)	Hedycaryol β-elemene γ-eudesmol

(*continued*)

TABLE 10.2 (Continued)
Comprehensive Overview of the *C. sativa* Terpene Synthase (TPS) Family

	Xu et al., 2024				Allen et al., 2019 (cv. Purple Kush)	
New Name	cv. CBDRx	Gene Name	Expression pattern explored by RNA-seq	Expression validation in qPCR	gene name	Expression pattern explored by RNA-seq
TPS27	XM_ 030622956.1	TPS27	highly expressed at flower, bract, and leaf	highly expressed at bract, and also expressed at flower	TPS8	highly expressed at flower, and also expressed at shoot
TPS28	XM_ 030622957.1	TPS28	highly expressed at flower		TPS14, TPS8-like	expressed at flower and shoot
TPS29	XM_ 030622624.1	TPS29	highly expressed at bract		TPS58	
TPS30	XM_ 030622631.1	TPS30			TPS23, TPS59, TPS12-like	expressed at flower and shoot
TPS31	XM_ 030622632.1	TPS31			TPS60	highly expressed at shoot
TPS32	XM_ 030654009.1	TPS32	highly expressed at bract		TPS55	expressed at flower and shoot
TPS33	XM_ 030622635.1	TPS33	highly expressed at bract			
TPS34	XM_ 030653866.1	TPS34			TPS4-like, TPS9-like1, TPS9-like2	TPS4-like is highly expressed at shoot, and also expressed at flower; TPS9-like2 are expressed at flower and shoot
TPS35	XM_ 030623148.1	TPS35			TPS48	expressed at flower and shoot
TPS36	XM_ 030623146.1	TPS36	highly expressed at bract and root		TPS47	expressed at flower, shoot and root
TPS37	XM_ 030624715.1	TPS37				
TPS38	XM_ 030624716.1	TPS38			TPS46	
TPS39	XM_ 030628902.1	TPS39	highly expressed at flower		TPS43	highly expressed at stem, and also expressed at flower
TPS40	XM_ 030626238.1	TPS40	highly expressed at root	highly expressed at bract and leaf	TPS64	highly expressed at root

This Study, Pfam Search				
Data source: NCBI (GCF_900626175)	Data source: SuperCan			
cv. CBDRx	cv. CBDRx	cv.Purple Kush	Genbank Accession Number	Main Product
XP_030478816.1		CsPK_09G0016400		
XP_030478817.1, XP_ 030478818.1		CsPK_09G0016420, CsPK_ 09G0026020	MK801762 (cv. Canna Tsu, Zager et al., 2019)	Hedycaryol
XP_030478484.1	CsCBD_ 08G0027890	CsPK_02G0000710		
XP_030478491.1	CsCBD_ 08G0028010	CsPK_06G0044410, CsPK_ 06G0044440, CsPK_ 02G0034230, CsPK_ 02G0034220, CsPK_ 02G0034240		
XP_030478492.1		CsPK_06G0044420, CsPK_ 02G0034260, CsPK_ 02G0034280		
XP_030509869.1		CsPK_06G0044430		
XP_030478495.1				
XP_030509727.1, XP_ 030509726.1	CsCBD_ 08G0028720	CsPK_00G0052360, CsPK_ 00G0103040, CsPK_ 00G0074910, CsPK_ 00G0052370, CsPK_ 00G0074920, CsPK_ 00G0074900	KY014564 (cv. Finola, Booth et al., 2017) MN967479 (cv. Purple Kush, Booth et al., 2020)	Alloaroma-dendrene No products found
XP_030479008.1, XP_ 030479011.1, XP_ 030479010.1	CsCBD_ 09G0000550	CsPK_00G0055110, CsPK_ 00G0082000		
XP_030479006.1, XP_ 030479007.1	CsCBD_ 09G0000560	CsPK_00G0034020, CsPK_ 00G0055120		
XP_030480575.1		CsPK_00G0055130		
XP_030480576.1	CsCBD_ 09G0000570	CsPK_00G0055140, CsPK_ 00G0034030		
XP_030484762.1	CsCBD_ 10G0019290	CsPK_08G0022890	MN967468 (cv. Purple Kush, Booth et al., 2020)	Linalool
XP_030482098.1	CsCBD_ 10G0019420	CsPK_08G0022580, CsPK_ 00G0023400, CsPK_ 08G0022590		

(continued)

TABLE 10.2 (Continued)
Comprehensive Overview of the *C. sativa* Terpene Synthase (TPS) Family

	Xu et al., 2024				Allen et al., 2019 (cv. Purple Kush)	
New Name	cv. CBDRx	Gene Name	Expression pattern explored by RNA-seq	Expression validation in qPCR	gene name	Expression pattern explored by RNA-seq
TPS41	XM_030629488.1	TPS41	highly expressed at flower and bract	highly expressed at bract	TPS53, TPS51	expressed at flower and shoot
TSP42					TPS2	
TPS43					TPS3	
TPS44						
TPS45					TPS32	expressed at flower and root
TPS46					TPS18	highly expressed at flower, and also expressed at shoot
TPS47					TPS13-like1, TPS30-like	TPS30-like is highly expressed at flower, and also expressed at shoot and stem; TPS13-like1 is expressed at flower and shoot
TPS48					TPS41	highly expressed at root
TPS49					TPS40	
TPS50						
TPS51						

This Study, Pfam Search				
Data source: NCBI (GCF_900626175)	Data source: SuperCan			
cv. CBDRx	cv. CBDRx	cv.Purple Kush	Genbank Accession Number	Main Product
XP_030485350.1, XP_030485348.1, XP_030485349.1	CsCBD_00G0000810		MN967470 (cv. Afghan Kush, Booth et al., 2020)	Myrcene
			MK614216 (cv. Finola, Livingston et al., 2019)	Terpinolene
			ABI21838.1 (cv. Skunk, Gunnewich et al., 2007)	(+)-α-pinene
	CsCBD_05G0002150	CsPK_01G0002580	MK801765 (cv. Canna Tsu, Zager et al., 2019)	Myrcene
			KY014561 (cv. Finola, Booth et al., 2017)	
		CsPK_09G0016410, CsPK_09G0026050, CsPK_09G0026040	MN967477 (cv. Purple Kush, Booth et al., 2020)	Himachalane
			KY014554 (cv. Finola, Booth et al., 2017)	δ-selinene
	CsCBD_08G0026830	CsPK_03G0020040, CsPK_03G0020020	MN967472 (cv. Lemon Skunk, Booth et al., 2020)	(E)-β-farnesene
		CsPK_04G0001180	MK131289 (cv. Cherry Chem, Zager et al., 2019)	Germacrene B
			MN967471 (cv. Purple Kush, Booth et al., 2020)	
			MN967480 (cv. Chocolope, Booth et al., 2020)	Myrcene
			MK614217 (cv. Finola, Livingston et al., 2019)	(E)-β-ocimene
	CsCBD_08G0028670			
	CsCBD_01G0013340			

patterns supported by transcriptomics data and/or qPCR. This is the first cannabis TPS catalog available.

10.4.3 Transcriptional Responses to Abiotic Stress

Abiotic stress is the main environmental factor that limits cannabis production (Gao, 2018). Drought and low irrigation regimes significantly reduce hemp growth, fiber quality, and overall productivity (Amaducci, 2008). Moreover, environmental transition is intensifying the frequency and severity of these abiotic stressors. Thus, it is imperative to develop innovative agricultural management strategies to increase cannabis resilience to suboptimal environments, optimize resource use efficiency, and adopt practices that minimize environmental impact (Mittler & Blumwald, 2010). Abiotic stress induces different types of damage at the tissue, cell, and molecular levels in plants, leading to altered growth patterns, reduced photosynthetic efficiency, and compromised reproductive success, among other abnormalities (Pugnaire et al., 2019; O'Neill et al., 2021). For instance, water scarcity induces stomatal closure, limiting CO_2 uptake and leaf temperature regulation, whereas salinity stress disrupts ion homeostasis, leading to toxic ion accumulation and osmotic imbalances. Such physiological disruptions have direct repercussions on crop yield and quality. Moreover, the use of abusive watering or fertilizers to counteract abiotic stress factors is also a major cause of environmental degradation (Ramankutty et al., 2018).

Understanding the mechanisms that regulate abiotic stress tolerance in cannabis is important not only for breeding programs, but also because of its direct connection with the biosynthesis of specialized metabolites with antioxidant properties, such as terpenes and flavonoids (Alonso 2015; Murcia 2017). Gao et al. (2018) carried out a transcriptomic study on gene responses to drought using the drought sensitive hemp cv. "Yunma 1". A total number of 1292 DEGs were identified after 7 days of drought, 51 of which were TFs: 20 been up-regulated (i.e. NACs, MYBs, and HD-Zips) and 31 down-regulated (i.e. B3s, KNOXs, C2H2Ls, and bHLHs). These TFs are not only related with drought stress, but also may be connected with the synthesis of specialized metabolites, some with pharmacological applications. On the other hand, DEGs showed a significant biological pathway GO enrichment in plant hormone signal transduction, phenylpropanoid biosynthesis, and carbon fixation in photosynthetic organisms, among 11 significant categories. Regarding plant hormone signal transduction pathways, genes related to abscisic acid (ABA) (8 DEGs) were the most represented. Some of these DEGs were validated by qRT-PCR and analyzed under ABA treatment (100 μM for 3h and 6h). As a result, the expression of ABA related PP2Cs and SAPK3 were up-regulated, and by contrast the expression of the ABA receptor PYL4-2 was down-regulated. Authors conclude that ABA signaling must be a key player in hemp tolerance to drought stress, but more research on this topic is required.

10.4.4 Multi Omics Data Integration Helps Identify Alternative Botanical Sources of Cannabinoids

Non-THC cannabinoids are gaining interest in the medical community for their therapeutic potential in neurological disorders, cancer, and other illnesses (Anokwuru et al., 2022; Singh et al., 2023). Nonetheless, the isolation of these compounds from cannabis is subjected to legal constraints related to the *in planta* co-occurrence of narcotic compounds such as Δ^9-THC. Although genetic engineering seems the most feasible strategy to obtain Δ^9-THC-free plants, stable transformation in cannabis is restricted by its low susceptibility to *Agrobacterium* infection and recalcitrant regeneration of most genotypes (Shiels et al., 2022). These limitations have motivated the discovery of alternative botanical sources with naturally occurring cannabinoid-like compounds (Bohlmann & Hoffmann, 1979). So far, the Asteraceae plant *Helichrysum umbraculigerum* is the first and only known plant species outside of the Cannabaceae from which CBGA, CBD, and their corresponding analogues have been isolated (Pollastro et al., 2018). However, a later study could not corroborate the existence of phytocannabinoids in this species, pointing out the need for further research to clarify such contradictions (Pollastro et al., 2017). After exhaustive chemical profiling of leaf and flower tissues, Berman et al. (2023) reported that cannabinoid production in trichomes of aerial organs of *H. umbraculigerum* was equivalent to maximum levels obtained in cannabis inflorescences. This work not only demonstrated the presence of cannabinoids in this species, but further validated the use of integrative data to prove it, including metabolomics and unambiguous chemical structure annotation, *de novo* whole-genome sequencing, short-read transcriptome analysis (including a protocol for trichome enrichment), and metabolic engineering. CBGA accumulated up to 4.3% of dry weight in *H. umbraculigerum* leaves, making it a promising alternative source of Δ^9-THC-free bioactive cannabinoids. Besides canonical and previously uncharacterized cannabinoids, leaf GTs also stored a diversity of other specialized metabolites, including amorfrutins, prenyl-acylphloroglucinoids, prenylchalcones and prenylflavanones. This rich metabolite profile suggests the existence of primary intermediates that enter parallel pathways to synthesize all compounds found in *H. umbraculigerum* tissues. Genome gene co-expression analysis of transcriptomic data followed by *in vitro* characterization of protein candidates led to the identification of three core cannabinoid biosynthetic enzymes and two novel enzymes that glycosylate and *O*-acylate cannabinoids (Berman et al., 2023). This clearly illustrates a case of parallel evolution in *C. sativa* and *H. umbraculigerum*, where the core cannabinoid pathway consists of paralogue enzymes that evolved autonomously in the two phylogenetically distant plant lineages. Together with a better understanding of cannabinoid synthesis in plants, these findings highlight the potential of *H. umbraculigerum* as an unexplored source for the production of bioactive cannabinoids.

10.5 ONLINE RESOURCES AND LITERATURE/OMICS DATABASES

Knowledge of cannabis molecular processes has grown vigorously in the past years. Several cannabis genome sequencing projects, as well as transcriptomic, proteomic and metabolomic studies are available (e.g., Livingston et al., 2020; Vincent et al., 2019; Zager et al., 2019, revised in this chapter). These releases have been followed by the raising of functional genomic databases that facilitate the examination and use of such data. A good example is CannabisGDB (https://gdb.supercann.net; Cai et al., 2021) that includes more than 5 Gb of genome sequences of eight different cannabis cultivars ('Jamaican Lion DASH', 'Finola', 'Purple Kush', 'CBCRx-18', 'Pineapple Banana Bubba Kush', 'LA Confidential', 'Chemdog91' and 'Cannatonic'). Moreover, CannabisGDB hosts more than 200 Gb of RNA-seq data from 17 varieties, as well as proteomics and cannabinoid targeted metabolomics data. This database counts with different tools, such as blast tools and a genome browser, that shows different annotation tracks with functional annotation, repeated sequences and mapped Iso-seq and short-read sequencing data.

CannabisGDB is divided into four modules: *varieties*, *gene loci*, *metabolites*, and *proteins*. In the *varieties* module it is possible to find information about phenotype and genome assembly of eight cannabis cultivars, as well as global statistics on genome assembly and the functional annotation for protein-coding genes. In the *gene loci* section, all genes identified to date in cannabis (more than 28,000 genes) can be searched, together with functional annotations, gene orthogroups, and their coordinates within genomes. Besides, it is also possible to obtain CDS, protein, cDNA, and gene sequences. In the *metabolites* module we find chemical phenotype information of 210 cannabis cultivars, together with charts to visualize cannabinoids contents in flower buds, cured flowers, vegetative leaf, and whole inflorescence. Finally, the *proteins* section shows all the proteins identified in seeds, trichomes, mature flowers, and apical flower buds. Certainly, CannabisGDB is a valuable resource for the research community that boosts basic, translational and applied science on cannabis.

10.5.1 The Cannabis Visualization Platform (CannaViz)

The cannabis visualization platform (CannaViz; available at https://plantaeviz.tomsbiolab.com/cannaviz) is a web-based interface that aims to offer genomic and transcriptomic derived resources to the wider community whilst providing a set of online tools to extract biological meaning from the gathered data. The raw data consists of public transcriptomic studies that have been processed into normalized expression data, differential expression analyses, and whole genome co-expression networks. The available tools allow exploring the data through various means, which include a user-friendly interface and the ability to extract custom data subsets, subjecting these for *on the go* analyses.

CannaViz showcases public data available at the Sequence Read Archive (SRA). Since it is not quite clear yet whether *indica* and *ruderalis* are species or subspecies of *Cannabis*, the search for public transcriptomic data in the SRA was carried out only using the term of the genus. The query below aimed to retrieve all Illumina-based transcriptomic data from any variety of cannabis, resulting in a total of 676 SRA runs (to date May 2024). Most of the search terms were placed to filter out any other sequencing strategies that did not correspond to mRNA transcriptomic studies:

*("**cannabis** "[Organism] AND ILLUMINA[Platform]) NOT (RIP-Seq[Strategy] OR OTHER[Strategy] OR ChIP-Seq[Source] OR OTHER[Source] OR METATRANSCRIPTOMIC[Source] OR Bisulfite-Seq[Strategy] OR GENOMIC[Source] OR METAGENOMIC[Source] OR DNase-Hypersensitivity[Strategy] OR Bisulfite-Seq[Strategy] OR WGS*[Strategy] OR ncRNA-Seq[Strategy] OR WCS[Strategy] OR degradome OR miRNA-Seq[Strategy] OR small RNA[Title] OR sRNA[Title])*

The transcriptomic data used in CannaViz has been aligned to three different chromosome-scaled genomes corresponding to 'Purple Kush', 'Jamaican Lion' and 'CBDRx-18' cultivars, representing chemotypes I, II and III, respectively. All tools and resources in CannaViz are triplicated in this manner given the key differences in cannabinoid related genes between the studied cultivars. Another important aspect of these re-analyzed transcriptomic data was the curation of its metadata. For this purpose, an ontology was devised to facilitate the classification of transcriptomic runs based on five categories; organism, tissue, developmental stage, sex, and cultivar. Figure 10.3 shows the hierarchical relationship of terms within each classification category. Although the metadata allowed the classification into 24 drug and 14 hemp cultivars, these were not shown in the diagram due to space constraints.

A set of scripts has been used to automatically classify each run according to the categories described above, through the use of regular expressions examined in key metadata columns. This classification has been used to improve the quality of the offered resources (Table 10.3), for instance, allowing tissue-specific expression across all SRA samples, or building hemp and drug specific co-expression networks. The hierarchical nature of the terms provides flexibility, allowing both general and more specific classification.

A gene catalog pooling together gene names/symbols and assigned functions given by the cannabis research community has also been put together after manual revision of the literature. A functional validation level was also provided as developed for grapevine in Navarro-Payá et al. (2022). This classification reflects the degree of certainty in the described function, based on *in silico* or experimental evidence. The incorporation of gene-specific information into the platform further enhances data interpretation as biological meaning is often lost when indicating just gene ids. A suite of web-based applications is offered within the CannaViz platform, which include the *GeneCards*, *GCNs* (gene co-expression networks) and *expression atlases* apps. The *GeneCards* app provides an overview of a gene of interest which can be searched for using its gene id or catalog gene symbol. The overview includes gene catalog columns such as full name, validation level, gene name synonyms and described function together with the corresponding paper(s). Moreover, expression is shown across all processed 676 SRA runs in the form of log(TPM+1) expression visualized

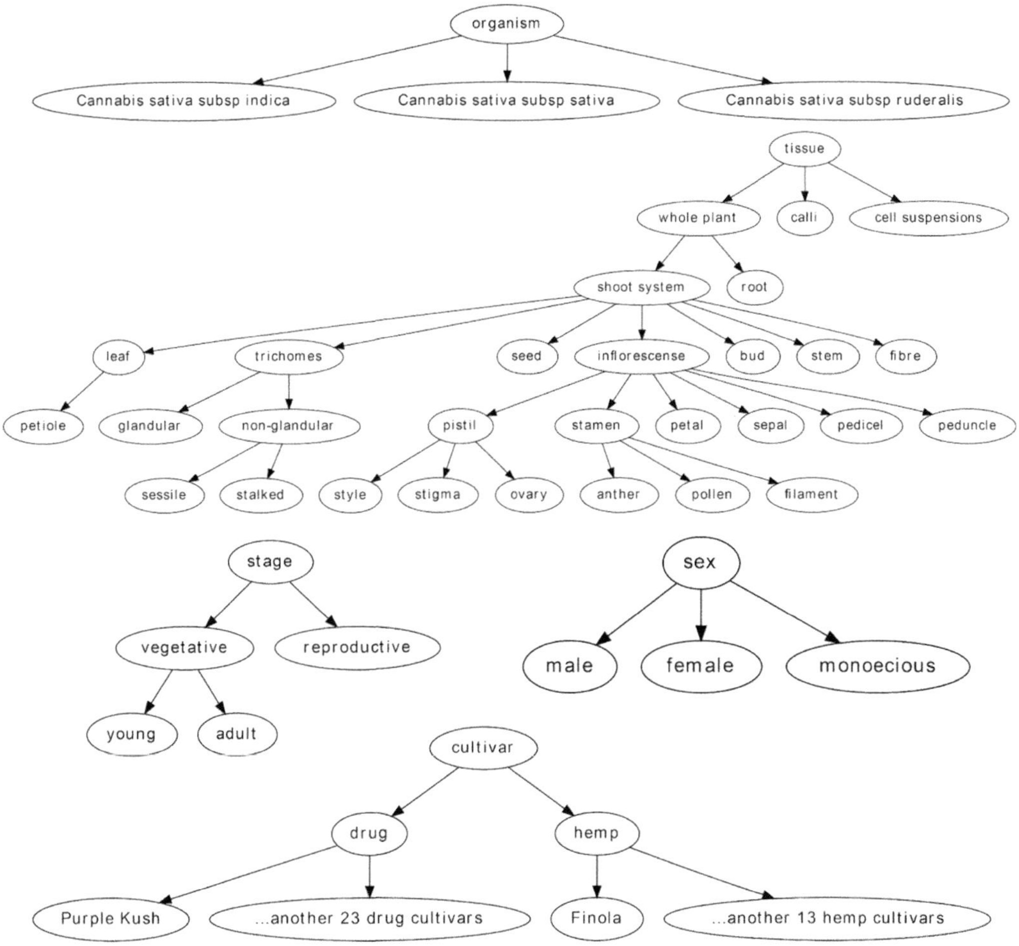

FIGURE 10.3 Ontology terms used to classify *Cannabis* species, tissues, developmental stage, sex, and cultivars.

TABLE 10.3
Metadata Classification According to the Above-Mentioned Ontology

Tissue # Runs		Cultivar Type # Runs		Species/ Subspecies # Runs		Sex # Runs	
leaf	89	hemp	308	sativa	653	female	109
root	32	drug	210	indica	6	male	31
bud	12	**unassigned**	**141**			monoecious	26
stem	5					**unassigned**	**493**
fiber	60						
trichomes	125						
shoot system	137						
whole plant	166						
unassigned	**26**						

as individual violin plots per tissue term. Its interactive nature allows hovering over each point to find out the corresponding run id, opening up the possibility to explore specific factors, which may explain particular expression values. The *expression atlases* app allows the visualization of normalized expression data for custom groups of genes within different pre-arranged heatmaps with optional row clustering. The pre-arranged heatmaps have involved selection of samples to allow the examination of gene expression across a developmental time-course and the evaluation of cultivar-specific differences. Finally, the *GCNs* app provides the means to examine whole genome networks built specifically for the platform using highest reciprocal ranking of Pearson's correlation coefficients (as described in Orduña et al. 2023). A single whole genome network built with this method will typically contain millions of edges for an average plant genome, making visualization on a genome scale unsuitable. Therefore, the app provides both a gene centered view of co-expressed genes for a gene of interest (Individual GCN tab) and a group-level view of network interactions where a custom set of genes is explored (Interaction network tab). The interaction network uses the *networkD3* R library to display all network edges amongst a group of genes in an interactive 3D plot which defines topology based on node connectivity. Custom node colors based on user defined categories can help evaluate node clusters in a visual manner for pathways or biological processes of interest.

The CannaViz effort is hoped to provide an innovative portal that serves for data exploration and analysis of manually curated genomic and transcriptomic studies in cannabis following F.A.I.R. (findable, accessible, interoperable and reusable) data principles.

REFERENCES

Abot, A., Bonnafous, C., Touchard, F. (2012). Effects of cultural conditions on the hemp (*Cannabis sativa*) phloem fibres: Biological development and mechanical properties. Journal of Composite Materials, 47(8): 1067–1077.

Aizpurua-Olaizola, O., Soydaner, U., Öztürk, E., Schibano, D., Simsir, Y., Navarro, P., Etxebarria, N., & Usobiaga, A. (2016). Evolution of the cannabinoid and terpene content during the growth of *Cannabis sativa* plants from different chemotypes. Journal of Natural Products, 79(2): 324–331. https://doi.org/10.1021/acs.jnatprod.5b00949

Allen, K. D., McKernan, K., Pauli, C., Roe, J., Torres, A., & Gaudino R. (2019). Genomic characterization of the complete terpene synthase gene family from *Cannabis sativa*. PLOS ONE, 14(9): e0222363. https://doi.org/10.1371/journal.pone.0222363

Alonso, R., Berli, F. J., Bottini, R., & Piccoli, P. (2015). Acclimation mechanisms elicited by sprayed abscisic acid, solar UV-B and water deficit in leaf tissues of field-grown grapevines. Plant Physiology and Biochemistry, 91: 56–60.

Amaducci S., Zatta, A., Pelatti, F., & Venturi, G. (2008). Influence of agronomic factors on yield and quality of hemp (*Cannabis sativa* L.) fibre and implication for an innovative production system. Field Crops Research, 107(2): 161–169.

Andre, C. M., Hausman, J. F., & Guerriero, G. (2016). *Cannabis sativa*: The plant of the thousand and one molecules. Frontiers in Plant Science, 7: 174167. https://doi.org/10.3389/FPLS.2016.00019

Anokwuru, C. P., Makolo, F. L., Sandasi, M., Tankeu, S. Y., Elisha, I. L., Agoni, C., Combrinck, S., & Viljoen, A. (2022). Cannabigerol: A bibliometric overview and review of research on an important phytocannabinoid. Phytochemistry Reviews, 21(5): 1523–1547. https://doi.org/10.1007/S11101-021-09794-W

Balant, M., Vitales, D., Wang, Z., Barina, Z., Fu, L., Gao, T., Garnatje, T., Gras, A., Hayat, M.Q., Oganesian, M., Pellicer, J., Salami, A.S., Seregin, A.P., Stepanyan-Gandilyan, N., Sultana, N., Tsooj, S., Urgamal, M., Vallès, J., van Velzen, R., & Pokorny, L. 2024. Integrating target capture with whole genome sequencing of recent and natural history collections to explain the phylogeography of wild-growing and cultivated Cannabis. bioRxiv 2024.10.17.618884; doi: https://doi.org/10.1101/2024.10.17.618884

Berman, P., de Haro, L. A., Jozwiak, A., Panda, S., Pinkas, Z., Dong, Y., Cveticanin, J., Barbole, R., Livne, R., Scherf, T., Shimoni, E., Levin-Zaidman, S., Dezorella, N., Petrovich-Kopitman, E., Meir, S., Rogachev, I., Sonawane, P. D., & Aharoni, A. (2023). Parallel evolution of cannabinoid biosynthesis. Nature Plants, 9(5). https://doi.org/10.1038/s41477-023-01402-3

Bohlmann, F., & Hoffmann, E. (1979). Cannabigerol-ähnliche verbindungen aus *Helichrysum umbraculigerum.* Phytochemistry, 18(8). https://doi.org/10.1016/0031-9422(79)83025-3

Booth, J. K., Yuen, M. M. S., Jancsik, S., Madilao, L. L., Page, J. E., & Bohlmann, J. (2020). Terpene synthases and terpene variation in *Cannabis sativa.* Plant Physiology, 184(1): 130–147. https://doi.org/10.1104/pp.20.00593

Braich, S., Baillie, R. C., Jewell, L. S., Spangenberg, G. C., & Cogan, N. O. I. (2019). Generation of a comprehensive transcriptome atlas and transcriptome dynamics in medicinal cannabis. Scientific Reports, 9(1): 1–12. https://doi.org/10.1038/s41598-019-53023-6

Cai, S., Zhang, Z., Huang, S., Bai, X., Huang, Z., Zhang, Y. J., ... & Liu, Y. (2021). CannabisGDB: A comprehensive genomic database for Cannabis Sativa L. *Plant Biotechnology Journal*, 19(5), 857–859.

Chalvin, C., Drevensek, S., Dron, M., Bendahmane, A., & Boualem, A. (2020). Genetic Control of Glandular Trichome Development. Trends in Plant Science, 25(5): 477–487. https://doi.org/10.1016/J.TPLANTS.2019.12.025

Chang, J., Yu, T., Yang, Q., Li, C., Xiong, C., Gao, S., Xie, Q., Zheng, F., Li, H., Tian, Z., Yang, C., & Ye, Z. (2018). Hair, encoding a single C_2H_2 zinc-finger protein, regulates multicellular trichome formation in tomato. Plant Journal, 96(1). https://doi.org/10.1111/tpj.14018

De Meijer, E. P. M., Hammond, K. M., & Sutton, A. (2009). The inheritance of chemical phenotype in *Cannabis sativa* L.(IV): cannabinoid-free plants. *Euphytica*, 168: 95–112.

Di Marzo, V. (2006). A brief history of cannabinoid and endocannabinoid pharmacology as inspired by the work of British scientists. Trends in Pharmacological Science, 27: 134–140.

Feng, Z., Bartholomew, E. S., Liu, Z., Cui, Y., Dong, Y., Li, S., Wu, H., Ren, H., & Liu, X. (2021). Glandular trichomes: New focus on horticultural crops. Horticulture Research, 8(1). https://doi.org/10.1038/s41438-021-00592-1

Fischedick, J. T., Hazekamp, A., Erkelens, T., Choi, Y. H., & Verpoorte, R. (2010). Metabolic fingerprinting of *Cannabis sativa* L., cannabinoids and terpenoids for chemotaxonomic and drug standardization purposes. Phytochemistry, 71(17): 2058–2073. https://doi.org/10.1016/j.phytochem.2010.10.001

Fischedick, J. T. (2017). Identification of terpenoid chemotypes Among high (−)- *trans*-Δ9−tetrahydrocannabinol-producing *Cannabis sativa* L. Cultivars. Cannabis and Cannabinoid Research, 2(1): 34–47. https://doi.org/10.1089/can.2016.0040

Fournier, G., Richez-Dumanois, C., Duvezin, J., Mathieu, J.-P., & Paris, M. (1987). Identification of a new chemotype in*cannabis sativa*: Cannabigerol – dominant plants, biogenetic and agronomic prospects. *Planta Medica*, 53(03): 277–280. https://doi.org/10.1055/s-2006-962705

Gao, C., Cheng, C., Zhao, L., Yu, Y., Tang, Q., Xin, P., Liu, T., Yan, Z., Guo, Y., & Zang, G. (2018). Genome-wide expression profiles of hemp (*Cannabis sativa* L.) in response to drought stress. International Journal of Genomics, ID3057272.

Grassa, C. J., Wenger, J. P., Dabney, C., Poplawski, S. G., Motley, S. T., Michael, T. P., Schwartz, C. J., & Weiblen, G. D. (2018). A complete *Cannabis* chromosome assembly and adaptive admixture for elevated cannabidiol (CBD) content. *bioRxiv*, 458083. https://doi.org/10.1101/458083

Haiden, S. R., Apicella, P. V., Ma, Y., & Berkowitz, G. A. (2022). Overexpression of CsMIXTA, a transcription factor from *Cannabis sativa*, increases glandular trichome density in tobacco leaves. Plants (Basel, Switzerland), 11(11). https://doi.org/10.3390/PLANTS11111519

Hammond, C. T., & Mahlberg, P. G. (1973). Morphology of glandular hairs of *Cannabis sativa* from scanning electron microscopy. American Journal of Botany, 60(6): 524. https://doi.org/10.2307/2441375

Hillig, K. W. (2005). Genetic evidence for speciation in Cannabis (Cannabaceae). Genetic Resources and Crop Evolution, 52(2): 161–180. https://doi.org/10.1007/s10722-003-4452-y

House J. D., Neufeld, J., & Leson, G. (2010). Evaluating the quality of protein from hemp seed (*Cannabis sativa* L.) products through the use of the protein digestibility-corrected amino acid score method. Journal of Agricultural and Food Chemistry, 58(22): 11801–11807.

Hua, B., Chang, J., Wu, M., Xu, Z., Zhang, F., Yang, M., Xu, H., Wang, L. J., Chen, X. Y., & Wu, S. (2021). Mediation of JA signalling in glandular trichomes by the woolly/SlMYC1 regulatory module improves pest resistance in tomato. Plant Biotechnology Journal, 19(2). https://doi.org/10.1111/pbi.13473

Huang, M., Sanchez-Moreiras, A. M., Abel, C., Sohrabi, R., Lee, S., Gershenzon, J., & Tholl, D. (2012). The major volatile organic compound emitted from Arabidopsis thaliana flowers, the sesquiterpene

(E)-β-caryophyllene, is a defense against a bacterial pathogen. New Phytologist, 193(4): 997–1008. https://doi.org/10.1111/j.1469-8137.2011.04001.x

Huang, X., Chen, W., Zhao, Y., Chen, J., Ouyang, Y., Li, M., Gu, Y., Wu, Q., Cai, S., Guo, F., Zhu, P., Ao, D., You, S., Vasseur, L., & Liu, Y. (2024). Deep learning-based quantification and transcriptomic profiling reveal a methyl jasmonate-mediated glandular trichome formation pathway in *Cannabis sativa*. The Plant Journal. https://doi.org/10.1111/TPJ.16663

Huchelmann, A., Boutry, M., & Hachez, C. (2017). Plant glandular trichomes: Natural cell factories of high biotechnological interest. Plant Physiology, 175(1): 6–22. https://doi.org/10.1104/PP.17.00727

Jaffé, F. W., Tattersall, A., & Glover, B. J. (2007). A truncated MYB transcription factor from Antirrhinum majus regulates epidermal cell outgrowth. Journal of Experimental Botany, 58(6): 1515–1524. https://doi.org/10.1093/JXB/ERM020

Kobayashi, H., Yanaka, M., & Ikeda, T. M. (2010). Exogenous methyl jasmonate alters trichome density on leaf surfaces of Rhodes grass (*Chloris gayana* Kunth). Journal of Plant Growth Regulation, 29(4). https://doi.org/10.1007/s00344-010-9161-0

Laverty, K. U., Stout, J. M., Sullivan, M. J., Shah, H., Gill, N., Holbrook, L., Deikus, G., Sebra, R., Hughes, T. R., Page, J. E., & van Bakel, H. (2019). A physical and genetic map of *Cannabis sativa* identifies extensive rearrangements at the *THC/CBD acid synthase* loci. Genome Research, 29(1): 146–156. https://doi.org/10.1101/gr.242594.118

Livingston, S. J., Bae, E. J., Unda, F., Hahn, M. G., Mansfield, S. D., Page, J. E., & Lacey Samuels, A. (2021). Cannabis glandular trichome cell walls undergo remodeling to store specialized metabolites. Plant & Cell Physiology, 62(12): 1944–1962. https://doi.org/10.1093/PCP/PCAB127

Livingston, S. J., Chou, E. Y., Quilichini, T. D., Page, J. E., & Samuels, A. L. (2023). Overcoming the challenges of preserving lipid-rich *Cannabis sativa* L. glandular trichomes for transmission electron microscopy. Journal of Microscopy, 291(1): 119–127. https://doi.org/10.1111/JMI.13165

Livingston, S. J., Quilichini, T. D., Booth, J. K., Wong, D. C. J., Rensing, K. H., Laflamme-Yonkman, J., Castellarin, S. D., Bohlmann, J., Page, J. E., & Samuels, A. L. (2020). Cannabis glandular trichomes alter morphology and metabolite content during flower maturation. The Plant Journal: For Cell and Molecular Biology, 101(1): 37–56. https://doi.org/10.1111/TPJ.14516

Livingston, S. J., Rensing, K. H., Page, J. E., & Samuels, A. L. (2022). A polarized supercell produces specialized metabolites in cannabis trichomes. Current Biology: CB, 32(18): 4040–4047.e4. https://doi.org/10.1016/J.CUB.2022.07.014

Luo, K., Jin, Y., Wen, S., Li, Y., Rong, J., & Ding, M. (2023). Detection and quantification of cotton trichomes by deep learning algorithm. Computers and Electronics in Agriculture, 210: 107936. https://doi.org/10.1016/J.COMPAG.2023.107936

Maes, L., Inzé, D., & Goossens, A. (2008). Functional specialization of the Transparent Testa GLABRA1 network allows differential hormonal control of laminal and marginal trichome initiation in Arabidopsis rosette leaves. Plant Physiology, 148(3). https://doi.org/10.1104/pp.108.125385

Mahlberg, P. G., & Eun, S. K. (2004). Accumulation of cannabinoids in glandular trichomes of Cannabis (Cannabaceae). Journal of Industrial Hemp, 9(1): 15–36. https://doi.org/10.1300/J237V09N01_04

Matías-Hernández, L., Jiang, W., Yang, K., Tang, K., Brodelius, P. E., & Pelaz, S. (2017). AaMYB1 and its orthologue AtMYB61 affect terpene metabolism and trichome development in *Artemisia annua* and *Arabidopsis thaliana*. Plant Journal, 90(3). https://doi.org/10.1111/tpj.13509

McKernan, K. J., Helbert, Y., Kane, L. T., Ebling, H., Zhang, L., Liu, B., Eaton, Z., McLaughlin, S., Kingan, S., Baybayan, P., Concepcion, G., Jordan, M., Riva, A., Barbazuk, W., & Harkins, T. (2020). Sequence and annotation of 42 cannabis genomes reveals extensive copy number variation in cannabinoid synthesis and pathogen resistance genes. *bioRxiv*, 2020.01.03.894428. https://doi.org/10.1101/2020.01.03.894428

Mishra, P., Sadeh, R., Bino, E., Polder, G., Boer, M. P., Rutledge, D. N., & Herrmann, I. (2021). Complementary chemometrics and deep learning for semantic segmentation of tall and wide visible and near-infrared spectral images of plants. Computers and Electronics in Agriculture, 186. https://doi.org/10.1016/J.COMPAG.2021.106226

Mittler, R., & E. Blumwald. (2010). Genetic engineering for modern agriculture: Challenges and perspectives. Annual Review of Plant Biology, 2010(61): 443–462.

Murcia, G., Fontana, A., Pontin, M., Baraldi, R., Bertazza, G., & Piccoli, P. N. (2017). ABA and GA3 regulate the synthesis of primary and secondary metabolites related to alleviation from biotic and abiotic stresses in grapevine. Phytochemistry, 135: 34–52.

Navarro-Payá, D., Santiago, A., Orduña, L., Zhang, C., Amato, A., D'Inca, E., Fattorini, C., Pezzotti, M., Tornielli, G. B., Zenoni, S., Rustenholz, C., & Matus, J. T. (2022). The grape gene reference catalogue as a standard resource for gene selection and genetic improvement. Frontiers in Plant Science, 12. https://doi.org/10.3389/fpls.2021.803977

Nevavuori, P., Narra, N., & Lipping, T. (2019). Crop yield prediction with deep convolutional neural networks. Computers and Electronics in Agriculture, 163. https://doi.org/10.1016/j.compag.2019.104859

O'Neill, M., Lanigan, G. J., Forristal, P. D., & Osborne, B. A. (2021). Greenhouse gas emissions and crop yields from winter oilseed rape cropping systems are unaffected by management practices. *Frontiers in Environmental Science*, 9, 716636.

Orduña, L., Santiago A., Navarro-Payá, D., Zhang, C., Wong, D. C. J., Matus, J. T. (2023). Aggregated gene co-expression networks predict transcription factor regulatory landscapes in grapevine. Journal of Experimental Botany, 74(21), 6522–6540. https://doi.org/10.1093/jxb/erad344

Oshima, Y., Shikata, M., Koyama, T., Ohtsubo, N., Mitsuda, N., & Ohme-Takagi, M. (2013). MIXTA-like transcription factors and WAX INDUCER1/SHINE1 coordinately regulate cuticle development in *Arabidopsis* and *Torenia fournieri*. The Plant Cell, 25(5): 1609–1624. https://doi.org/10.1105/TPC.113.110783

Pollastro, F., De Petrocellis, L., Schiano-Moriello, A., Chianese, G., Heyman, H., Appendino, G., & Taglialatela-Scafati, O. (2017). Amorfrutin-type phytocannabinoids from *Helichrysum umbraculigerum*. Fitoterapia, 123: 13–17. https://doi.org/10.1016/J.FITOTE.2017.09.010

Pollastro F., De Petrocellis L, Schiano-Moriello A, Chianese G, Heyman H, Appendino G, Taglialatela-Scafati O. Reprint of: Amorfrutin-type phytocannabinoids from *Helichrysum umbraculigerum*. Fitoterapia. 2018 Apr;126:35-39. doi: 10.1016/j.fitote.2018.04.002. Epub 2018 Apr 11. PMID: 29655820

Pugnaire, F. I., Morillo, J. A., Peñuelas, J., Reich, P. B., Bardgett, R. D., Gaxiola, A., Wardle, D. A., & van der Putten, W. H. (2019). Climate change effects on plant-soil feedbacks and consequences for biodiversity and functioning of terrestrial ecosystems. Science Advances, 5(11): eaaz1834.

Punja, Z. K., Sutton, D. B., & Kim, T. (2023). Glandular trichome development, morphology, and maturation are influenced by plant age and genotype in high THC-containing cannabis (*Cannabis sativa* L.) inflorescences. Journal of Cannabis Research, 5(1): 1–26. https://doi.org/10.1186/S42238-023-00178-9

Ramankutty, N., Mehrabi, Z., Waha, K., Jarvis, L., Kremen, C., Herrero, M., & Rieseberg, L. H. (2018). Trends in global agricultural land use: Implications for environmental health and food security. Annual Review of Plant Biology, 69: 789–815

Ren, G., Zhang, X., Li, Y., Ridout, K., Serrano-Serrano, M. L., Yang, Y., Liu, A., Ravikanth, G., Nawaz, M. A., Mumtaz, A. S., Salamin, N., & Fumagalli, L. (2021). Large-scale whole-genome resequencing unravels the domestication history of *Cannabis sativa*. Science Advances, 7(29). https://doi.org/10.1126/sciadv.abg2286

Richins, R. D., Rodriguez-Uribe, L., Lowe, K., Ferral, R., & O'Connell, M. A. Accumulation of bioactive metabolites in cultivated medical *Cannabis*. PLoS ONE, 13(7): e0201119. https://doi.org/10.1371/journal.pone.0201119

Romero, D., Johansson, M., Hermansson, U., & Lindner, P. (2021). Impact of users' attitudes toward anonymous internet interventions for cannabis vs. Alcohol use: A secondary analysis of data from two clinical trials. Frontiers in Psychiatry, 12. https://doi.org/10.3389/fpsyt.2021.730153

Schellmann, S., Schnittger, A., Kirik, V., Wada, T., Okada, K., Beerman, A., Thumfahrt, J., Jürgens, G., & Hülskamp, M. (2002). TRIPTYCHON and CAPRICE mediate lateral inhibition during trichome and root hair patterning in Arabidopsis. EMBO Journal, 21(19). https://doi.org/10.1093/emboj/cdf524

Schuurink, R., & Tissier, A. (2020). Glandular trichomes: Micro-organs with model status? New Phytologist, 225(6): 2251 2266. https://doi.org/10.1111/NPH.16283

Shi, P., Fu, X., Shen, Q., Liu, M., Pan, Q., Tang, Y., Jiang, W., Lv, Z., Yan, T., Ma, Y., Chen, M., Hao, X., Liu, P., Li, L., Sun, X., & Tang, K. (2018). The roles of AaMIXTA1 in regulating the initiation of glandular trichomes and cuticle biosynthesis in Artemisia annua. New Phytologist, 217(1). https://doi.org/10.1111/nph.14789

Shiels, D., Prestwich, B. D., Koo, O., Kanchiswamy, C. N., O'Halloran, R., & Badmi, R. (2022). Hemp genome editing-challenges and opportunities. Frontiers in Genome Editing, 4. https://doi.org/10.3389/FGEED.2022.823486

Singh, K., Bhushan, B., Chanchal, D. K., Sharma, S. K., Rani, K., Yadav, M. K., Porwal, P., Kumar, S., Sharma, A., Virmani, T., Kumar, G., & Noman, A. Al. (2023). Emerging therapeutic potential of Cannabidiol (CBD) in neurological disorders: A comprehensive review. Behavioural Neurology. https://doi.org/10.1155/2023/8825358

Small, E., & Beckstead, H. D. (1973). Cannabinoid phenotypes in cannabis sativa. Nature, 245(5421): 147–148. https://doi.org/10.1038/245147a0

Sutton, D. B., Punja, Z. K., & Hamarneh, G. (2023). Characterization of trichome phenotypes to assess maturation and flower development in *Cannabis sativa* L. (cannabis) by automatic trichome gland analysis. Smart Agricultural Technology, 3: 100111. https://doi.org/10.1016/J.ATECH.2022.100111

Tassis, L. M., Tozzi de Souza, J. E., & Krohling, R. A. (2021). A deep learning approach combining instance and semantic segmentation to identify diseases and pests of coffee leaves from in-field images. Computers and Electronics in Agriculture, 186. https://doi.org/10.1016/j.compag.2021.106191

van Bakel, H., Stout, J. M., Cote, A. G., Tallon, C. M., Sharpe, A. G., Hughes, T. R., & Page, J. E. (2011). The draft genome and transcriptome of *Cannabis sativa.* Genome Biology, 12(10): R102. https://doi.org/10.1186/gb-2011-12-10-r102

Vincent, D., Rochfort, S., & Spangenberg, G. (2019). Optimisation of protein extraction from medicinal cannabis mature buds for bottom-up proteomics. Molecules, 24: 659.

Wang, X., Shen, C., Meng, P., Tan, G., & Lv, L. (2021). Analysis and review of trichomes in plants. BMC Plant Biology, 21(1). https://doi.org/10.1186/S12870-021-02840-X

Wenger, J. P., Dabney, C. J., III, ElSohly, M. A., Chandra, S., Radwan, M. M., Majumdar, C. G., & Weiblen, G. D. (2020). Validating a predictive model of cannabinoid inheritance with feral, clinical, and industrial *Cannabis sativa.* American Journal of Botany, 107(10): 1423–1432. https://doi.org/10.1002/ajb2.1550

Xu, J., Kong, L., Ren, W., Wang, Z., Tang, L., Wu, W., Liu, X., Ma, W., & Zhang, S. Identification and expression analysis of TPS family gene in *Cannabis sativa* L. Heliyon, 10(6): e27817. https://doi.org/10.1016/j.heliyon.2024.e27817

Xu, J., Van Herwijnen, Z. O., Dräger, D. B., Sui, C., Haring, M. A., & Schuurink, R. C. (2018). SlMYC1 regulates type VI glandular trichome formation and terpene biosynthesis in tomato glandular cells. Plant Cell, 30(12). https://doi.org/10.1105/tpc.18.00571

Yamaguchi, T., Tanaka, Y., Imachi, Y., Yamashita, M., & Katsura, K. (2020). Feasibility of combining deep learning and RGB images obtained by unmanned aerial vehicle for leaf area index estimation in rice. Remote Sensing, 13(1). https://doi.org/10.3390/rs13010084

Yan, T., Chen, M., Shen, Q., Li, L., Fu, X., Pan, Q., Tang, Y., Shi, P., Lv, Z., Jiang, W., Ma, Y. N., Hao, X., Sun, X., & Tang, K. (2017). HOMEODOMAIN PROTEIN 1 is required for jasmonate-mediated glandular trichome initiation in Artemisia annua. New Phytologist, 213(3). https://doi.org/10.1111/nph.14205

Yang, C., Li, H., Zhang, J., Luo, Z., Gong, P., Zhang, C., Li, J., Wang, T., Zhang, Y., Lu, Y., & Ye, Z. (2011). A regulatory gene induces trichome formation and embryo lethality in tomato. Proceedings of the National Academy of Sciences of the United States of America, 108(29). https://doi.org/10.1073/pnas.1100532108

Yeo, H. C., Reddy, V. A., Mun, B.-G., Leong, S. H., Dhandapani, S., Rajani, S., & Jang, I.-C. (2022). Comparative transcriptome analysis reveals coordinated transcriptional regulation of central and secondary metabolism in the trichomes of Cannabis cultivars. International Journal of Molecular Sciences, 23(15): 8310. https://doi.org/10.3390/ijms23158310

Ying, S., Su, M., Wu, Y., Zhou, L., Fu, R., Li, Y., Guo, H., Luo, J., Wang, S., & Zhang, Y. (2020). Trichome regulator SlMIXTA-like directly manipulates primary metabolism in tomato fruit. Plant Biotechnology Journal, 18(2): 354–363. https://doi.org/10.1111/PBI.13202

Zager, J. J., Lange, I., Srividya, N., Smith, A., & Lange, B. M. (2019). Gene networks underlying cannabinoid and terpenoid accumulation in cannabis. Plant Physiology, 180: 1877–1897.

Zhang, H., Li, W., Niu, D., Wang, Z., Yan, X., Yang, X., Yang, Y., & Cui, H. (2019). Tobacco transcription repressors NtJAZ: Potential involvement in abiotic stress response and glandular trichome induction. Plant Physiology and Biochemistry, 141. https://doi.org/10.1016/j.plaphy.2019.06.021

11 Metabolomics in Cannabis

Tiziana Maria Sirangelo, Simona Felletti, Tatiana Chenet, Martina Catani and Natasha Damiana Spadafora

11.1 INTRODUCTION

Cannabis (*Cannabis sativa L.*) originated from Central Asia and today is distributed to a lesser extent all over the world. It is a highly adaptable species, growing in a wide variety of habitats and climatic conditions (Clarke and Merlin, 2016).

Cannabis has been cultivated for thousands of years and it is also one of the oldest multipurpose crops. It can be classified, according to its main purposes, as fiber crop (hemp), for rope and textile practices, or drug crop (medicinal cannabis) when it is used for treating specific medical conditions (Andre et al., 2016). The understanding of its therapeutic potential is continuously improving, and the number of works investigating cannabis as a source of bioactive compounds for medicinal preparations has increased significantly in recent years (Procaccia et al., 2022).

Cannabis is a complex matrix, containing more than 500 compounds, including terpenes, waxes, and cannabinoids. Cannabinoids are a group of terpenophenolic compounds naturally occurring in different amounts in flowers and leaves. Tetrahydrocannabinol (THC) and cannabidiol (CBD) are the most important secondary metabolites in cannabis, among the ~130 identified (Morales et al., 2017; Kovalchuk et al., 2020), and they are the main focus of breeding programs aimed at obtaining useful varieties for pharmaceutical purposes (Andre et al., 2016; Piluzza et al., 2013) (see Sec. 2). Hemp contains a percentage of THC lower than 0,3%, but medicinal cannabis is characterized by a greater amount, reaching up to 17–28% and in some varieties exceeding the 30% of the dry weight (Schwabe et al., 2023).

CBD and THC are each other isomers and they share the same molecular formula with 21 carbon atoms, 30 hydrogen atoms, and two oxygen atoms. Their molecular structure is also similar, since they both have a n-pentyl chain structure, a limonene, and a benzene ring but the structure of CBD consists of two 6-membered rings whereas THC has an additional 6-membered ring (Lian et al., 2022). In spite of their alike chemical structures, THC and CBD have distinct chemical properties and show very different mechanisms of action and effects on brain function (Andre et al., 2016). THC is mainly associated with psychoactive effects, furthermore its potential activities as anti-inflammatory, anti-cancer and muscle relaxant have been investigated in previous research (De Petrocellis et al., 2011). However, THC has also shown anxiolytic or anxiogenic effects (Phan et al., 2008; Rubino et al., 2008), while CBD is often associated with beneficial effects on depression and anxiety and is employed for therapeutic uses, such as against epilepsy, chronic pain, and Alzheimer's disease (Andre et al., 2016; Boyaji et al., 2020; Raucci et al., 2020; Trojan et al., 2023).

Several studies suggest that the interaction or synergy between these main cannabinoids and other secondary metabolites leads to more marked therapeutic effects than those produced by the single compound (Russo, 2019; Maayah et al., 2020; Goerl et al., 2021).

DOI: 10.1201/9781003386438-11

Despite the small size of the cannabis genus, the exact number of species is still not well defined. According to previous classifications, there are three cannabis species characterized by distinct phenotypic differences: *C. sativa* L., *C. indica Lam* (Lamarck), and *C. ruderalis* (Small et al., 1976; Henry et al., 2020). However, to date, the scientific community defines *C. sativa* and *C. indica* species only. Crosses (hybrids) between the two species show highly variable phenotypes, intermediate between those of the parents (Vergara et al., 2016). However, these distinctions do not represent the real evolutionary relationships (Vergara et al., 2021). A significant phenotypic variation among cannabis species can be found, especially about cannabinoids (McPartland and Russo, 2001; Russo et al., 2008) and terpenoids levels (De la Fuente et al., 2020; Reimann-Philipp et al., 2020). Relevant differences among genotypes can be detected as well (Small et al., 1976; Lynch et al., 2016; Vergara et al., 2016), together with visible features like branch and internode length and flowering time (Clarke and Merlin, 2016). *C. sativa* phenotype is characterized by a tall appearance, less dense buds, narrow leaves, and higher levels of THC. Conversely, *C. indica Lam* is shorter, has denser buds and broad leaves, and shows higher levels of both CBD and THC (McPartland and Russo, 2001; Clarke and Merlin, 2016; Vergara et al., 2016). In each of the previously mentioned cannabis species, cultivars are further grouped on the basis of their chemotype, from I to V, according to the ratio and number of key cannabinoids (Salami et al., 2020). Chemotype I is defined "drug type" cannabis, due to its high THC to CBD ratio (THC content > 0.3–20% and CBD < 0.5%). The second chemotype has both the two main cannabinoids, THC and CBD occur in a content ratio close to one, but generally with a slight prevalence of CBD. The third one, named "fiber" type (chemotype III), has mainly CBD and thus a high CBD/THC ratio (THC < 0.3% or undetectable). Chemotype IV has cannabigerol (CBG) as the predominant cannabinoid, and CBD at trace levels; chemotype V is used to classify cultivars with an undetectable amount of any cannabinoids.

Cannabis growth is characterized by four distinct phases: germination, seedling, vegetative and flowering stages. Each of them has its nutritional and climatic needs, as well as a different photoperiod (Green, 2017). In the wild, cannabis germinates, grows, reaches sexual maturity, reproduces and dies in just one year. Cannabis male and female plants are distinct (dioecious), but hermaphrodites have often been documented (Cervantes, 2007; Green, 2017). However, males and females cannot be morphologically distinguished until the second week of the flowering cycle (Cervantes, 2007; Green, 2017). Since only female inflorescences contain the commonly extracted bioactive compounds, male plants are usually discarded. The vegetative phase shows the greatest overall growth and increase in biomass. In this step, roots increase their length considerably and their architecture becomes more complex, leaves become larger in order to increase their photosynthetic area, transpiration rate rises, to promote water intake. The reproductive phase involves great photoperiod-induced hormonal changes, which can be further enhanced by increasing red and far-red light irradiation intensity (Cervantes, 2007; Green, 2017). In this phase, a first major increase in cannabinoid levels can be observed in female inflorescences (Cervantes, 2007; Green, 2017).

Due to the strict legislation regulating cannabis uses and cultivation, the cannabinoid biosynthetic pathway has only recently been investigated, especially at genetic and molecular level (Hurgobin et al., 2021).

However, the recent relaxation of the pre-existing norms (Pacula and Smart, 2017; Cox, 2018), as well as the availability of the completely assembled cannabis reference sequence, consisting of a genome of 843 Mb for male and 818 Mb for female plants, respectively (van Bakel et al., 2011), is now making research easier. The legalization of cannabis for medical and recreational purposes in many countries has requested the implementation of advanced bioanalytical tools as well as the development of innovative metabolomics applications to identify non-conformities (Aliferis and Bernard-Perron, 2020). Furthermore, the control on cannabis products has led to the development of novel analytical methods to monitor cannabinoid content (García-Valverde et al., 2022).

Combined -omic methods, based on the relationship among genomic, transcriptomic and metabolomic approaches, which have been developed for a wide set of crop species, finding synergies among these heterogeneous data, now can be applied also for cannabis research. This is crucial for metabolomics studies in cannabis. In fact, the particularly complex chemical profile of cannabis, including many lipophilic cannabinoids and terpenoids, along with other primary and secondary metabolites, makes metabolic analyses extremely challenging (Aliferis and Bernard-Perron, 2020) requiring the use of novel, integrated approaches.

11.2 METABOLOMIC PROFILE OF CANNABIS

Cannabis is a polymorphic plant species producing a large set of chemical compounds (~530) (Bonini et al., 2018). The most important active compounds are the cannabinoids, covering ~20% of the total secondary metabolites, followed by terpenoids, also highly abundant, especially isoprenes, monoterpenes, and sesquiterpenes (Chandra et al., 2017). Flavonoids are also high in number, approximately 26, and share a wide set of health benefits with cannabinoids and terpenoids including anti-inflammatory, anti-cancer, and neuroprotective effects (Jin et al., 2020). The adaptive value of cannabinoid production in plants is not well-known, though several hypotheses have been reported including the protection from ultraviolet radiation, pathogens, and herbivores. About the latter, there is evidence that the cannabinoid production could have been an adaptation in response to the herbivore expansion (McPartland, 2018). Terpenes are responsible for the characteristic aroma of cannabis and have a key role in the defense system, against pests, fungi, and bacteria (Russo and Marcu, 2017). Finally, many flavonoids possess protective functions for plants, and their production is responsive to environmental factors, which is also observed in *C. sativa*. For example, cannflavin A production depends on temperature, solar radiation, rainfall, and humidity (Calzolari et al., 2017).

11.2.1 Cannabinoids

The major cannabinoids produced by cannabis varieties are Δ9-tetrahydrocannabinolic acid (Δ9-THCA), cannabidiolic acid (CBDA), cannabichromenic acid (CBCA), and their precursor cannabigerolic acid (CBGA). When the acidic form loses the carboxyl group the neutral decarboxylated compounds are formed (Raharjo and Verpoorte, 2004). Cannabidiol (CBD) and Δ9-tetrahydrocannabinol (Δ9-THC) are the two decarboxylated cannabinoids with the highest reported bioactivity (McPartland et al., 2017).

Along with CBD and THC, other cannabinoids investigated for their potential medical benefits include cannabinol (CBN), cannabichromene (CBC), cannabigerol (CBG), and tetrahydrocannabivarin (THCV). CBN, a by-product of THC, was the first cannabinoid to be isolated and identified from *C. sativa*, found after prolonged storage time, especially at high temperature (Zamengo et al., 2019). It may be an effective treatment for insomnia disorder (Lavender et al., 2023). CBC shows anti-inflammatory (Gojani et al., 2023) and analgesic activities (Raup-Konsavage et al., 2024). Furthermore, it shows cytotoxicity against cancer cells (Ligresti et al., 2006). CBG shows relevant antibiotic effects and could decrease intraocular pressure, and like CBD, it has analgesic and anti-inflammatory properties (Kowal et al., 2016). THCV is structurally similar to THC, except for a shortened propyl ($-C_3$) chain structure, and recent studies suggest its therapeutic potential to fight cerebellar dysfunction and hyperexcitability (Kowal et al., 2016). Cannabinoids are mainly produced in the glandular trichomes of female flowers (Livingston et al., 2020). They contain resin storage cells and the composition of cannabinoids in the resin is not constant, reaching the highest value at flower maturity (Livingston et al., 2020). This content is lower in bulbous trichomes compared to other trichome types, such as stalked or sessile ones (Tahir et al., 2021).

Cannabinoid content is influenced by several factors including temperature of growth, soil composition, and light (De Prato et al., 2022). High temperatures seem to affect CBD and THC content in a variety dependent manner (Sikora et al., 2011). The sensitivity of cannabinoids metabolism to mineral nutrition varies between cannabinoids and plant tissues, for instance, treatments with phosphorus do not influence THC, CBD, CBN, and CBG concentrations in cannabis flower, however the inflorescence leaves show ~15% reduction of THC (Bernstein et al., 2019).

Cannabinoid yield is also affected by UV radiation (Eichhorn Bilodeau et al., 2019). A recent study suggests that using UV radiation as a production approach does not lead to relevant benefits to cannabis yield or inflorescence metabolite composition (Rodriguez-Morrison et al., 2021).

11.2.2 Terpenoids

The cannabis genus contains a wide range of terpenes and terpenoids, showing therapeutic effects either alone and/or as synergic compounds with other terpenes, terpenoids, and/or cannabinoids (Jin et al., 2020). More specifically, the combination of different terpenes/terpenoids with cannabinoids is believed to increase the range of therapeutic properties, this characteristic goes under the name of "entourage effect" (Jin et al., 2020). In cannabis, terpenes share the biosynthesis and accumulation districts with cannabinoids. In fact, both compound classes are biosynthesized in the glandular trichomes of leaves and flowers and are mainly accumulated in the exuded resin (Tanney et al., 2021). Terpenoids are also responsible for cannabis distinctive aroma, and most of them are Volatile Organic Compounds (VOCs) (Chandra et al., 2017). About 200 compounds have been reported to contribute to the complex scent of the genus (Oswald et al., 2021). The terpenoids found in higher amounts in most cannabis varieties are the two isomers of pinene, namely α-pinene and β-pinene; α-terpineol; β-myrcene and d-limonene (Pavlovic et al., 2019; Oswald et al., 2021). The isomer α-pinene is characterized by apoptotic, antimetastatic and antibiotic properties (Pandey et al., 2022). The second isomer, β-pinene, has shown antimicrobial and antidepressant properties (Salem et al., 2023). The monoterpenoid alcohol, α-terpineol, with its delicate floral and lilac-like sweet smell, is widely used in flavors and perfumes production (Saeed et al., 2023). The acyclic monoterpene, β-myrcene, enhancing the anxiolytic and sedative effects of CBD, is the most abundant VOC in most hemp cultivars (do Vale et al., 2002). D-limonene, a precursor in other monoterpes biosynthesis, may have significant health applications, showing anti-inflammatory, anti-cancer and neuroprotective effects, as well as helping mitigate psychoactive adverse events resulting from THC (Russo, 2011).

Dioecious female plants release more VOCs compared to male plants and monoecious cultivars (Kumeroa et al., 2022), and the diversity of these volatile compounds is increasing with the development of new purpose-bred cultivars.

11.2.3 Flavonoids

Flavonoids are polyphenolic compounds with bioactive properties found in cannabis in sufficient amounts to show pharmacological activity. They have been reported as potential prospective neuroprotectants and investigated for their possible therapeutic effects against neurological disorders (Ayaz et al., 2019).

Flavonoids, which are produced primarily in cannabis leaves, include flavones (luteolin, apigenin, orientin, vitexin, and isovitexin) and flavonols (quercetin and kaempferol), and exist as free aglycones or as conjugated O-glycosides or C-glycosides (Flores-Sanchez and Verpoorte, 2008; Bautista et al., 2021). Flavonoid metabolism in *C. sativa* and their synergistic effects to enhance the bioactivities of phytocannabinoids, are currently not clear (Bautista et al., 2021). Several studies report on medicinal potential and anti-inflammatory activity of cannflavins, and particularly on cannflavin A and B, which are prenylated flavonoids that are unique to cannabis, but little attention has been focused on their biosynthesis within *C. sativa* (Andre et al., 2016). The group of Rea (2019) investigated this topic using a combination of phylo-genomic and biochemical approaches. They identified an aromatic prenyltransferase (CsPT3) that catalyzes either geranyl diphosphate (GPP) or dimethylallyl diphosphate (DMAPP) to the methylated flavone, chrysoeriol, to produce cannflavins A and B, respectively (Rea et al., 2019).

11.2.4 Cannabinoids, Terpenoids, and Flavonoids Biosynthesis

Figure 11.1 shows the main steps of THC, CBD, terpenoids, and flavonoids biosynthesis.

THC and CBD are produced from two different metabolic pathways: the polyketide and the methylerythritol phosphate (MEP), resulting in olivetolic acid (OA) and geranyl diphosphate (GPP), respectively (Kovalchuk et al., 2020; Dehnavi et al., 2022). OA and GPP synthesize the cannabigerolic acid (CBGA), whose pentyl side chain produces the acidic precursors of THC (THCA), of CBD (CBDA) and of CBC (CBCA) (Weiblen et al., 2015; Gülck and Møller, 2020).

Terpenoids are produced by two metabolic pathways: the dimethylallyl diphosphate (DMAPP) and isopentenyl diphosphate (IPP) pathway, which share a GPP precursor with cannabinoids biosynthesis (Grof, 2018).

Terpenoids also result from the mevalonate (MVA) pathway. The MEP pathway provides GPP to form monoterpenoids (C_{10}) while the MVA pathway produces farnesyl diphosphate (FPP) for sesquiterpenoids (C_{15}) and squalene, which are precursors for triterpenoids (C_{30}) and sterols (Flores-Sanchez and Verpoorte, 2008).

The phenylpropanoid pathway produces *p*-coumaroyl-CoA from phenylalanine. In conjunction, malonyl-CoA, *p*-coumaroyl-CoA produces naringenin, which constitutes the substrate for flavone and flavonol biosynthesis (Flores-Sanchez and Verpoorte, 2008).

Overall, the cannabis metabolites biosynthesis has not been completely understood at molecular level (Tahir et al., 2021). Furthermore, new and innovative investigations could be crucial for healthcare applications, where less abundant cannabinoids could cover a significant role (Gallily et al., 2018).

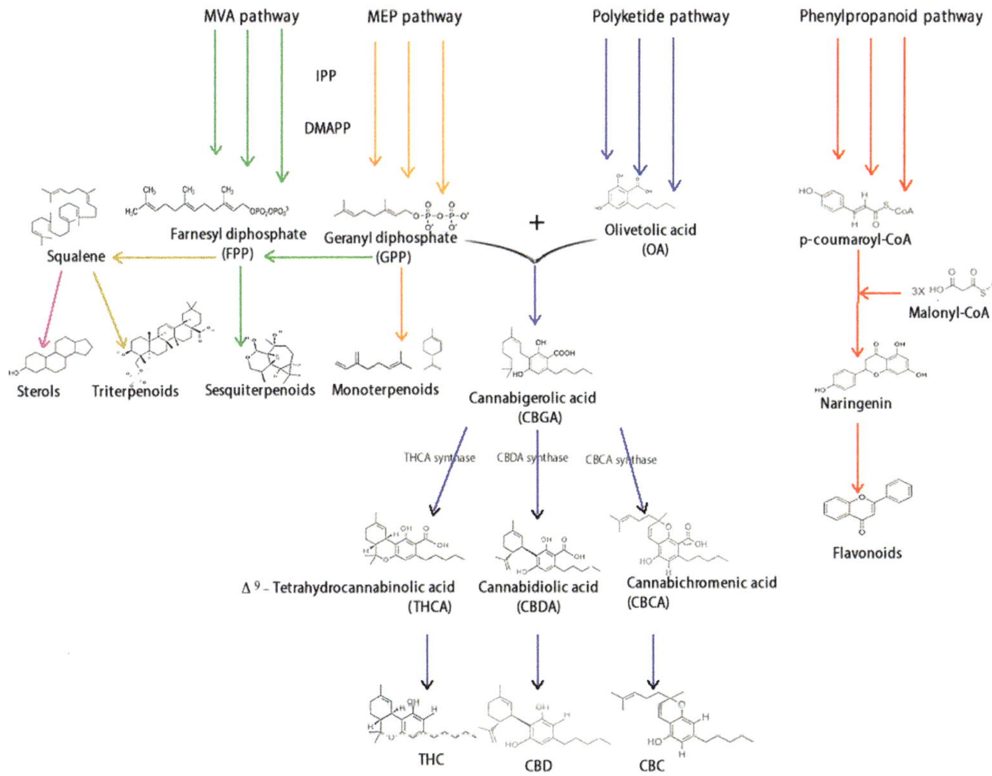

FIGURE 11.1 Main steps of cannabinoids (in blue), terpenoids (green and yellow) and flavonoids (red) biosynthesis in *C. sativa*.

11.3 CANNABIS METABOLITE PROFILING TECHNIQUES AND APPLICATIONS

11.3.1 Cannabinoids Analysis Techniques

The metabolic fingerprint is crucial to explore differences among cannabis cultivars (Zager et al., 2019).

A wide set of techniques have been employed to extract and analyze cannabis compounds, and their choice depends on the compound family to analyze either cannabinoids or terpenes (Leghissa et al., 2018a).

The most commonly used platforms to analyze cannabinoids are Gas Chromatography (GC) and Liquid Chromatography (LC), coupled with Mass Spectrometry (MS) (Fuhrer and Zamboni, 2015; Leghissa et al., 2018a). LC is usually used to investigate non-volatile and thermally unstable compounds (such as acidic forms), while GC makes it possible to analyze more thermally stable molecules, often with the aid of derivatization agents. GC-MS often relies on Electron Ionization (EI) to disrupt the analytes in smaller fragments, while the process to generate ions in LC-MS is less drastic. Being the metabolomic matrix of cannabinoids quite complex, investigating less abundant and novel cannabinoids is usually quite demanding (Leghissa et al., 2018a). Laboratories often prefer to rely on LC for cannabinoid analysis, due to simpler protocols and sample preparation, making it possible to skip the derivatization step required for GC-based methods (Leghissa et al., 2018a).

GC-MS was adopted by Magagnini et al. (2018) to study the effects of lighting on cannabinoid metabolism. More specifically, cannabis crops treated with three different light spectra, high-pressure sodium (HPS), AP673L (LED), and NS1 (LED) (Magagnini et al., 2018) showed changes in CBG, CBD, and THC content and cannabis morphology. In particular, LED lights resulted in greater plant growth and improved the cannabinoid profile when compared to HPS. However, how the LED lighting manipulation exactly affects cannabis extract quality and yield is still not clear (Eichhorn Bilodeau et al., 2019). Gas Chromatography Electron Impact Mass Spectrometry (GC-EI-MS) approach has recently been adopted for the identification and analysis of cannabinoids from lipid extracts from seeds of *C. sativa,* resulting in the identification of over 40 cannabinoids, 16 of them never reported before (Basas-Jaumandreu and De Las Heras, 2020). Other approaches applied to cannabis quantitative metabolite analysis include GC coupled with triple-quadrupole MS methods, known as QQQ-MS, where the first and third quadrupoles work as mass filters and the second fragments the chemicals (Ramirez et al., 2018). The methods are characterized by high selectivity and sensitivity, thus, they may be suitable to evaluate the abundance of less abundant cannabinoids in different parts of the crop, which may not be detected with other platforms (Leghissa et al., 2018b).

High Performance Liquid Chromatography (HPLC) has also been extensively used to investigate cannabis metabolites with more than 100 available methods (Felletti et al., 2021). For instance, a study analyzing a sub-set of cannabis varieties, representative of all chemotypes (Fulvio et al., 2021) reported the total yields of the major cannabinoids CBD, CBG and THC in female inflorescences, showing significant differences among cultivars and chemotypes. The varieties characterized by the greatest CBD content were "CS" and "Carmagnola" (chemotype III), while the lowest amount of CBD was detected in "Santhica 27" (chemotype IV). Conversely, "Bernabeo" showed the highest CBG content and, among the medical varieties "CINBOL" (chemotype I) was characterized by very high THC content.

High-resolution mass spectrometry (HRMS) is another powerful technique used to investigate the cannabis metabolite profiles, allowing a more precise identification of compounds with the same mass, due to a more reliable calculation of mass to charge ratio (m/z) to several decimal places compared to MS. The technique allowed a better definition of cannabis chemovars (Wang et al., 2018), as proved by one recent study (Vásquez-Ocmín et al., 2021) where, by using data from 20 varieties and a combined LC-HRMS platform, metabolites were reliably mapped and annotated, and cannabis peculiar markers identified. The results obtained from this approach were compared with those based solely on major cannabinoid quantification, highlighting how minor compounds were highly reliable markers to differentiate cannabis varieties. Therefore, a comprehensive knowledge of the profile of these minor secondary metabolites is relevant to establish and to assure the composition and quality of a given *C. sativa* chemovar (Jin et al., 2021; Puri et al., 2021).

Nuclear Magnetic Resonance spectroscopy (NMR) (Markley et al., 2017) is a versatile technique even if it is characterized by a lower sensitivity than MS. NMR provides a more reliable metabolite structure and does not require any destructive sample processing. Furthermore, NMR makes it possible to simultaneously identify multiple analytes. Recently, the cannabinoid and untargeted metabolomic profiles from inflorescences of two *C. sativa* varieties with either high THCA or high CBDA contents harvested at different times were investigated (Fernández et al., 2023) by means of HPLC and NMR approaches. The two cultivars showed expected variations in cannabinoid profiles. In addition, the untargeted metabolomic approach allowed the distinction between harvests of the same variety. Furthermore, variations in the metabolome were correlated with the presence of powdery mildew, which is a well-known fungal infection in cannabis. These findings led to the identification of potential molecular markers associated with a pathogen, helping to close the lack of knowledge in the field (Sirangelo, 2023a; Sirangelo et al., 2023b). However, despite its potential, NMR has rarely been applied to investigate cannabinoids, since it is still a quite expensive technique.

11.3.2 Terpenoids Analysis Techniques

Methods to quantify terpenes in cannabis have become increasingly available in the literature, especially with the increased interest in medicinal cannabis and the contribution of terpenes to the "entourage effect" (Giese et al., 2015; Fischedick, 2017; Ibrahim et al., 2019).

They are usually investigated by GC, coupled with different types of detectors, such as Flame Ionization Detection (FID), the most commonly used detector type, due to its low cost and reliable quantification power, and MS, highly used for compound identification. The terpenoid profile of 30 cannabis cultivars quantified by GC-FID and confirmed by GC-MS was used to develop classification model based multivariate statistical analyses. The results highlighted how several cultivars could be associated with specific terpenoid chemotype classes, while other cultivars were not well discriminated based on their terpenoid compounds identified and quantified in their profiles (Fischedick, 2017). Recently, these technologies were used to analyze terpenes commonly present in cannabis samples. For instance, a straightforward solid phase microextraction-gas chromatography-flame ionization detection method (SPME-GC-FID) was used for the quantification of major terpenes and terpenoids in two cannabis varieties reporting high concentrations of β-myrcene, α-pinene, β-pinene, limonene, (E)-β-ocimene, β-caryophyllene, α-humulene, (E)-nerolidol, and linalool on both varieties (Giovannoni et al., 2023).

Furthermore, a GC-MS method was developed and validated for the quantification of terpenes in cannabis, including the major terpenes (α-pinene, β-pinene, β-myrcene, limonene, terpinolene, linalool, α-terpineol, β-caryophyllene) (Ibrahim et al., 2019). The method, optimized according to the Association of Official Analytical Chemists guidelines, was applied to the simultaneous analysis of the main terpenes, and showed excellent accuracy, repeatability, and intermediate precision.

Further method development efforts have been applied for a high-throughput GC-MS method able to facilitate terpene quantitation, fast profiling separation (30-min) and microscale sampling (40 mg sample) (Krill et al., 2020). The optimized method enabled the identification of over 45 terpenes and terpenoids and showed similar validation results to other studies with the advantage of a shorter time and a simple extraction protocol (Krill et al., 2020).

Considering the complexity of cannabis terpene profile, the use of comprehensive two-dimensional gas chromatograph time-of-flight mass spectrometry (GC × GC-TOFMS) has come in aid of researchers allowing the separation identification of a large number of terpenes and terpenoids in the dried flowers of six cannabis strains (Kaur et al., 2023). A total of 146 terpenes and terpenoids were detected. Chemical differences were observed in the number and relative abundance of monoterpenes, monoterpenoids, sesquiterpenes, and sesquiterpenoids in all six strains, reporting four new terpenoids for the first time.

11.3.3 Flavonoids Analysis Techniques

Flavonoids have been mainly determined using spectrophotometric or chromatographic techniques. Fundamentally, spectrophotometry has been used for a simple, fast and non-selective total flavonoids estimation, while chromatography has been employed for an individual flavonoid determination with high sensitivity and accuracy (Sammani et al., 2021). To date, the most common analytical technique used for the analysis of flavonoids is liquid chromatography (Sammani et al., 2021).

For example, an innovative study aimed at the profiling of non-psychoactive cannabinoids and flavonoids was recently carried out by means of HPLC-DAD-UV (HPLC coupled to Diode-Array Ultraviolet Detector) and ESI-MS (Electrospray ionization and MS) and MS^2 (the ions are selected and fragmented) detection, showing the flavonoid cannflavin A as the main compound present in almost all samples (Pellati et al., 2018).

However, several attempts to analyze cannflavins in extracts using UV absorbance detection led to low or not quantifiable amounts of this compound. In fact, despite almost all flavonoids having adsorption power, this method was found to be non-selective (Sammani et al., 2021). Instead, a validated approach based on chromatographic separation and positive mode electrospray mass spectrometry has been recently carried out for the detection of cannflavins in hemp extracts (O'Croinin et al., 2023). The LC-MS method proved to be simple and sensitive and can be adapted to other cannabis compounds.

11.3.4 Combined Techniques

In cannabis, due to the complexity of the metabolite matrix, the combination of several analytical approaches usually gives the most comprehensive profile picture (Andre et al., 2016). For instance, LC/QQQ/MS and NMR metabolomics studies highlighted the presence of several cannabinoids in cell extracts of capitate-sessile and stalked trichomes as well (Happyana et al., 2013). Extracting and analyzing the chemical profile of specific trichome lines in such a precise way is promising for future multi-omics experiments.

The combined approach, relying on LC-DAD and GC-FID technologies, allowed further classifications of cultivars based on cannabinoid and terpenoid profiles (Zager et al., 2019) confirming that the comprehensive chemical composition of secondary metabolites greatly changes according to the cultivar.

Recently, two complementary techniques, SPME-GC-MS and proton-transfer-reaction time-of-flight mass spectrometer (PTR-ToF-MS), were used to characterize the volatile fraction of the chemical profile of pollen deriving from three different cultivars of *Cannabis sativa*, while the cannabinoid content was determined by GC-MS (Taiti et al., 2022). SPME-GC-MS and PTR-ToF-MS require small amounts of material to be investigated, without pre-treatment and with a short extraction time. PTR-ToF-MS produces the whole volatile spectrum (including low-molecular-weight compounds) in real time and with high sensitivity, with few misidentifications. While SPME-GC-MS combines a simple sample preparation with a versatile and highly sensitive detection (Vitalini et al., 2021).

Generally, it is preferable not to use PTR-MS to obtain the separation of chemical compounds, which is achieved using the GC-MS technique. Therefore, using a combined approach, it is possible to join the advantages of each method obtaining a better comprehension of the molecular composition of cannabis pollen. In the study of Taiati (2022) the use of the combined approaches allowed to show a high content of volatile compounds differing in the pollen of the three cultivars. Furthermore, the pollen of one of the three cultivars was particularly rich in cannabinoid content compared to the others (Taiati et al., 2022).

These findings confirm the applicability of innovative combined analytical techniques to obtain a comprehensive cannabis profile.

11.4 METABOLOMICS COMBINED WITH OTHER OMICS APPROACHES IN CANNABIS

Next Generation Sequencing (NGS) technologies have revolutionized plant biology and contributed to elucidate molecular mechanisms in several crops, including cannabis. Their applications did not just include metabolomics but also combined methods, on the basis of the relationship among metabolomics, genomic, and transcriptomic approaches, in order to find synergies among heterogeneous data (Pandita et al., 2021; Sirangelo et al., 2022).

Many investigations focused on trichomes, being the primary storage organ for cannabinoids (Zager et al., 2019; Liu et al., 2021; Tahir et al., 2021). Among them, one study was about the changes occurring during flower development, the main phase of metabolite accumulation, and was based on metabolomics and transcriptomics (Livingston et al., 2020). SPME, GC-MS, and LC-MS methods for carrying out the quantitative analysis of terpenes and cannabinoids were used, while RNA-Seq gene co-expression networks were developed. Results demonstrated how medicinal cannabis trichomes undergo changes to their morphology and metabolite profile during flower development. Furthermore, they demonstrated a high expression of cannabinoid and terpene related genes, as well as genes involved in CBDA synthase, suggesting that the glandular trichomes represent the heart of the cannabinoids and terpenes production.

In the same year, a metabolome and transcriptome analysis of several cannabis cultivars, which was based on WGCNA approach (Zager et al., 2019), demonstrated that subsets of cannabis terpene synthase gene family are expressed in all plant tissues, and include a set of root specific monoterpene synthases. These findings underline the role played by terpenes in plant interaction with the soil biome, a role that could be further investigated due to its impact in agronomic programs.

Quite recently, a multi-omics study was aimed at investigating genes involved in terpene synthase, by detecting about 30 different *CsTPS* genes, as well as further genes involved in isoprenoid and cannabinoid biosynthesis (Booth et al., 2020). Despite these recent findings, the relationship between terpene and cannabinoid biosynthesis to date has not been fully elucidated.

The genetic controls of complex agronomic and biochemical traits in cannabis were also investigated, and research exploring how genes involved in cannabinoid synthase pathway affect the THC:CBD ratio and the overall cannabinoids amount, has been increasing in the latest years (Hurgobin et al., 2021). Metabolomics, genomics, and transcriptomics approaches were employed to study the relationships between cannabinoid content and cannabinoid synthase Copy Number Variations (CNVs) (Vergara et al., 2019). More specifically, the cannabinoid profiles of ~70 cannabis cultivars of different lineages were assessed using the HPLC technique. A whole genome shotgun data set from a diversity of cultivars and a RNA-Seq method was also applied. Results showed a positive correlation between the accumulation of specific cannabinoids, including THCA, and the CNVs of certain synthase paralogs. These results were further confirmed in (Grassa et al. (2021). Table 11.1 summarizes some of the most significant recent cannabis metabolomics studies.

TABLE 11.1
A Summary of Cannabis Metabolomics Studies

Omic Technologies	Description	Reference
Metabolomics/GC-Cannabinoid synthase genotyping, linkage mapping, and QTL analysis	Study about cannabinoids and terpenoids biosynthesis in cannabis	(Weiblen et al., 2015) (Laverty et al., 2019)
Metabolomics/Analytical methods available for cannabinoids analysis	Review—Cannabis metabolites	(Chandra et al., 2017)
Metabolomics/GC-FID method	Definition of cannabis chemovars based on their terpenoid profile. The Effect of Light Spectrum on cannabis morphology	(Magagnini et al., 2018) (Fischedick, 2017)
Metabolomics/Supercritical CO_2 extraction of the cannabis inflorescence	Study of cannabinoids and terpenoids biosynthesis in cannabis	(Walsh et al., 2021)
Metabolomics/GC-MS/LC-MS	Overview about methods for the chemical characterization of cannabis	(Leghissa et al., 2018a)
Metabolomics/HRMS—LC-HRMS	Potential in the definition of cannabis chemovars of HRMS techniques	(Wang et al., 2018) (Vásquez-Ocmín et al., 2021)
Metabolomics/SFE/SPE	Isolation of tetrahydrocannabinol from cannabis	(Gallo-Molina et al., 2019)
Metabolomics/SPME/DLLME/LC-QQQ-MS	Physicochemical characterization of cannabis	(Ramirez et al., 2018)
Metabolomics/Analytical methods available for cannabinoids and terpenoids analysis	Review—Cannabinomics: Metabolomics applications in cannabis. Terpenoids properties	(Aliferis and Bernad-Perron., 2020; Gallily et al., 2018)
Metabolomics/Structural classification of phytocannabinoids	Study about phytocannabinoids	(Gülck and Møller, 2020)
Metabolomics/GC-EIMS	Investigation about lipids extracted from cannabis seeds	(Basas-Jaumandreu and De Las Heras, 2020)
Metabolomics/HPLC	Investigation about the major cannabinoids: CBD, CBG, and THC	(Fulvio et al., 2021)
Metabolomics/GC, Transcriptomics/RNA-Seq e PCR	Quantification of the transcript levels of different cannabinoid synthase genes	(Allen et al., 2019) (Booth et al., 2020)
Metabolomics/HPLC, Genomics/WGS and GBS	Investigation about cannabis ancestry	(Lynch et al., 2016)
Metabolomics/SPME GC-MS, and LC-MS, Transcriptomics/RNA-Seq	Investigation about trichomes changes during flower maturation in cannabis	(Livingston et al., 2020)
Metabolomics/HPLC/GC, Transcriptomics/RNA-Seq	Study about nine cannabis cultivars having different basic characteristics	(Zager et al., 2019)
Metabolomics/HPLC, Genomics/WGS, Transcriptomics/RNA-Seq	Study about the relation between CNVs and cannabinoids profile	(Vergara et al., 2019)
Metabolomics, Genomics, and Transcriptomics data are collected and structured, and available on a web site for researchers	"CannabisGDB," a comprehensive multi-omics database	(Cai et al., 2021)
Metabolomics, Genomics, and Transcriptomics platforms. WGS approach in medical cannabis is discussed	Review on medicinal plants multi-omics applications	(Pandita et al., 2021)
Metabolomics/GC, Genomics/WGS/Linkage mapping and QTL analysis	Comparison between Carmagnola and USO31 cannabis cultivars	(Woods et al., 2021)
Metabolomics/HPLC, Genomics/Nanopore technology/Linkage mapping and QTL analysis, Transcriptomics/RNA-Seq	Study about the CBDRx cannabis cultivar	(Grassa et al., 2021)

REFERENCES

Aliferis, K.A., Bernard-Perron, D. 2020. Cannabinomics: Application of metabolomics in Cannabis (*Cannabis sativa* L.) research and development. *Front. Plant. Sci.* 11:554.

Allen, K.D., McKernan, K., Pauli, C., Roe, J., Torres, A., Gaudino, R. 2019. Genomic characterization of the complete terpene synthase gene family from *Cannabis sativa*. *PLoS ONE* 14:e0222363.

Andre, C.M., Hausman, J.F., Guerriero, G. 2016. *Cannabis sativa*: The plant of the thousand and one molecules. *Front. Plant Sci.* 7:19.

Ayaz, M., Sadiq, A., Junaid, M., Ullah, F., Ovais, M., Ullah, I., Ahmed, J., Shahid, M. 2019. Flavonoids as prospective neuroprotectants and their therapeutic propensity in aging associated neurological disorders. *Front Aging Neurosci.* 11:155.

Basas-Jaumandreu, J., De las Heras, F.X.C. 2020. GC-MS metabolite profile and identification of unusual homologous cannabinoids in high potency *Cannabis sativa*. *Planta Med.* *86*:338–347.

Bautista, J. L., Yu, S., Tian, L. 2021. Flavonoids in *Cannabis sativa*: Biosynthesis, bioactivities, and biotechnology. *ACS Omega* 6:5119–5123.

Bernstein, N., Gorelick, J., Zerahia, R., Koch, S. 2019. Impact of N, P, K, and humic acid supplementation on the chemical profile of medical cannabis (*Cannabis sativa* L.). *Front. Plant Sci.* 10:736.

Bonini, S.A., Premoli, M., Tambaro, S., Kumar, A., Maccarinelli, G., Memo, M., Mastinu, A. 2018. *Cannabis sativa:* A comprehensive ethnopharmacological review of a medicinal plant with a long history. *J. Ethnopharmacol.* 227:300–315.

Booth, J.K., Yuen, M.M., Jancsik, S., Madilao, L., Page, J., Bohlmann, J. 2020. Terpene synthases and terpene variation in *Cannabis sativa*. *Plant Physiol.* 184:130–147.

Boyaji, S., Merkow, J., Elman, R.N.M., Kaye, A.D., Yong, R.J., Urman, R.D. 2020. The role of Cannabidiol (CBD) in chronic pain management: An assessment of current evidence. *Curr. Pain Headache Rep.* 24:4.

Cai, S., Zhang, Z., Huang, S., Bai, X., Huang, Z., Zhang, Y.J., Huang, L., Tang, W., Haughn, G., You, S., et al. 2021. CannabisGDB: A comprehensive genomic database for *Cannabis sativa* L. *Plant Biotechnol. J.* 19:857–859.

Calzolari, D., Magagnini, G., Lucini, L., Grassi, G., Appendino, G.B., Amaducci, S. 2017. High added-value compounds from cannabis threshing residues. *Ind. Crops Prod.* 108:558–563.

Cervantes, J. 2007. *Marijuana Horticulture: The Indoor/Outdoor Medical Grower's Bible*. Van Patten Publishing: Vancouver, WA, USA.

Chandra, S., Lata, H., Khan, I.A., Elsohly, M.A. 2017. *Cannabis sativa* L.: Botany and horticulture. In *Cannabis sativa L.—Botany and Biotechnology*, ed. Chandra, S., Lata, H., Elsohly, M.A, 79–100. Springer: Cham Switzerland.

Clarke, R.C., Merlin, M.D. 2016. Cannabis domestication, breeding history, present day genetic diversity, and future prospects. *Crit. Rev. Plant Sci.* 35:293–327.

Cox, C. 2018. The Canadian Cannabis Act legalizes and regulates recreational cannabis use in 2018. *Health Policy* 122:205–209.

De la Fuente, A., Zamberlan, F., Ferrán, A.S., Carrillo, F., Tagliazucchi, E., Pallavicini, C. 2020. Relationship among subjective responses, flavor, and chemical composition across more than 800 commercial cannabis varieties. *J. Cannabis Res.* 2:21.

Dehnavi, M., Ebadi, A., Peirovi, A., Taylor, G., Salami, S.A. 2022. THC and CBD fingerprinting of an elite cannabis collection from Iran: Quantifying diversity to underpin future cannabis breeding. *Plants* 11:129.

De Petrocellis, L., Ligresti, A., Moriello, A. S., Allarà, M., Bisogno, T., and Petrosino, S. 2011. Effects of cannabinoids and cannabinoid-enriched cannabis extracts on TRP channels and endocannabinoid metabolic enzymes. *Br. J. Pharmacol.* 163:1479–1494.

De Prato, L., Ansari, O., St.J. Hardy, G. E., Howieson, J., O'Hara, G., Ruthrof, K.X. 2022. The cannabinoid profile and growth of hemp (*Cannabis sativa L.*) is influenced by tropical daylengths and temperatures, genotype and nitrogen nutrition. *Ind. Crops Prod.* 178:114605.

do Vale, T.G., Furtado, E.C., Santos Jr, J.G., Viana G.S.B. 2002. Central effects of citral, myrcene and limonene, constituents of essential oil chemotypes from *Lippia alba* (Mill.) n.e. Brown. *Phytomedicine* 9:709–14.

Eichhorn Bilodeau, S., Wu, B.S., Rufyikiri, A.S., MacPherson, S., Lefsrud, M. 2019. An update on plant photobiology and implications for Cannabis production. *Front. Plant Sci.* 10:296.

Felletti, S., De Luca, C., Buratti, A., Bozza, D., Cerrato, A., Capriotti, A.L., Laganà, A., Cavazzini, A., Catani, M. 2021. Potency testing of cannabinoids by liquid and supercritical fluid chromatography: Where we are, what we need. *J. Chromatogr. A.* 1651:462304.

Fernández, S., Castro, R., Lopez-Radcenco, A., Rodriguez, P., Carrera, I., Garcia-Carnelli, C., Moyna, G. 2023. Beyond cannabinoids: Application of NMR-based metabolomics for the assessment of *Cannabis sativa* L. crop health. *Front. Plant Sci.* 14:1025932.

Fischedick, J.T. 2017. Identification of terpenoid chemotypes among High ()-*trans*-Δ^9-Tetrahydrocannabinol-Producing *Cannabis sativa* L. Cultivars. *Cannabis Cannabinoid Res.* 2: 34–47.

Flores-Sanchez, I. J., Verpoorte, R. 2008. Secondary metabolism in cannabis. *Phytochem. Rev.* 7:615–639.

Fuhrer, T., Zamboni, N. 2015. High-throughput discovery metabolomics. *Curr. Opin. Biotechnol.* 31:73–78.

Fulvio, F., Paris, R., Montanari, M., Citti, C., Cilento, V., Bassolino, L., Moschella, A., Alberti, I., Pecchioni, N., Cannazza, G., et al. 2021. Analysis of sequence variability and transcriptional profile of cannabinoid synthase genes in *Cannabis sativa* L. Chemotypes with a focus on cannabichromenic acid synthase. *Plants* 10:1857.

Gallily, R., Yekhtin, Z., Hanuš, L.O. 2018. The anti-inflammatory properties of terpenoids from cannabis. *Cannabis Cannabinoid Res.* 3:282–290.

Gallo-Molina, A.C., Castro-Vargas, H.I., Garzón-Méndez, W.F., Ramírez, J.A.M., Monroy, Z.J.R., King, J.W., Parada-Alfonso, F. 2019. Extraction, isolation and purification of tetrahydrocannabinol from the *Cannabis sativa* L. plant using supercritical fluid extraction and solid phase extraction. *J. Supercrit. Fluids* 146:208–216.

García-Valverde, M.T., Sánchez-Carnerero, C.C, Díaz-Liñán, M.C., Sánchez de Medina, V., Hidalgo-García, J., Nadal, X., Hanuš, L., Ferreiro-Vera, C. 2022. Effect of temperature in the degradation of cannabinoids: From a brief residence in the gas chromatography inlet port to a longer period in thermal treatments. *Front Chem.* 10:1038729.

Giese, M.W., Lewis, M.A., Giese, L., Smith, K.M. 2015. Development and validation of a reliable and robust method for the analysis of cannabinoids and terpenes in Cannabis. *J. AOAC Int.* 98:1503–1522.

Giovannoni, S., Lancioni, C., Vaccarini, C., Sedan, D., Andrinolo, D., Castells, C. 2023. Determination of variability of terpenes and terpenoids in *Cannabis sativa* by gas chromatography-flame ionization detection and gas chromatography-mass spectrometry. *J. Chromatogr. A* 1687:463669.

Goerl, B., Watkins, S., Metcalf, C., Smith, M., Beenhakker, M. 2021. Cannabidiolic acid exhibits entourage-like improvements of anticonvulsant activity in an acute rat model of seizures. *Epilepsy Res.* 169:106525.

Gojani, E.G., Wang, B., Li, D.P., Kovalchuk, O., Kovalchuk, I. 2023. Anti-Inflammatory effects of minor cannabinoids CBC, THCV, and CBN in human macrophages. *Molecules* 28:6487.

Grassa, C.J., Weiblen, G.D., Wenger, J.P., Dabney, C., Poplawski, S.G., Motley, S.T., Michael, T.P., Schwartz, C.J. 2021. A new Cannabis genome assembly associates elevated cannabidiol (CBD) with hemp introgressed into marijuana. *New Phytol.* 230:1665–1679.

Green, G. 2017. *The Cannabis Grow Bible: The Definitive Guide to Growing Marijuana for Recreational and Medicinal Use.* Green Candy Press: San Francisco, CA, USA.

Grof, C.P. 2018. Cannabis, from plant to pill. *Br. J. Clin. Pharmacol.* 84:2463–2467.

Gülck, T., Møller, B.L. 2020. Phytocannabinoids: Origins and biosynthesis. *Trends Plant Sci.* 25:985–1004.

Happyana, N., Agnolet, S., Muntendam, R., Van Dam, A., Schneider, B., Kayser, O. 2013. Analysis of cannabinoids in laser- microdissected trichomes of medicinal *Cannabis sativa* using LCMS and cryogenic NMR. *Phytochemistry* 87:51–59.

Henry, P., Khatodia, S., Kapoor, K., Gonzales, B., Middleton, A., Hong, K., Hilyard, A., Johnson, S., Allen, D., Chester, Z.A., et al. 2020. Single Nucleotide Polymorphism assay sheds light on the extent and distribution of genetic diversity, population structure and functional basis of key traits in cultivated North American cannabis. *J. Cannabis Res.* 2:26.

Hurgobin, B., Tamiru-Oli, M., Welling, M.T., Doblin, M.S., Bacic, A., Whelan, J.A., Lewsey, M.G. 2021. Recent advances in *Cannabis sativa* genomics research. *New Phytol.* 230:73–89.

Ibrahim, E.A., Wang, M., Radwan, M.M., Wanas, A.S., Majumdar, C.G., Avula, B., Wang, Y.H., Khan, I.A., Chandra, S., Lata, H., et al. 2019. Analysis of terpenes in *Cannabis sativa* L. using GC/MS: method development, validation, and application. *Planta Med.* 85:431–438.

Jin, D., Dai, K., Xie, Z., et al. 2020. Secondary metabolites profiled in Cannabis inflorescences, leaves, stem barks, and roots for medicinal purposes. *Sci. Rep.* 10:3309.

Jin, D., Henry, P., Shan, J., Chen, J. 2021. Identification of chemotypic markers in three chemotype categories of cannabis using secondary metabolites profiled in inflorescences, leaves, stem bark, and roots. *Front. Plant Sci.* 12:699530.

Kaur, J., Sun, N., Hill, J.E. 2023. Comprehensive profiling of terpenes and terpenoids in different Cannabis strains using GC × GC-TOFMS. *Separations* 10:500.

Kovalchuk, I., Pellino, M., Rigault, P., van Velzen, R., Ebersbach, J., Ashnest, J.R., Mau, M., Schranz, M., Alcorn, J., Laprairie, R. 2020. The genomics of cannabis and its close relatives. *Annu. Rev. Plant Biol.* 71:713–739.

Kowal, M.A., Hazekamp, A., Grotenhermen, F. 2016. Review on clinical studies with cannabis and cannabinoids 2010-2014. *Cannabinoids* 11:1–18.

Krill, C., Rochfort, S., Spangenberg, G.A. 2020. High-Throughput method for the comprehensive analysis of terpenes and terpenoids in medicinal Cannabis biomass. *Metabolites* 10:276.

Kumeroa, F., Komahan, S., Sofkova-Bobcheva, S., Clavijo McCormick, A. 2022. Characterization of the volatile profiles of six industrial Hemp (*Cannabis sativa* L.) Cultivars. *Agronomy* 12:2651.

Lavender, I., McCartney, D., Marshall, N., Suraev, A., Irwin, C., D'Rozario, A.L., Gordon, C.J., Saini, B., Grunstein, R.R., Yee, B., McGregor, I., Hoyos, C.M. 2023. Cannabinol (CBN, 30 and 300 mg) effects on sleep and next-day function in insomnia disorder ('CUPID' study): protocol for a randomised, double-blind, placebo-controlled, cross-over, three-arm, proof-of-concept trial. *BMJ Open* 13:e071148.

Laverty, K.U., Stout, J.M., Sullivan, M.J., Shah, H., Gill, N., Holbrook, L., Deikus, G., Sebra, R., Hughes, T.R., Page, J.E. 2019. A physical and genetic map of *Cannabis sativa* identifies extensive rearrangements at the THC/CBD acid synthase loci. *Genome Res.* 29:146–156.

Leghissa, A., Hildenbrand, Z.L., Schug, K.A. 2018a. A review of methods for the chemical characterization of cannabis natural products. *J. Sep. Sci.* 41:398–415.

Leghissa, A., Hildenbrand, Z.L., Foss, F.W., Schug, K.A. 2018b. Determination of cannabinoids from a surrogate hops matrix using multiple reaction monitoring gas chromatography with triple quadrupole mass spectrometry. *J. Sep. Sci.* 41:459–468.

Lian, S., Gao, X., Song, C., Li, H., Chen, A., Lin, J. 2022. The characteristics of Raman spectroscopy of isomer CBD- and THC-Au nanoparticles using the density functional theory. *Spectrochim. Acta Part A: Mol. Biomol. Spectrosc.* 268:120682.

Ligresti, A., Moriello, A.S., Starowicz, K., Matias, I., Pisanti, S., De Petrocellis, L., et al. 2006. Antitumor activity of plant cannabinoids with emphasis on the effect of cannabidiol on human breast carcinoma. *J. Pharmacol. Exp. Ther.* 318:1375–1387.

Liu, Y., Zhu, P., Cai, S., Haughn, G., Page, J.E. 2021. Three novel transcription factors involved in cannabinoid biosynthesis in *Cannabis sativa* L. *Plant Mol. Biol.* 106:49–65.

Livingston, S., Quilichini, T., Booth, J., Wong, D., Rensing, K., Laflamme-Yonkman, J., Castellarin, S., Bohlmann, J., Page, J., Samuels, A.L. 2020. Cannabis glandular trichomes alter morphology and metabolite content during flower maturation. *Plant J.* 101:37–56.

Lynch, R.C., Vergara, D., Tittes, S., White, K., Schwartz, C.J., Gibbs, M.J., Ruthenburg, T.C., deCesare, K., Land, D.P., Kane, N.C. 2016. Genomic and Chemical Diversity in Cannabis. *Crit. Rev. Plant Sci.* 35:349–363.

Maayah, Z.H., Takahara, S., Ferdaoussi, M., Dyck, J.R.B. 2020. The antiinflammatory and analgesic effects of formulated full-spectrum cannabis extract in the treatment of neuropathic pain associated with multiple sclerosis. *Inflamm. Res.* 69:549–558.

Magagnini, G., Grassi, G., Kotiranta, S. 2018. The Effect of Light Spectrum on the Morphology and Cannabinoid Content of *Cannabis sativa* L. *Med. Cannabis Cannabinoids* 1:19–27.

Markley, J.L., Brüschweiler, R., Edison, A.S., Eghbalnia, H.R., Powers, R., Raftery, D., Wishart, D.S. 2017. The future of NMR-based metabolomics. *Curr. Opin. Biotechnol.* 43:34–40.

McPartland, J.M. and Russo, E.B. 2001. Cannabis and cannabis extracts: Greater than the sum of their parts? *J. Cannabis Ther.* 1:103–132.

McPartland, J.M., MacDonald, C., Young, M., Grant, P.S., Furkert, D.P., Glass, M. 2017. Affinity and efficacy studies of tetrahydrocannabinolic acid A at cannabinoid receptor types one and two. *Cannabis Cannabinoid Res.* 2:87–95.

McPartland, JM. 2018. Cannabis systematics at the levels of family, genus, and species. *Cannabis Cannabinoid Res.* 3:203–12

Morales, P., Hurst, D.P., Reggio, P.H. 2017. Molecular targets of the phytocannabinoids: A complex picture. *Prog. Chem. Org. Nat. Prod.* 103:103–131.

O'Croinin, C., Le, T.S., Doschak, M., Löbenberg, R., Davies, N.M. 2023. A validated method for detection of cannflavins in hemp extracts. *J. Pharm. Biomed. Anal.* 235:115631.

Oswald, I.W.H., Ojeda, M.A., Pobanz, R.J., Koby, K.A., Buchanan, A.J., Del Rosso, J., Guzman, M.A., Martin, T.J. 2021. Identification of a new family of prenylated volatile sulfur compounds in Cannabis revealed by comprehensive two-dimensional gas chromatography. *ACS Omega* 6:31667–31676.

Pacula, R.L.. Smart, R. 2017. Medical Marijuana and Marijuana Legalization. *Ann. Rev. Clin. Psychol.* 13:397–419.

Pandey, V.K., Rayees, I.U.L., Shams, R., Dar, A.H. 2022. A comprehensive review on the application of essential oils as bioactive compounds in Nano-emulsion based edible coatings of fruits and vegetables. *Appl. Food Res.* 2:100042.

Pandita, D., Pandita, A., Wani, S.H., Abdelmohsen, S.A.M., Alyousef, H.A., Abdelbacki, A.M.M., Al-Yafrasi, M.A., Al-Mana, F.A., Elansary, H.O. 2021. Crosstalk of multi-omics platforms with plants of therapeutic importance. *Cells* 10:1296.

Pavlovic, R., Panseri, S., Giupponi, L., Leoni, V., Citti, C., Cattaneo, C., Cavaletto, M., Giorgi, A. 2019. Phytochemical and ecological analysis of two varieties of hemp (*Cannabis sativa L.*) grown in a mountain environment of Italian Alps. *Front. Plant Sci.* 10:1265.

Pellati, F., Brighenti, V., Sperlea, J., Marchetti, L., Bertelli, D., Benvenuti, S. 2018. New methods for the comprehensive analysis of bioactive compounds in *Cannabis sativa* L. (hemp). *Molecules* 23:2639.

Phan, K.L., Angstadt, M., Golden, J., Onyewuenyi, I., Popovska, A., de Wit, H. 2008. Cannabinoid modulation of amygdala reactivity to social signals of threat in humans. *J Neurosci.* 28:2313–9.

Piluzza, G., Delogu, G., Cabras, A., Marceddu, S., Bullitta, S. 2013. Differentiation between fiber and drugs of hemp (*Cannabis sativa* L.) from a collection of wild and domesticated accessions. *Genet. Resour. Crop Evol.* 60:2331–2342.

Procaccia, S., Lewitus, G.M., Lipson Feder, C., Shapira, A., Berman, P., Meiri, D. 2022. Cannabis for medical use: Versatile plant rather than a single drug. *Front. Pharmacol.* 13:894960.

Puri, S., Sahal, D., Sharma, U. 2021. A conversation between hyphenated spectroscopic techniques and phytometabolites from medicinal plants. *Anal. Sci. Adv.* 2:579–593.

Raharjo, T.J., Verpoorte, R. 2004. Methods for the analysis of cannabinoids in biological materials: a review. *Phytochem. Anal.* 15:79–94.

Ramirez, C.L., Fanovich, M.A., Churio, M.S. 2018. Cannabinoids: Extraction methods, analysis, and physicochemical characterization. *Stud. Nat. Prod. Chem.* 61:143–173.

Raucci, U., Pietrafusa, N., Paolino, M.C., Di Nardo, G., Villa, M.P., Pavone, P., Terrin, G., Specchio, N., Striano, P., Parisi, P. 2020. Cannabidiol treatment for refractory epilepsies in pediatrics. *Front. Pharmacol.* 11:586110.

Raup-Konsavage, W.M., Sepulveda, D.E., Wang, J., Dokholyan, N.V., Vrana, K.E., Graziane, N.M. 2024. Antinociceptive effects of cannabichromene (CBC) in mice: Insights from von frey, tail-flick, formalin, and acetone tests. *Biomedicines* 12:83.

Rea, K.A., Casaretto, J.A., Al-Abdul-Wahid, M.S., Sukumaran, A., Geddes-McAlister, J., Rothstein, S.J., Akhtar, T.A. 2019. Biosynthesis of cannflavins A and B from *Cannabis sativa* L. *Phytochemistry* 164:162–171.

Reimann-Philipp, U., Speck, M., Orser, C., Johnson, S., Hilyard, A., Turner, H., Stokes, A.J., Small-Howard, A.L. 2020. Cannabis chemovar nomenclature misrepresents chemical and genetic diversity, survey of variations in chemical profiles and genetic markers in Nevada Medical Cannabis samples. *Cannabis Cannabinoid Res.* 5:215–230.

Rodriguez-Morrison, V., Llewellyn, D., Zheng, Y. 2021. Cannabis inflorescence yield and cannabinoid concentration are not increased with exposure to short-wavelength ultraviolet-B radiation. *Front Plant Sci.* 12:725078.

Rubino, T., Guidali, C., Vigano, D., Realini, N., Valenti, M., Massi, P., et al. 2008. CB1 receptor stimulation in specific brain areas differently modulate anxiety-related behaviour. *Neuropharmacology* 54:151–60.

Russo, E.B. 2011. Taming THC: potential cannabis synergy and phytocannabinoid-terpenoid entourage effects. *Br. J. Pharmacol.* 163:1344–64.

Russo, E.B., Marcu, J. 2017. Cannabis pharmacology: The usual suspects and a few promising leads. *Adv. Pharmacol.* 80:67–134.

Russo, E.B. 2019. The case for the entourage effect and conventional breeding of clinical cannabis: No "Strain," no gain. *Front. Plant Sci.* 9:1969.

Russo, E.B., Jiang, H.E., Li, X., Sutton, A., Carboni, A., Del Bianco, F., Mandolino, G., Potter, D.J., Zhao, Y.X., Bera, S., et al. 2008. Phytochemical and genetic analyses of ancient cannabis from Central Asia. *J. Exp. Bot.* 59:4171–4182.

Saeed, F., Afzaal, M., Raza, M.A., Rasheed, A., Hussain, M., Nayik, G.A., Ansari, M.J. 2023. Lavender essential oil: Nutritional, compositional, and therapeutic insights. In *Essential Oils*, ed. Nayik, G.A., Ansari, M.J., 85–101. Academic Press: New York.

Salami, S.A., Martinelli, F., Giovino, A., Bachari, A., Arad, N., Mantri, N. 2020. It is our turn to get cannabis high: Put cannabinoids in food and health baskets. *Molecules* 25:4036.

Salem, S.S., Elsayed. H.E., Shabana, S., Khazaal, M.T., Moharram, F.A. 2023. Phytochemical profile and antimicrobial activity of essential oils from two Syzygium species against selected oral pathogens. *BMC Complement. Med.* 23:448.

Sammani, M.S., Clavijo, S., Cerdà, V. 2021. Recent, advanced sample pretreatments and analytical methods for flavonoids determination in different samples. *TrAC Trends Analyt. Chem.* 138:116220.

Schwabe, A.L., Johnson, V., Harrelson, J., McGlaughlin, M.E. 2023. Uncomfortably high: Testing reveals inflated THC potency on retail Cannabis labels. *PLoS ONE* 18:e0282396.

Sikora, V., Berenji, J., Latković, D. 2011. Influence of agroclimatic conditions on content of main cannabinoids in industrial hemp (*Cannabis sativa L.*). *Genetika* 43:449–456.

Sirangelo, T.M., Ludlow, R.A., Spadafora, N.D. 2022. Multi-Omics approaches to study molecular mechanisms in *Cannabis sativa. Plants* 11:2182.

Sirangelo, T.M., Ludlow, R.A., Spadafora, N.D. 2023a. Molecular mechanisms underlying potential pathogen resistance in *Cannabis sativa. Plants* 12:2764.

Sirangelo, T.M. 2023b. NLR- and *mlo*-based resistance mechanisms against powdery mildew in *Cannabis sativa. Plants* 13:105.

Small, E., Cronquist, A. 1976. A practical and natural taxonomy for Cannabis. *Taxon* 4:405–435.

Tahir, M.N., Shahbazi, F., Rondeau-Gagné, S., Trant, J.F. 2021. The biosynthesis of the cannabinoids. *J. Cannabis Res.* 3:7.

Taiti, C., Masi, E., Cicaloni, V., Vinciguerra, V., Salvini, L., Garzoli, S. 2022. SPME-GC-MS and PTR-ToF-MS techniques for the profiling of the metabolomic pattern of VOCs and GC-MS for the determination of the cannabinoid content of three cultivars of *Cannabis sativa L.* Pollen. *Molecules* 27:8739.

Tanney, C.A.S., Backer, R., Geitmann, A., Smith, D.L. 2021. Cannabis glandular trichomes: A cellular metabolite factory. *Front. Plant Sci.* 12:721986.

Trojan, V., Landa, L., Šulcová, A., Slíva, J., Hřib, R. 2023. The main therapeutic applications of cannabidiol (CBD) and its ptential effects on aging with respect to Alzheimer's disease. *Biomolecules* 13:1446.

Van Bakel, H., Stout, J., Cote, A., Tallon, C., Sharpe, A., Hughes, T., Page, J. 2011. The draft genome and transcriptome of *Cannabis sativa. Genome Biol.* 12:102.

Vásquez-Ocmín, P.G., Marti, G., Bonhomme, M., Mathis, F., Fournier, S., Bertani, S., Maciuk, A. 2021. Cannabinoids vs. whole metabolome: Relevance of cannabinomics in analyzing Cannabis varieties. *Anal. Chim. Acta* 1184:339020.

Vergara, D., Baker, H., Clancy, K., Keepers, K.G., Mendieta, J.P., Pauli, C.S., Tittes, S.B., White, K.H., Kane, N.C. 2016. Genetic and genomic tools for *Cannabis sativa. Crit. Rev. Plant Sci.* 35:364–377.

Vergara, D., Huscher, E.L., Keepers, K.G., Givens, R.M., Cizek, C.G., Torres, A., Gaudino, R., Kane, N.C.J.B. 2019. Gene copy number is associated with phytochemistry in Cannabis sativa. *AoB Plants* 11:plz074.

Vergara, D., Feathers, C.; Huscher, E.L., Holmes, B., Haas, J.A. Kane, N.C. 2021. Widely assumed phenotypic associations in *Cannabis sativa* lack a shared genetic basis. *PeerJ* 9:e10672.

Vitalini, S., Iriti, M., Vinciguerra, V., Garzoli, S. 2021. A comparative study of the chemical composition by SPME-GC/MS and antiradical activity of less common citrus species. *Molecules* 26:5378.

Walsh, K.B., McKinney, A.E., Holmes, A.E. 2021. Minor cannabinoids: Biosynthesis, molecular pharmacology and potential therapeutic uses. *Front. Pharmacol.* 12:777804.

Wang, X., Harrington, P.D.B., Baugh, S.F. 2018. Effect of preprocessing high-resolution mass spectra on the pattern recognition of Cannabis, hemp, and liquor. *Talanta* 180:229–238.

Weiblen, G.D., Wenger, J.P., Craft, K.J., ElSohly, M.A., Mehmedic, Z., Treiber, E.L., Marks, M.D. 2015. Gene duplication and divergence affecting drug content in *Cannabis sativa. New Phytol.* 208:1241–1250.

Woods, P., Campbell, B.J., Nicodemus, T.J., Cahoon, E.B., Mullen, J.L., McKay, J.K. 2021. Quantitative trait loci controlling agronomic and biochemical traits in *Cannabis sativa. Genetics* 219:iyab099.

Zager, J.J., Lange, I., Srividya, N., Smith, A., Lange, B.M. 2019. Gene networks underlying cannabinoid and terpenoid accumulation in Cannabis. *Plant Physiol.* 180:1877–1897.

Zamengo, L., Bettin, C., Badocco, D., Di Marco, V., Miolo, G., Frison, G. 2019. The role of time and storage conditions on the composition of hashish and marijuana samples: A four-year study. *Forensic Sci. Int.* 298:131–137.

12 Genome Engineering and Gene Editing of Cannabis for Crop Improvement

Chandrakanth Emani, Arturo Ortega, Julie Duong, Marcus Parungao, Angel Fernandez Marti and Ajith Anand

12.1 INTRODUCTION

Cannabis sativa L., commonly known as marijuana or hemp, is an annual primarily dioecious flowering plant that has diverse agricultural and pharmaceutical uses. The plant is native to Central Asia and was first cultivated in Asia and Europe before expanding as one of the popular globally cultivated crops (Schultes et al., 1974). The cultivation of industrial hemp is now a global profitable business venture generating a diverse array of economically important products ranging from pharmaceutical cannabinoids, nutritional seeds, and seed oil, fiber, to textile materials, construction materials, and biocomposites. The pharmaceutical importance of the crop is indicated by the recent spurt in health products derived from many hemp and cannabis varieties.

Hemp as a fiber crop was domesticated by humans 12,000 years ago (Ren et al., 2021) and is a multipurpose plant grown for thousands of years for its fiber, oil, and nutritional value (Hillig, 2005; Clarke & Merlin, 2013; Clarke & Merlin, 2016; Ren et al., 2021). Hemp can grow in a wide range of climates in most regions of the world and can be cultivated on farmlands that would otherwise be unsuitable for other crops, and is known to contribute to soil replenishment by eliminating heavy metals and other impurities. The seeds of hemp are high in vitamins, minerals, and nutrients and an excellent source of proteins, providing numerous health benefits. Cannabis is a biofactory of bioactive compounds which is illustrated by the wide range of secondary metabolites, of which the phytocannabinoids are the most valuable from a pharmaceutical perspective (Radwan et al., 2009; Fischedick et al., 2010). Almost a thousand compounds fall into different categories including, 278 cannabinoids, 174 terpenes, 221 terpenoids, 19 flavonoids, 63 flavonoid glycosides, 46 polyphenols, 92 steroids have been identified (Gould, 2015; Radwan et al., 2009). Nearly 278 of these compounds are cannabinoids and classified as phytocannabinoids (plant-based) to distinguish them from endocannabinoids (non-plant). The major cannabinoids in cannabis include Δ9-tetrahydrocannabinol (THC), cannabidiol (CBD), and cannabichromene (CBC), their precursor cannabigerol (CBG) and cannabinol (CBN) (Flores-Sanchez & Verpoorte, 2008). Cannabigerolic acid (CBGA), a CBG-type cannabinoid, is the central precursor for the biosynthesis of psychoactive THC, non-psychoactive CBD, and CBC (ElSohly & Slade, 2005; Gould, 2015; Radwan et al., 2009). The classification of cannabis has been reported using the biochemical composition of the cannabinoid content, in particular the THC/CBD ratio, in several publications. Chemical phenotypes (chemotypes) can be used to define cannabis varieties with different chemical variants and different morphological features to classify *C. sativa* into three principal classes differing in their THC/CBD ratios: chemotype I (drug type plants which exhibit a THC/CBD ratio well beyond 1.0 especially due to high THC content); chemotype II (an intermediate ratio of 0.5–2.0); and chemotype III (fiber-type plants that have a ratio well below 1.0, with a prevalence of CBD) [Hurgobin et al., 2021]. The

DOI: 10.1201/9781003386438-12

accumulation of cannabinoids is dependent on the age and tissue type of the plant with the glandular trichomes of the female flowers and the tiny leaves grouped within the inflorescence containing the highest quantities (Kim & Mahlberg, 1997). Cannabinoids and terpenes are among the many valuable secondary metabolites found in cannabis. Cannabinoids with documented medical properties include CBD, CBG, CBN, CBC, cannabidivarin (CBDV), cannabicyclol (CBL), cannabivarin (CBV), cannabiichromevarin (CBCV), and cannabielsoin (CBE) (Shiels et al., 2022).

With the legalization of cannabis in many countries, the demand for advanced genetic enhancement technology in cannabis is gradually expanding (Hesami et al., 2021). The growing applications in cannabis plants necessitate the need for development of novel cultivars with enhanced desirable traits and classical breeding approaches have proven to be time-consuming and labor intensive. Molecular breeding strategies, especially with marker-assisted selection, have catalyzed the traditional breeding approaches, but the development of such strategies has faced contextual challenges with cannabis. Moreover, the cannabinoid and terpene profile can be compromised while transferring agronomically important traits through conventional breeding (Ingvardsen & Brinch-Pedersen, 2023). The recent advances made in molecular disease diagnostics, marker development, plant tissue culture, and genome modification technologies have revolutionized the plant biotechnology industry and can prove effective for the cannabis industry (Adams et al., 2021; Simiyu et al., 2022; Ingvardsen & Brinch-Pedersen, 2023). The more recent biotechnological innovations of plant genetic transformation and genome editing technologies such as CRISPR (Clustered Regularly Interspaced Short Palindromic Repeats) gene editing potentially offer precise and effective tools to develop cultivars with targeted traits more efficiently while advancing the cannabis plant breeding program to the next level. Gene editing also overcomes the biggest challenge of breeding strategies where hemp's dioecious nature and inability to self-pollinate prevents many breeding approaches to obtain homozygous plants and stable quality seeds. The ability to modify genes on both alleles within a plant with gene editing technologies enables researchers to produce homozygous lines in a single generation (Deguchi et al., 2020). The traits of interest and targets of gene editing in cannabis include genes involved in cannabinoid pathways, fiber, disease resistance, nutritional and oil quality. The present review is an overview of genome modification of cannabis with a focus on the multiple opportunities, potential challenges, and strategies to overcome the limitations.

12.2 CANNABIS GENOMICS

12.2.1 The Cannabis Genome

Cannabis is a dioecious plant with a diploid genome (2n=20) containing nine pairs of autosomes and a pair of sex (XX and XY) chromosomes (Sakamoto et al., 1998). The Y chromosome is larger than the X chromosome with the female haploid genome measuring about 818 Mb and a male haploid genome of 843 Mb (Sakamoto et al. 1998; Ming et al., 2011). The first high-quality reference genome of wild *C. sativa* China varieties measured approximately 808 Mb with scaffold and contig N50 sizes of 83Mb and 513.57kb (Gao et al., 2020). The genome showcases 38,828 protein-coding genes with 74.75% repetitive elements (Gao et al., 2020). A more recent comprehensive draft genome sequence assembly of ~900Mb was generated from medical cannabis strain Cannobio-2, which produces a balanced ratio of cannabidiol and Δ9-tetrahydrocannabinol (Braich et al., 2020). The total predicted features from this version of a final gene set were 109,686 with 36,632 genes, 37,107 mRNA, and 35,947 proteins (Braich et al., 2020).

12.3 GENETIC ENGINEERING AND GENE EDITING

Recombinant DNA technology, or genetic engineering, refers to methodologies that isolate genetic material, particularly DNA from various biological species, and then introduce the resulting recombinant or transgenic DNA into an organism to create novel heritable genetic material. On the other hand, genome or gene editing is the precise and predictable indigenous modification of a specific genome. Numerous genes could be altered in various cell types and organisms using nucleases that provide targeted alterations (Pabo et al., 2001). NHEJ (Non-Homologous End Joining) random insertions or deletions in the coding region result in frameshift mutations (indels, deletions, insertions), resulting in gene knockouts or non-functional proteins. The application of gene editing tools such as CRISPR technology opens up the potential for analyzing gene expression, loss-of-function, and gain-of-function. CRISPR has numerous applications in plant biology and can be easily used to create high-quality agriculturally sustainable products (Puchta, 2005).

12.3.1 Genetic Transformation

Genetic transformation has the potential to advance cannabis breeding significantly. This has been observed in other plant species, where these techniques have resulted in genetically improved varieties exhibiting resistance to both biotic and abiotic stresses, improved nutritional and processing qualities, and increased yields, among other desirable traits (Gosal & Wani, 2018). Several well-established genetic transformation techniques have been demonstrated to effectively and durably insert novel genes into the nuclear genomes of various plant species (Feng et al., 2023; Su et al., 2023). Genome engineering has successfully modified and improved various traits in important agronomic crops (Li et al., 2022). The shape and color or grain and fruit have been improved in wheat, rice, and tomato by knocking out various negative regulators using Non-homologous end Joining (NHEJ) (Liu et al., 2002; Xu et al., 2016; Rodrigez-Leal et al., 2017; Wang et al., 2019; Yuyu et al., 2020). Similarly, the biofortification of several crop plants including increasing B-carotene content in rice, banana, and tomato (Li et al., 2018; Dong et al., 2020; Kaur et al., 2020), and micronutrients in rice and wheat was achieved (Che et al., 2021; Sun et al., 2021).

Notwithstanding the considerable progress made in technology over several decades, the lack of an efficient, robust, and reliable plant transformation and regeneration continues to pose a significant challenge to the genetic improvement of cannabis and hemp crops (Ramkumar et al., 2020). The first crucial step in developing a robust transformation system is to develop an efficient tissue culture, DNA-delivery, and regeneration protocol using explants that are amenable to tissue culture and transformation. While the micropropagation system has enabled researchers and producers in the rapid propagation of insect, disease, and virus-free clonal plants that aid both germplasm storage and provide ideal material for biotechnological manipulations (Monthony et al., 2021), these protocols are not developed for plant transformation or gene editing in cannabis. Several academic labs and industries have successfully developed micropropagation protocols by tweaking factors such as media composition (MS and DKW media), explant type (shoot tips, inflorescences, florets, axillary node, hypocotyls, cotyledons, leaves, and roots), and genotype (Monthony et al., 2021). There are reports of establishing reliable tissue culture and plant regeneration protocols using various explants such as leaf, hypocotyl, and cotyledon (Farag, 2014; Movahedi et al., 2015; Chaohua et al., 2016; Galan-Avila et al., 2020; Holmes et al., 2021; Monthony et al., 2021), which has opened the possibility for genetic modification of cannabis and hemp.

Transformation protocols for cannabis by *Agrobacterium*-mediated methodology examined various parameters for transient gene expression in various explants (Mohammad et al., 2024), and stable transformation methods using hypocotyls and cotyledons were reported (Galan-Avila et al., 2021; Petersen et al., 2021). Other genetic transformation methods include a transient transformation system for a protoplast-based method (Beard et al., 2021) and particle bombardment (Petersen et al., 2021). We have recently established a protocol for a rapid *Agrobacterium*-mediated transformation using mature seeds (Figure 12.1; provisional patent filed) for multiple hemp cultivars, which is highly efficient producing stably transformed and gene-edited events in less than 60 days.

Figure: The different steps involved in the gene editing process in cannabis. (a) Germinating uncoated seedlings co-cultivated as the choice explant;

FIGURE 12.1 Genetic transformation through direct organogenesis from hypocotyl and cotyledonary explants of hemp and cannabis.

(b) Germinating seedlings on cocultivation medium after infection with Agrobacterium harboring the gene editing cassette; (c) Putative transgenic shoots regenerating on a selection medium; (d) Putative transgenic plant in a culture tub on a selection medium; (e) PCR detection of gene-edited events which are positive for Cas9 (top row) and free of Agrobacterium (bottom row). The PCR was performed with gene-specific primers including Cas9 (gene cassette) and VirD2 (Agrobacterium).

12.3.2 Gene Editing in Cannabis

Unlike traditional breeding methods, targeted gene editing allows for precise modifications of plant genomes without changing the overall genetic makeup of the plant. Zinc finger nucleases (ZFNs) were one of the first programmable tools used for gene editing in various species, including plants. Transcription activator-like effector nuclease (TALEN) then became the 2nd generation of gene editors which allowed for more precision and less cytotoxicity. The discovery of the CRISPR/Cas (Clustered Regularly Interspaced Short Palindromic Repeats-CRISPR associated protein) system pushed science into the next generation of gene editing technologies and CRISPR/Cas quickly became the most ubiquitous programmable nuclease in gene editing platforms. The CRISPR/Cas system was discovered in bacteria in 1987 (Ishino Y et al., 1987), and was originally characterized as an endogenous adaptive immune system used by bacteria against bacteriophages. CRISPR/Cas-based gene editing technologies have revolutionized plant science and agriculture, facilitating the modification of genomes and enhancement of crop traits. Advancements in gene editing in plants give the potential for application in *Cannabis sativa*, a medicinal and fiber crop of significant value worldwide.

CRISPR-Cas9 is currently the most ubiquitous gene editing tool used across different crop species due to its precision, ease of use, and versatility. It employs a single guide RNA (sgRNA) sequence to direct the Cas9 enzyme, a specific class of Cas protein, to a specific DNA sequence, where it cleaves the DNA leaving a double-strand break (Westra et al., 2012). Following this, endogenous plant DNA repair mechanisms repair the break via molecular pathways like non-homologous end joining (NHEJ) or homology-directed repair (HDR). NHEJ repair pathways are error prone and may lead to mutations whereas the HDR pathway is more precise due to its employment of another uncleaved DNA strand as a repair template. If a precise gene edit is the intended target for a plant genome, typically a DNA donor template with the desired gene edit is introduced along with CRISPR-Cas9.

Other RNA-guided nuclease complexes such as CRISPR-Cas12 and CRISPR-Cas13 have shown promise in plants, offering mechanisms and specificities that can be tailored to different editing requirements (Schindele P et al., 2018). These CRISPR-Cas enzymes have allowed for a range of genetic modifications, from gene knockouts to the introduction of precise gene edits in crops (Wada N et al., 2022). As of yet, their employment in cannabis to enable the fine-tuning of cannabis traits has been explored limitedly.

Previously, CRISPR-Cas has been used across many plant species for the modifications of genes related to abiotic stress to improve crop performance against various environmental stressors including salinity, drought, temperature, heavy metal, etc (Kumar, 2023; Shi, 2017). Many of the orthologs of genes involved in abiotic stress previously discovered in other plants are present in cannabis genomes and thus may be potential gene editing targets to confer resistance to sources of abiotic stress. As for biotic stresses, CRISPR/Cas has also been used to edit plant genes involved in disease resistance/susceptibility to confer resistance to bacteria, fungi, insects, and viruses (Wang, 2016; Paul et al., 2021). Regarding cannabis, the establishment of gene-edited cannabis plants with resistance to pathogens like powdery mildew or hop latent viroid has the potential to improve crop loss and reduce the use of chemical intervention, making cannabis agricultural practices more sustainable. Traits such as flowering time, plant height, and biomass production can be optimized using CRISPR-Cas tools. For example, editing genes involved in photoperiod sensitivity can help develop cannabis varieties that flower independently of light cycles, making them suitable for various growing environments as well as indoor growing. Additionally, improvement of biomass production may allow for cannabis to be a target crop for carbon sequestration. Lastly, gene editing can significantly shorten breeding cycles, accelerating the development of unlimited iterations of cannabis cultivars with discerning traits, stable genetics, and high-quality seeds. Thus reducing the industry's reliance on cloning mother plants, micropropagation, and currently employed breeding practices for trait discovery and product development.

One of the most promising applications of gene editing in cannabis is the modification of cannabinoid profiles in specific cultivars. CRISPR-Cas can target and modify genes involved in cannabinoid biosynthesis pathways. This could allow for the development of cultivars with tailored levels of THC, CBD, and other cannabinoids, optimizing them for medicinal and recreational use. Beyond cannabinoids, cannabis produces a variety of secondary metabolites, including terpenes and flavonoids. These metabolites contribute to the plant's aroma and therapeutic properties, which are sought after in medicinal and recreational contexts. Gene editing can upregulate or downregulate the production of specific terpenes and flavonoids in certain cultivars, allowing for the customization of flavor and medicinal properties.

To facilitate accuracy during the gene editing process, the precise genetic sequence of the cultivar of interest is needed to design sgRNAs. This entails full sequencing of the region spanning the gene to be targeted by CRISPR-Cas. The relatively limited genomic information available for cannabis cultivars compared to other crops is an issue for identifying target genes. Further, the availability of the full genomic sequence for a certain cultivar can offer the opportunity to analyze for possible off-target effects (regions of the genome that are unintended targets). Future endeavors should include expanding genomic resources for specific cannabis cultivars.

The availability of a high-quality de novo reference genome of cannabis (Gao et al., 2020) and using the current transformation system (Sorokin et al., 2020; Ahmed et al., 2021) one could attempt to edit agronomically important traits in cannabis. Gene editing of the phytoene desaturase (PDS) gene to produce albino cannabis plantlets has been reported (Zhang, 2021). The same group also reported improvements in cannabis transformation with the introduction of GRF3-GIF1 developmental genes. Due to the redundancy of the THCAS gene in the cannabis genome, it would be prudent to use CRISPR/Cas9 technology to eliminate THCAS gene copies via multiple gene knockout with a single gene editing step to produce THC-free varieties. Glandular trichomes are where the majority of cannabinoids are generated and stored by cannabis plants (Tanney et al., 2021). Transgenic *C. sativa* overexpressing *Artemisia annua* MYB1 gene showed a higher number of trichomes than control plants. The homologs of AaMYB1 from the cannabis genome can be edited to increase the number of trichomes that will, in turn, increase the amount of cannabinoids (Sayre et al., 2022; Yin et al., 2022).

Scientists have known for a long time that THC and CBD are different derivatives of the same precursor cannabigerolic acid (CBGA). So it's always been a keen interest for researchers to study the pathways underlying these processes leading to the formation of different compounds. The enzymes that turn CBGA into tetrahydrocannabinol or cannabidiolic acid, which are then changed into THC or CBD upon heating, were encoded by either one gene with two variations or by two closely related yet separate genes (De Meijer et al., 2003). The genome sequence, of strain Purple Kush, had a functional THC synthase gene, as was to be expected. However, it also contained numerous copies of the CBD synthase gene that were inactive due to premature stop signals or other mutations (Van Bakel et al., 2011). Hemp-based bioplastics have shown good potential and may outperform traditional polymers in some ways. They also provide a more sustainable and environmentally friendly alternative to petroleum-based plastics (Fike, 2016).

Cannabis produced indoors is vulnerable to various microbes and insect pests, thus, effective disease management techniques are needed to prevent crop loss (Punja, 2021). While indoor cultivation under strictly regulated circumstances can enhance yields, it is one of the most energy-intensive industries with extremely high operational expenses (Warren et al., 2019). The viable option is to grow outdoors in fields that can reduce production costs but pose various environmental stresses to plants. Plants respond to stress in various ways, including changes in gene expression, physiology, organ and tissue architecture, and changes in primary and secondary metabolism. The complexity of this reaction is greatly determined by the duration and intensity of the stress, the simultaneous occurrence of many stressors, the plant genotype, and the developmental period at which the stress is applied (Cramer et al., 2011; Suzuki et al., 2014; He et al., 2018; Zhang et al., 2020).

In cannabis, biotic and abiotic stress factors affect the production and accumulation of secondary metabolites. For instance, drought increases the production of CBG while significantly reduces the production of other cannabinoids like THC and CDB in a high CBD chemovar (Park et al., 2022). High salinity and UV stress reduces the accumulation of cannabinoids in cannabis (Rodriguez-Morrison et al., 2021; Yep & Zheng, 2021). A few genes have been explored for potential use for enhancement of abiotic stress tolerance in cannabis. A group studied the expression of cannabis NAC transcription factors in tobacco under high salinity stress (Liu et al., 2022). Their results showed that over-expression of CsNAC1, CsNAC2, and CsNAC3 in tobacco plants enhanced salinity tolerance. A genome-wide expression profiling showed increased transcript abundance of various genes including expansin, B3, POD, and NAC under the drought stress conditions in cannabis (Gao et al., 2018).

A knockout of the THC acid synthase gene via genome editing is a potential method of producing THC-free, high-CBD plants, which would be extremely valuable in countries with laws restricting THC content. Over-expression of the Glabrous Inflorescence (GIS) gene increased the number of trichomes per unit area, ultimately increasing THC and other cannabinoids (Roscow, 2019). In hemp, various potential gene targets for bast fiber production, disease resistance, and seed oil profile can be manipulated using CRISPR/Cas-based genome editing (Behr et al., 2016; Mihalyov et al., 2021).

12.4 CONCLUSION

Despite its promise, genome modification and gene editing in cannabis faces several challenges. The current protocols are not efficient and are not developed for commercial scale production of genome modified crops as is the case of other agronomically important crops. The current legal status of the crop and the lack of investment in science and technology have hindered advancement in genomics, and adaptation of cutting technologies for breeding, crop improvement and public acceptance of a crop that is of immense value. Although CRISPR-Cas gene-edited crops are not considered genetically modified organisms in the United States, regulatory hurdles must be overcome and public perception of cannabis must be addressed to garner wider acceptance. In other regions, gene-edited crops are subject to strict regulations and may require extensive safety assessments and approval processes before market entry. This highlights the importance of addressing regulatory considerations to expedite the adoption of gene-edited cannabis varieties by growers and industry stakeholders.

Gene editing, particularly with CRISPR-Cas9 and other RNA-guided nucleases, holds immense promise for the advancement of cannabis breeding, trait development, and crop improvement. By enabling precise genetic modifications, CRISPR-Cas technologies can help establish new cannabis varieties with enhanced cannabinoid profiles, improved disease resistance, and optimized agronomic traits.

REFERENCES

Adams, T. K., Masondo, N. A., Malatsi, P., & Makunga, N. P. (2021). *Cannabis sativa*: from therapeutic uses to micropropagation and beyond. *Plants*, *10*(10), 2078.

Adesina, I., Bhowmik, A., Sharma, H., & Shahbazi, A. (2020). A review on the current state of knowledge of growing conditions, agronomic soil health practices and utilities of hemp in the United States. *Agriculture*, *10*(4), 129.

Adkar-Purushothama, C. R., Sano, T., & Perreault, J. P. (2023). Hop latent viroid: a hidden threat to the cannabis industry. *Viruses*, *15*(3), 681.

Ahmed, S., Gao, X., Jahan, M. A., Adams, M., Wu, N., & Kovinich, N. (2021). Nanoparticle-based genetic transformation of *Cannabis sativa*. *Journal of Biotechnology*, *326*, 48–51.

Alghanim, H. J., & Almirall, J. R. (2003). Development of microsatellite markers in *Cannabis sativa* for DNA typing and genetic relatedness analyses. *Analytical and Bioanalytical Chemistry*, *376*, 1225–1233.

Aliekperova, N., Kosyachenko, K., & Kaniura, O. (2020). Perspectives on the formation of the medical cannabis market in Ukraine based on a holistic approach. *Journal of Cannabis Research*, *2*, 1–19.

Andre, C. M., Hausman, J. F., & Guerriero, G. (2016). *Cannabis sativa*: the plant of the thousand and one molecules. *Frontiers in Plant Science*, *7*, 19.

Atallah, O. O., Yassin, S. M., & Verchot, J. (2023). New insights into hop latent viroid detection, infectivity, host range, and transmission. *Viruses*, *16*(1), 30.

Barcaccia, G., Palumbo, F., Scariolo, F., Vannozzi, A., Borin, M., & Bona, S. (2020). Potentials and challenges of genomics for breeding cannabis cultivars. *Frontiers in Plant Science*, *11*, 573299.

Beard, K. M., Boling, A. W., & Bargmann, B. O. (2021). Protoplast isolation, transient transformation, and flow-cytometric analysis of reporter-gene activation in *Cannabis sativa* L. *Industrial Crops and Products*, *164*, 113360.

Behr, M., Legay, S., Žižková, E., Motyka, V., Dobrev, P. I., Hausman, J. F., ... & Guerriero, G. (2016). Studying secondary growth and bast fiber development: the hemp hypocotyl peeks behind the wall. *Frontiers in Plant Science*, *7*, 1733.

Borin, M., Palumbo, F., Vannozzi, A., Scariolo, F., Sacilotto, G. B., Gazzola, M., & Barcaccia, G. (2021). Developing and testing molecular markers in *Cannabis sativa* (Hemp) for their use in variety and dioecy assessments. *Plants*, *10*(10), 2174.

Braich, S., Baillie, R. C., Spangenberg, G. C., & Cogan, N. O. (2020). A new and improved genome sequence of *Cannabis sativa*. *Gigabyte*, *2020*, gigabyte10.

Bridgeman, M. B., & Abazia, D. T. (2017). Medicinal cannabis: history, pharmacology, and implications for the acute care setting. *Pharmacy and Therapeutics*, *42*(3), 180.

Cascini, F., Farcomeni, A., Migliorini, D., Baldassarri, L., Boschi, I., Martello, S., ... & Bernardi, J. (2019). Highly predictive genetic markers distinguish drug-type from fiber-type *Cannabis sativa* L. *Plants*, *8*(11), 496.

Chaohua, C., Gonggu, Z., Lining, Z., Chunsheng, G., Qing, T., Jianhua, C., ... & Jianguang, S. (2016). A rapid shoot regeneration protocol from the cotyledons of hemp (*Cannabis sativa* L.). *Industrial Crops and Products*, *83*, 61–65.

Che, J., Yamaji, N., & Ma, J. F. (2021). Role of a vacuolar iron transporter OsVIT2 in the distribution of iron to rice grains. *New Phytologist*, *230*(3), 1049–1062.

Chiginsky, J., Langemeier, K., MacWilliams, J., Albrecht, T., Cranshaw, W., Fulladolsa, A. C., ... & Nachappa, P. (2021). First insights into the virus and viroid communities in hemp (*Cannabis sativa*). *Frontiers in Agronomy*, *3*, 778433.

Clarke, R. C., & Merlin, M. D. (2013). *Cannabis: Evolution and Ethnobotany*. University of California Press.

Clarke, R. C., & Merlin, M. D. (2016). *Cannabis: Evolution and Ethnobotany*. University of California Press.

Cramer G.R., Urano K., Delrot S., Pezzotti M., Shinozaki K. 2011. Effects of abiotic stress on plants: a systems biology perspective. *BMC Plant Biology*, *11*(1), 163.

Deguchi, M., Kane, S., Potlakayala, S., George, H., Proano, R., Sheri, V., ... & Rudrabhatla, S. (2020). Metabolic engineering strategies of industrial hemp (*Cannabis sativa* L.): A brief review of the advances and challenges. *Frontiers in Plant Science*, *11*, 580621.

Dehnavi, M. M., Damerum, A., Taheri, S., Ebadi, A., Panahi, S., Hodgin, G., ... & Taylor, G. (2024). Population genomics of a natural *Cannabis sativa* L. collection from Iran identifies novel genetic loci for flowering time, morphology, sex and chemotyping. *bioRxiv*, 2024-05.

De Meijer, E. P., Bagatta, M., Carboni, A., Crucitti, P., Moliterni, V. C., Ranalli, P., & Mandolino, G. (2003). The inheritance of chemical phenotype in *Cannabis sativa* L. *Genetics*, *163*(1), 335–346.

Dong, O. X., Yu, S., Jain, R., Zhang, N., Duong, P. Q., Butler, C., ... & Ronald, P. C. (2020). Marker-free carotenoid-enriched rice generated through targeted gene insertion using CRISPR-Cas9. *Nature Communications*, *11*(1), 1178.

ElSohly, M. A., & Slade, D. (2005). Chemical constituents of *Cannabis sativa*. In *Cannabis and Cannabinoids: Pharmacology, Toxicology, and Therapeutic Potential*, M. A. ElSohly (Ed.) (pp. 27–36). Haworth Press.

Farag, S. (2014). Cannabinoids production in *Cannabis sativa* L.: an in vitro approach. https://eldorado.tu-dortmund.de/server/api/core/bitstreams/5b33f194-bd89-4f6e-8c8d-7fb7f15f8c02/content

Feng, J., Wang, N., Li, Y., Wang, H., Zhang, W., Wang, H., & Chai, S. (2023). Recent progress in genetic transformation and gene editing technology in cucurbit crops. *Agronomy*, *13*(3), 755.

Fernandez i Marti, A., Parungao, M., Hollin, J., Selimotic, B., Farrar, G., Seyler, T., Anand, A., & Ahmad, R. (2023). A novel, precise and high-throughput technology for viroid detection in cannabis (MFDetect™). *Viruses, 15*(7), 1487. https://doi.org/10.3390/v15071487

Fike, J. (2016). Industrial hemp: renewed opportunities for an ancient crop. *Critical Reviews in Plant Sciences*, *35*(5–6), 406–424.

Fischedick, J. T., Hazekamp, A., Erkelens, T., Choi, Y. H., & Verpoorte, R. (2010). Metabolic fingerprinting of *Cannabis sativa* L., cannabinoids and terpenoids for chemotaxonomic and drug standardization purposes. *Phytochemistry*, *71*(17), 2058–2073.

Flores-Sánchez, I. J., & Verpoorte, R. (2008). Secondary metabolism in Cannabis. *Phytochemistry Reviews*, *7*, 615–639. https://doi.org/10.1007/s11101-008-9094-4

Fulvio, F., Righetti, L., Minervini, M., Moschella, A., & Paris, R. (2023). The B1080/B1192 molecular marker identifies hemp plants with functional THCA synthase and total THC content above legal limit. *Gene*, *858*, 147198.

Galán-Ávila, A., García-Fortea, E., Prohens, J., & Herraiz, F. J. (2020). Development of a direct in vitro plant regeneration protocol from *Cannabis sativa* L. seedling explants: developmental morphology of shoot regeneration and ploidy level of regenerated plants. *Frontiers in Plant Science*, *11*, 534467.

Galán-Ávila, A., Gramazio, P., Ron, M., Prohens, J., & Herraiz, F. J. (2021). A novel and rapid method for agrobacterium-mediated production of stably transformed *Cannabis sativa* L. plants. *Industrial Crops and Products*, *170*, 113691.

Gao, C., Cheng, C., Zhao, L., Yu, Y., Tang, Q., Xin, P., ... & Zang, G. (2018). Genome-wide expression profiles of hemp (*Cannabis sativa* L.) in response to drought stress. *International Journal of Genomics*, *2018*, 3057272.

Gao, S., Wang, B., Xie, S., Xu, X., Zhang, J., Pei, L., ... & Zhang, Y. (2020). A high-quality reference genome of wild *Cannabis sativa*. *Horticulture Research*, *7*, 73–84.

Gilmore, S., Peakall, R., & Robertson, J. (2003). Short tandem repeat (STR) DNA markers are hypervariable and informative in *Cannabis sativa*: implications for forensic investigations. *Forensic Science International*, *131*(1), 65–74.

Gosal, S. S., & Wani, S. H. (2018). Plant genetic transformation and transgenic crops: methods and applications. In *Biotechnologies of Crop Improvement, Volume 2: Transgenic Approaches*, S. S. Gosal & S. H. Wani (Eds) (pp. 1–23). Springer.

Gould, J. (2015). The cannabis crop. *Nature*, *525*, S2–S3.

Grassa, C. J., Weiblen, G. D., Wenger, J. P., Dabney, C., Poplawski, S. G., Timothy Motley, S., ... & Schwartz, C. J. (2021). A new Cannabis genome assembly associates elevated cannabidiol (CBD) with hemp introgressed into marijuana. *New Phytologist*, *230*(4), 1665–1679.

Hadad, L., Luria, N., Smith, E., Sela, N., Lachman, O., & Dombrovsky, A. (2019). Lettuce chlorosis virus disease: a new threat to cannabis production. *Viruses*, *11*(9), 802.

Halpin-McCormick, A., Heyduk, K., Kantar, M. B., Batora, N. L., Masalia, R. R., Law, K. B., & Kuntz, E. J. (2024). Examining population structure across multiple collections of Cannabis. *Genetic Resources and Crop Evolution*, *71*(8), 4705–4722.

He, M., He, C.-Q., Ding, N.-Z. 2018. Abiotic stresses: general defenses of land plants and chances for engineering multistress tolerance. *Frontier Plant Science*, *9*, 1771.

Hesami, M., Baiton, A., Alizadeh, M., Pepe, M., Torkamaneh, D., & Jones, A. M. P. (2021). Advances and perspectives in tissue culture and genetic engineering of Cannabis. *International Journal of Molecular Sciences*, *22*(11), 5671.

Hesami, M., Pepe, M., Alizadeh, M., Rakei, A., Baiton, A., & Jones, A. M. P. (2020). Recent advances in cannabis biotechnology. *Industrial Crops and Products*, *158*, 113026.

Hillig, K. W. (2005). Genetic evidence for speciation in Cannabis (Cannabaceae). *Genetic Resources and Crop Evolution*, *52*, 161–180.

Holmes, J. E., Lung, S., Collyer, D., & Punja, Z. K. (2021). Variables affecting shoot growth and plantlet recovery in tissue cultures of drug-type *Cannabis sativa* L. *Frontiers in Plant Science*, *12*, 732344.

Hsieh, H. M., Hou, R. J., Tsai, L. C., Wei, C. S., Liu, S. W., Huang, L. H., ... & Lee, J. C. I. (2003). A highly polymorphic STR locus in *Cannabis sativa*. *Forensic Science International*, *131*(1), 53–58.

Hurgobin, B., Doblin, M. S., & Bacic, A. (2021). The genome of *Cannabis sativa*. *Nature Plants*, *7*, 1025–1031. https://doi.org/10.1038/s41477-021-00997-6

Ingvardsen, C. R., & Brinch-Pedersen, H. (2023). Challenges and potentials of new breeding techniques in *Cannabis sativa*. *Frontiers in Plant Science*, *14*, 1154332.

Ishino, Y., Shinagawa, H., Makino, K., Amemura, M., Nakata, A. (1987). Nucleotide sequence of the iap gene, responsible for alkaline phosphatase isozyme conversion in Escherichia coli, and identification of the gene product. *Journal of Bacteriology*, *169*, 5429–5433. doi: 10.1128/jb.169.12.5429-5433.1987.

Jindal, S. K., Dhaliwal, M. S., & Meena, O. P. (2020). Molecular advancements in male sterility systems of Capsicum: a review. *Plant Breeding*, *139*(1), 42–64.

Kaur, N., Alok, A., Kumar, P., Kaur, N., Awasthi, P., Chaturvedi, S., ... & Tiwari, S. (2020). CRISPR/Cas9-directed editing of lycopene epsilon-cyclase modulates metabolic flux for β-carotene biosynthesis in banana fruit. *Metabolic Engineering*, *59*, 76–86.

Kim, E-S., Mahlberg, P. G. (1997). Immunochemical localization of tetrahydrocannabinol (THC) in cryofixed glandular trichomes of Cannabis (Cannabaceae). *American Journal of Botany*, *84*(3), 336–342.

Kumar, M., Prusty, M. R., Pandey, M. K., Singh, P. K., Bohra, A., Guo, B., & Varshney, R. K. (2023). Application of CRISPR/Cas9-mediated gene editing for abiotic stress management in crop plants. *Frontiers in Plant Science*, *14*, 1157678.

Li, L., Yu, S., Chen, J., Cheng, C., Sun, J., Xu, Y., ... & Zhang, X. (2022). Releasing the full potential of cannabis through biotechnology. *Agronomy*, *12*(10), 2439.

Li, X., Wang, Y., Chen, S., Tian, H., Fu, D., Zhu, B., ... & Zhu, H. (2018). Lycopene is enriched in tomato fruit by CRISPR/Cas9-mediated multiplex genome editing. *Frontiers in Plant Science*, *9*, 559.

Liu, H., Hu, H., Tang, K., Rehman, M., Du, G., Huang, Y., & Liu, F. (2022). Overexpressing hemp salt stress induced transcription factor genes enhances tobacco salt tolerance. *Industrial Crops and Products*, *177*, 114497.

Liu, J., Van Eck, J., Cong, B., & Tanksley, S. D. (2002). A new class of regulatory genes underlying the cause of pear-shaped tomato fruit. *Proceedings of the National Academy of Sciences*, *99*(20), 13302–13306.

Lynch, R. C., Vergara, D., Tittes, S., White, K., Schwartz, C. J., Gibbs, M. J., ... & Kane, N. C. (2016). Genomic and chemical diversity in Cannabis. *Critical Reviews in Plant Sciences*, *35*(5–6), 349–363.

Mass, E. (2009). Hemp: the new, old fiber makes a comeback for clothes, fabrics, and home furnishings. *Nature Life*, *127*, 36.

Melgarejo, T. A., Chen, L. F., Rojas, M. R., Schilder, A., & Gilbertson, R. L. (2022). Curly top disease of hemp (*Cannabis sativa*) in California is caused by mild-type strains of Beet curly top virus often in mixed infection. *Plant Disease*, *106*(12), 3022–3026.

Mihalyov, P. D., & Garfinkel, A. R. (2021). Discovery and genetic mapping of PM1, a powdery mildew resistance gene in *Cannabis sativa* L. *Frontiers in Agronomy*, *3*, 720215.

Ming, R., Bendahmane, A. & Renner, S. S. (2011). Sex chromosomes in land plants. *Annual Review of Plant Biology,* 62, 485–514.

Miotti, N., Sukhikh, N., Laboureau, N., Casati, P., & Pooggin, M. M. (2023). Cannabis virome reconstruction and antiviral RNAi characterization through small RNA sequencing. *Plants*, *12*(23), 3925.

Mohammad, T., Ghogare, R., Morton, L. B., Dhingra, A., Potlakayala, S., Rudrabhatla, S., & Dhir, S. K. (2024). Evaluation of parameters affecting agrobacterium-mediated transient gene expression in industrial hemp (*Cannabis sativa* L.). *Plants*, *13*(5), 664.

Monthony, A. S., Page, S. R., Hesami, M., & Jones, A. M. P. (2021). The past, present and future of *Cannabis sativa* tissue culture. *Plants*, *10*(1), 185.

Mostafaei Dehnavi, M., Ebadi, A., Peirovi, A., Taylor, G., & Salami, S. A. (2022). THC and CBD fingerprinting of an elite cannabis collection from Iran: quantifying diversity to underpin future cannabis breeding. *Plants*, *11*(1), 129.

Movahedi, M., Ghasemi-Omran, V., & Torabi, S. (2015). The effect of different concentrations of TDZ and BA on in vitro regeneration of Iranian cannabis (*Cannabis sativa*) using cotyledon and epicotyl explants. *Journal of Plant Molecular Breeding*, *3*(2), 20–27.

Olivas-Aguirre, M., Torres-López, L., Villatoro-Gómez, K., Perez-Tapia, S. M., Pottosin, I., & Dobrovinskaya, O. (2022). Cannabidiol on the path from the lab to the cancer patient: opportunities and challenges. *Pharmaceuticals*, *15*(3), 366.

Pabo, C. O., Peisach, E., & Grant, R. A. (2001). Design and selection of novel Cys2His2 zinc finger proteins. *Annual Review of Biochemistry*, *70*(1), 313–340.

Palumbo, F., Galla, G., Vitulo, N., & Barcaccia, G. (2018). First draft genome sequencing of fennel (Foeniculum vulgare Mill.): identification of simple sequence repeats and their application in marker-assisted breeding. *Molecular Breeding*, *38*, 1–17.

Pan, G., Li, Z., Huang, S., Tao, J., Shi, Y., Chen, A., ... & Zhao, L. (2021). Genome-wide development of insertion-deletion (InDel) markers for Cannabis and its uses in genetic structure analysis of Chinese germplasm and sex-linked marker identification. *BMC Genomics*, *22*(1), 1–12.

Park, S.-H., Pauli, C. S., Gostin, E. L., Staples, S. K., Seifried, D., Kinney, C., & Vanden Heuvel, B. D. (2022). Effects of short-term environmental stresses on the onset of cannabinoid production in young immature flowers of industrial hemp (*Cannabis sativa* L.). *Journal of Cannabis Research*, *4*(1), 1.

Patella, A., Palumbo, F., Galla, G., & Barcaccia, G. (2019). The molecular determination of hybridity and homozygosity estimates in breeding populations of lettuce (Lactuca sativa L.). *Genes*, *10*(11), 916.

Patella, A., Scariolo, F., Palumbo, F., & Barcaccia, G. (2019). Genetic structure of cultivated varieties of radicchio (Cichorium intybus l.): a comparison between f1 hybrids and synthetics. *Plants*, *8*(7), 213.

Paul, N. C., Park, S. W., Liu, H., Choi, S., Ma, J., MacCready, J. S., ... & Sang, H. (2021). Plant and fungal genome editing to enhance plant disease resistance using the CRISPR/Cas9 system. *Frontiers in Plant Science*, *12*, 700925.

Pellati, F., Borgonetti, V., Brighenti, V., Biagi, M., Benvenuti, S., & Corsi, L. (2018). *Cannabis sativa* L. and nonpsychoactive cannabinoids: their chemistry and role against oxidative stress, inflammation, and cancer. *BioMed Research International*, 2018, 1691428.

Petersen, M. W., Williams, E. J., Harnish, R., Kaeppler, H. F., Martinell, B., Collier, R., McFarland, F., & Kaeppler, S. M. (2021). Methods of gene editing and transforming cannabis. U.S. Patent Application 16/930,936, filed March 11, 2021.

Pisanti, S., Malfitano, A. M., Ciaglia, E., Lamberti, A., Ranieri, R., Cuomo, G., ... & Bifulco, M. (2017). Cannabidiol: state of the art and new challenges for therapeutic applications. *Pharmacology & Therapeutics*, *175*, 133–150.

Puchta, H. (2005). The repair of double-strand breaks in plants: mechanisms and consequences for genome evolution. *Journal of Experimental Botany*, *56*(409), 1–14.

Punja, Z. K. (2018). Flower and foliage-infecting pathogens of marijuana (*Cannabis sativa* L.) plants. *Canadian Journal of Plant Pathology*, *40*(4), 514–527.

Punja, Z. K. (2021). Emerging diseases of *Cannabis sativa* and sustainable management. *Pest Management Science*, *77*(9), 3857–3870.

Punja, Z. K., Kahl, D., Reade, R., Xiang, Y., Munz, J., & Nachappa, P. (2023). Challenges to *Cannabis sativa* production from pathogens and microbes—the role of molecular diagnostics and bioinformatics. *International Journal of Molecular Sciences*, *25*(1), 14.

Radwan, M. M., ElSohly, M. A., Slade, D., Ahmed, S. A., Khan, I. A., Ross, S. A. (2009). Biologically active cannabinoids from high-potency *Cannabis sativa*. *Journal of Natural Products*, *72*(5), 906–911.

Ramkumar, T. R., Lenka, S. K., Arya, S. S., & Bansal, K. C. (2020). A short history and perspectives on plant genetic transformation. In *Biolistic DNA Delivery in Plants: Methods and Protocols*, S. Rustgi & H. Luo (Eds) (pp. 39–68). Springer.

Ranalli, P., & Venturi, G. (2004). Hemp as a raw material for industrial applications. *Euphytica*, *140*(1–2), 1–6.

Ren, G., Zhang, X., Li, Y., Ridout, K., Serrano-Serrano, M. L., Yang, Y., ... & Fumagalli, L. (2021). Large-scale whole-genome resequencing unravels the domestication history of *Cannabis sativa*. *Science Advances*, *7*(29), eabg2286.

Rode, J., In-Chol, K., Saal, B., Flachowsky, H., Kriese, U., & Weber, W. E. (2005). Sex-linked SSR markers in hemp. *Plant Breeding*, *124*(2), 167–170.

Rodríguez-Leal, D., Lemmon, Z. H., Man, J., Bartlett, M. E., & Lippman, Z. B. (2017). Engineering quantitative trait variation for crop improvement by genome editing. *Cell*, *171*(2), 470–480.

Rodriguez-Morrison, V., Llewellyn, D., & Zheng, Y. 2021. Cannabis inflorescence yield and cannabinoid concentration are not increased with exposure to short-wavelength ultraviolet-B radiation. *Frontier in Plant Science*, *12*, 725078.

Roscow, R. F. (2019). "Enhanced cannabis plants and methods of making and using the same." U.S. Patent Application No. 16/093,066.

Sakamoto, K., Akiyama, Y., Fukui, K., Kamada, H., & Satoh, S. (1998). Characterization;genome sizes and morphology of sex chromosomes in hemp (*Cannabis sativa* L.). *Cytologia*, 63, 459–464.

Sawler, J., Stout, J. M., Gardner, K. M., Hudson, D., Vidmar, J., Butler, L., ... & Myles, S. (2015). The genetic structure of marijuana and hemp. *PloS ONE*, *10*(8), e0133292.

Sayre, R., Soto-Aguilar, M., Zidenga, T., & Goncalves, E. C. (2022). *U.S. Patent No. 11,352,635*. Washington, DC: U.S. Patent and Trademark Office.

Schindele, P., Wolter, F., & Puchta, H. (2018). Transforming plant biology and breeding with CRISPR/Cas9, Cas12 and Cas13. *FEBS Letters, 592*(12), 1954–1967. doi: 10.1002/1873-3468.13073. PMID: 29710373.

Schultes, R. E., Klein, W. M., Plowman, T., & Lockwood, T. E. (1974). Cannabis: an example of taxonomic neglect. *Botany Mus. Leafl., Harvard University*, *23*, 337–367.

Shi, J., Gao, H., Wang, H., Lafitte, H. R., Archibald, R. L., Yang, M., Hakimi, S. M., Mo, H., Habben, J. E. (2017). ARGOS8 variants generated by CRISPR-Cas9 improve maize grain yield under field drought stress conditions. *Plant Biotechnology Journal,* 15(2), 207–216. doi: 10.1111/pbi.12603. Epub 2016 Aug 17. PMID: 27442592; PMCID: PMC5258859.

Shiels, D., Prestwich, B. D., Koo, O., Kanchiswamy, C. N., O'Halloran, R., & Badmi, R. (2022). Hemp genome editing—Challenges and opportunities. *Frontiers in Genome Editing*, 4, 823486.

Simiyu, D. C., Jang, J. H., & Lee, O. R. (2022). Understanding *Cannabis sativa* L.: current status of propagation, use, legalization, and haploid-inducer-mediated genetic engineering. *Plants*, *11*(9), 1236.

Soler, S., Gramazio, P., Figàs, M. R., Vilanova, S., Rosa, E., Llosa, E. R., ... & Prohens, J. (2017). Genetic structure of *Cannabis sativa* var. indica cultivars based on genomic SSR (gSSR) markers: implications for breeding and germplasm management. *Industrial Crops and Products*, *104*, 171–178.

Soorni, A., Fatahi, R., Haak, D. C., Salami, S. A., & Bombarely, A. (2017). Assessment of genetic diversity and population structure in Iranian Cannabis germplasm. *Scientific Report*, *7*, 15668.

Sorokin, A., Yadav, N. S., Gaudet, D., & Kovalchuk, I. (2020). Transient expression of the β-glucuronidase gene in *Cannabis sativa* varieties. *Plant Signaling & Behavior*, *15*(8), 1780037.

Struik, P. C., Amaducci, S., Bullard, M. J., Stutterheim, N. C., Venturi, G., & Cromack, H. T. H. (2000). Agronomy of fibre hemp (*Cannabis sativa* L.) in Europe. *Industrial Crops and Products*, *11*(2–3), 107–118.

Su, W., Xu, M., Radani, Y., & Yang, L. (2023). Technological development and application of plant genetic transformation. *International Journal of Molecular Sciences*, *24*(13), 10646.

Sun, S. K., Xu, X., Tang, Z., Tang, Z., Huang, X. Y., Wirtz, M., ... & Zhao, F. J. (2021). A molecular switch in sulfur metabolism to reduce arsenic and enrich selenium in rice grain. *Nature Communications*, *12*(1), 1392.

Sun, Y., Zhang, X., Wu, C., He, Y., Ma, Y., Hou, H., Guo, X., Du, W., Zhao, Y., & Xia, L. (2016). Engineering herbicide-resistant rice plants through CRISPR/Cas9-Mediated homologous recombination of acetolactate synthase. *Molecular Plant, 9*(4), 628–631. doi: 10.1016/j.molp.2016.01.001. Epub 2016 Jan 6. PMID: 26768120.

Suzuki, N., Rivero, R. M., Shulaev, V., Blumwald, E., & Mittler, R. (2014). Abiotic and biotic stress combinations. *New Phytologist, 203*(1), 32–43.

Tanney, C. A., Backer, R., Geitmann, A., & Smith, D. L. (2021). Cannabis glandular trichomes: a cellular metabolite factory. *Frontiers in Plant Science*, *12*, 721986.

Thomazella, D. P. D. T., Seong, K., Mackelprang, R., Dahlbeck, D., Geng, Y., Gill, U. S., ... & Staskawicz, B. (2021). Loss of function of a DMR6 ortholog in tomato confers broad-spectrum disease resistance. *Proceedings of the National Academy of Sciences*, *118*(27), e2026152118.

Tomppo, L., Heikkinen, J., Salami, A., Jokinen, N., & Lappalainen, R. (2023). Slow pyrolysis processing of industrial hemp by-products. In *Current Applications, Approaches, and Potential Perspectives for Hemp*, I. F. Garcia-Tejero & V. H. Duran-Zuazo (Eds.) (pp. 315–335). Academic Press.

Törjék, O., Bucherna, N., Kiss, E., Homoki, H., Finta-Korpelová, Z., Bócsa, I., ... & Heszky, L. E. (2002). Novel male-specific molecular markers (MADC5, MADC6) in hemp. *Euphytica*, *127*, 209–218.

Van Bakel, H., Stout, J. M., Cote, A. G., Tallon, C. M., Sharpe, A. G., Hughes, T. R., & Page, J. E. (2011). The draft genome and transcriptome of *Cannabis sativa*. *Genome Biology*, *12*(10), 1–18.

Wada, N., Osakabe, K., & Osakabe, Y. (2022). Expanding the plant genome editing toolbox with recently developed CRISPR-Cas systems. *Plant Physiology*, *188*(4), 1825–1837. doi: 10.1093/plphys/kiac027. PMID: 35099553; PMCID: PMC8968252.

Wang, F., Wang, C., Liu, P., Lei, C., Hao, W., Gao, Y., Liu, Y. G., & Zhao, K. (2016). Enhanced rice blast resistance by CRISPR/Cas9-Targeted mutagenesis of the ERF transcription factor gene OsERF922. *PLoS One*, *11*(4), e0154027. doi: 10.1371/journal.pone.0154027. PMID: 27116122; PMCID: PMC4846023.

Wang, W., Pan, Q., Tian, B., He, F., Chen, Y., Bai, G., ... & Akhunov, E. (2019). Gene editing of the wheat homologs of TONNEAU 1-recruiting motif encoding gene affects grain shape and weight in wheat. *The Plant Journal*, *100*(2), 251–264.

Warren, J. G., Mercado, J., & Grace, D. (2019). Occurrence of hop latent viroid causing disease in *Cannabis sativa* in California. *Plant Disease*, *103*(10), 2699–2699. Scientific Societies.

Westra, E. R., Swarts, D. C., Staals, R. H., Jore, M. M., Brouns, S. J., van der Oost, J. (2012). The CRISPRs, they are a-changin': how prokaryotes generate adaptive immunity. *Annual Review of Genetics*, *46*, 311–39. doi: 10.1146/annurev-genet-110711-155447. PMID: 23145983.

Xu, R., Yang, Y., Qin, R., Li, H., Qiu, C., Li, L., ... & Yang, J. (2016). Rapid improvement of grain weight via highly efficient CRISPR/Cas9-mediated multiplex genome editing in rice. *Journal of Genetics and Genomics= Yi chuan xue bao*, *43*(8), 529–532.

Yano, H., & Fu, W. (2023). Hemp: a sustainable plant with high industrial value in food processing. *Foods*, *12*(3), 651.

Yep, B., & Zheng, Y. (2021). Potassium and micronutrient fertilizer addition in a mock aquaponic system for drug-type *Cannabis sativa* L. cultivation. *Canadian Journal of Plant Science*, *101*(3), 341–352.

Yin, M., Pan, G., Tao, J., Doblin, M. S., Zeng, W., Pan, L., ... & Huang, S. (2022). Identification of MYB genes reveals their potential functions in cadmium stress response and the regulation of cannabinoid biosynthesis in hemp. *Industrial Crops and Products*, *180*, 114607.

Yuyu, C., Aike, Z., Pao, X., Xiaoxia, W., Yongrun, C., Beifang, W., ... & Yingxin, Z. (2020). Effects of GS3 and GL3. 1 for grain size editing by CRISPR/Cas9 in rice. *Rice Science*, *27*(5), 405–413.

Zhang, H., Zhao, Y., & Zhu, J.-K. (2020). Thriving under stress: how plants balance growth and the stress response. *Developmental Cell*, *55*(5), 529–543.

Zhang, X., Xu, G., Cheng, C., Lei, L., Sun, J., Xu, Y., ... & Su, J. (2021). Establishment of an Agrobacterium-mediated genetic transformation and CRISPR/Cas9-mediated targeted mutagenesis in Hemp (Cannabis Sativa L.). *Plant Biotechnology Journal*, *19*(10), 1979–1987.

13 Sex Determination and Reproductive Mechanisms in *Cannabis sativa* L.

Yousoon Baek and Daniela Vergara

13.1 INTRODUCTION

In most flowering plants, both male flower structures, and egg producing female structures are found on the same flowers, making it challenging to strictly designate them as male or female and are therefore called hermaphrodite. However, in rare species like *Cannabis sativa*, some plants bear male reproductive organs (stamens) that exclusively produce pollen, while others have female reproductive organs (carpels). It is customary to refer to the former as "males" or "staminate," and the latter as "females" or "pistillate." In these cases where male and female individuals are separate, they are referred to as dioecy. Dioecy is relatively rare, accounting for just 5–6% of all flowering plant species (Charlesworth 2002; Charlesworth et al. 2005; Charlesworth 2006).

Although uncommon, dioecy has independently evolved multiple times over hundreds of years in various flowering plants (Angiosperm) lineages. Examples include asparagus, persimmon, spinach, hops, cannabis, kiwifruit, papaya, and grape. Transitioning from hermaphroditism to dioecy involves diverse mechanisms that suppress the growth of reproductive organs. Studies on dioecy shed light on sexual trait factors, sex determination mechanisms, and evolutionary paths. Transition from hermaphrodite to unisexual forms may have occurred more than 100 times in angiosperm lineages, possibly through pathways involving monoecious stages. Phylogenetic evidence supports the co-occurrence of monoecy and dioecy, although the genetic mechanisms are not fully understood yet (Charlesworth and Charlesworth 1978; Irish and Nelson 1989; Renner and Ricklefs 1995; Charlesworth 2002, 2006; Lan et al. 2006; Picq et al. 2014; Akagi et al. 2016; Akagi and Charlesworth 2019; Zhou et al. 2019; Harkess et al. 2020; Renner and Müller 2022).

Dioecy's genetic control is influenced by sex chromosomes and environmental factors such as plant hormones. Genetic research has unveiled DNA polymorphisms in loci correlated with sex dimorphism, clarifying on possible mechanisms. This leads to a diverse range of sex determination mechanisms, making the evolutionary path of dioecy in flowering plants uncertain (Truta et al. 2007; Ming et al. 2011; Renner 2014; Charlesworth 2019).

13.2 FAMILY CANNABACEAE CHARACTERISTICS

In the Cannabaceae family, there are species with dioecious or both dioecious and monoecious reproductive strategies, including *Humulus* and *Cannabis* (Yang et al. 2013). Humulus, commonly known as hop, is the closest relative to *C. sativa* within the Cannabaceae family, with the two genera diverging approximately 25–27 million years ago (McPartland et al. 2019; Ren et al. 2021). Both genera are dioecious but also exhibit as monoecious individuals, where male and female flowers are present on the same individual, though not within the same flower (Shephard et al. 1999; Shephard et al. 2000).

 DOI: 10.1201/9781003386438-13

In hemp-type plants, dioecious varieties may excel in fiber production, while monoecious varieties offer more uniformity in both seed and fiber production (Salentijn et al. 2019). Males, though taller and faster-growing, may be more prone to pests (Salentijn et al. 2019). Conversely, marijuana-type plants require careful sex determination to avoid fertilization, as seeded flowers produce lower-quality compounds compared to unfertilized ones (Punja and Holmes 2020). While metabolic differences exist between male and female flowers, THC levels can vary within and between sexes (Welling et al. 2021).

13.3 REPRODUCTIVE ORGANS OF *CANNABIS SATIVA* L.

Phenotypic differences play a crucial role in identifying sexes early in *C. sativa* plants. However, determining the sexual identity of a plant before maturity is challenging due to the absence of gender-specific features in young plants. Further, depending solely on traits like plant height and flowering time may not be reliable due to variations among varieties (Rahimi et al. 2019). While efforts to pinpoint specific sex traits have been made, early differentiation remains challenging (Rahimi et al. 2019).

After germination, *C. sativa* typically undergoes a vegetative phase for a few months. Male and female *C. sativa* plants are nearly identical throughout the early growth phase, and the plant generates leafy branches early in its seasonal life cycle. Sex expression varies between dioecious and monoecious plants across hemp varieties, with monoecious plants exhibiting shorter hypocotyl length and seedling height (Campbell et al. 2021). Additional distinctions between sexes include males exhibiting higher susceptibility to pests (Salentijn et al. 2019), and in the case of *C. sativa* hemp-type, they tend to possess a finer fiber, advantageous for textile production (Salentijn et al. 2019). Furthermore, male plants are often noted to be taller than their female counterparts (Onofri and Mandolino 2017; Rahimi et al. 2019; Punja and Holmes 2020) have shorter life cycles (Rahimi et al. 2019), exhibit faster growth rates (Punja and Holmes 2020), and they generally appear more slender (Onofri and Mandolino 2017).

Male plant morphology (Figure 13.2) in *C. sativa* is adapted for efficient pollen dispersal by wind, and their inflorescences are characterized by loose, axillary arrangements exhibiting a relatively diffuse structure (Small 2016; Punja and Holmes 2020). Male flowers are pedicellate, each bearing individual stalks, and are distinguished by five greenish or whitish tepals (distinct petals and sepals are absent). Five stamens with flaccid filaments are positioned opposite the tepal. After anthesis, or the shedding of pollen, male flowers detach from the plant (Small 2016; Punja and Holmes 2020).

FIGURE 13.1 Female plant (left), monoecious inflorescence (center) and male plant (right).

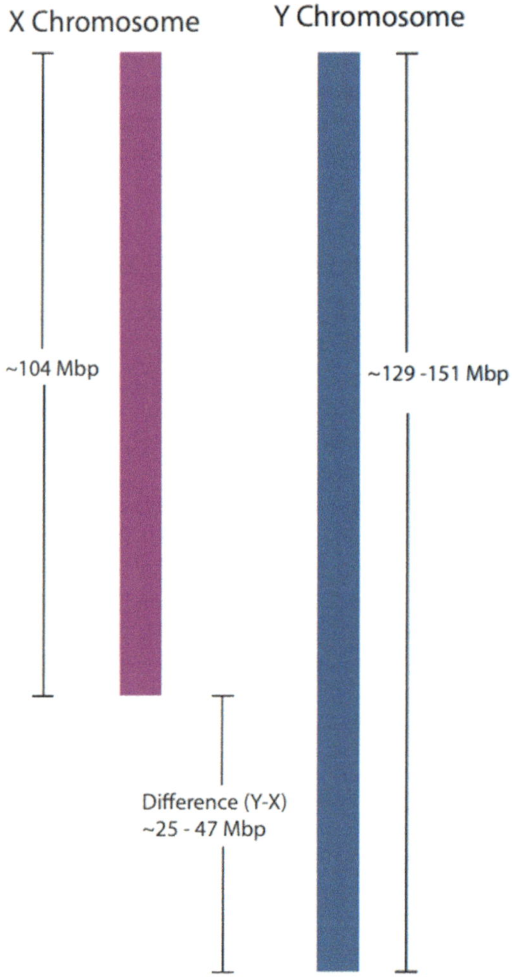

FIGURE 13.2 Size of the X chromosome and estimated size of the Y chromosome.

Female flowers (Figure 13.2) of *C. sativa* are characterized by their absence of sepals and petals as well as stamens. Instead, they are surrounded by perigonal bracts, which are essential to the synthesis of cannabinoids. In botany, a "bract" is a modified or specialized leaf that is frequently connected to flowers and these bracts in *C. sativa* have a single leaflet and resemble tiny unifoliate leaves. The female flowers are enclosed in a perigonal bract that resembles a cup. This bract grows larger and thicker with time, containing many tiny secretory glands that are in the role of synthesizing cannabinoid (Small and Naraine 2016; Punja et al. 2023a). Female flowers are found in axillary inflorescences that are smaller and relatively inconspicuous compared to male flowers. Each female flower comprises a one-celled ovary and a short apical style adorned with two long filiform stigmatic branches. In contrast to the male flowers, female flowers are essentially sessile, lacking individual stalks (Small 2016).

Degrees of monoecy exist (Figure 13.1), and the onset of flowering tends to feature male flowers initially (Faux et al. 2013). Monoecy's genetic basis suggests an association between early flowering and maleness, suggesting that flowering response genes may influence sex determination (Faux et al. 2014).

Understanding these sex expression differences is crucial for cultivation practices and product quality in both hemp and marijuana-type varieties. Particularly, sex and sex determination in marijuana-type *C. sativa* is very important for cannabinoid and terpene production. Females typically generate elevated levels of cannabinoids because of the abundance of trichomes on their flowers (Sirikantaramas et al. 2005; Gagne et al. 2012). Although there are different types of trichomes, most of the cannabinoid production is found in the stalked capitate trichomes, and unpollinated females contain the highest concentrations of these compounds (Livingston et al. 2020; Punja et al. 2023b). Fertilization and seed development are not desirable for cannabinoid production applications (Romanov et al. 2022). When females produce male flowers the allocation of resources to pollen followed by seed production can result in disproportionately lower levels of terpene and other compounds compared to unfertilized female flowers (Meier and Mediavilla 1998).

13.4 SEX CHROMOSOMES IN *C. SATIVA*

Cannabis sativa L. possesses a diploid genome with 20 chromosomes, including one pair of sex chromosomes (Sakamoto et al. 1998; Divashuk et al. 2014; Kovalchuk et al. 2020). Sex in *C. sativa* is determined by an XY sex chromosome system. Female and monoecious plants carry two X chromosomes (XX), while males have one X and one Y chromosome (Sakamoto et al. 1998; Divashuk et al. 2014; Kovalchuk et al. 2020). Despite its relatively small genome size compared to other species, *C. sativa* exhibits significant differences in genome size between males and females, primarily due to the larger Y chromosome in males (Balant et al. 2022).

In *C. sativa*, the Y chromosome is markedly heterochromatic, containing repetitive sequences and gene-poor regions (Moliterni et al. 2004; Divashuk et al. 2014). Conversely, the X chromosome is less condensed and gene rich (Divashuk et al. 2014). Because the complete assembly of the Y chromosome is still incomplete, precise estimations of its size and understanding of its role in sex determination have not yet been established (van Bakel et al. 2011; Laverty et al. 2019; Grassa et al. 2021).

The sex chromosomes in *C. sativa* appear to be among the oldest in the plant kingdom, with a Y chromosome showing signs of degeneration, indicating that about 70% of it no longer recombines with the X chromosome (Divashuk et al. 2014; Baránková et al. 2020). This degeneration has led to a significant loss of genes, much more than in other species (Prentout et al. 2020). Studies estimate that the X and Y chromosomes diverged around 12–29 million years ago, coinciding with the split from *Humulus* sp. (McPartland et al. 2019; Prentout et al. 2020; Ren et al. 2021). However, the specific mechanisms underlying sex determination, including the role of the Y chromosome and its interaction with environmental factors, remain subjects of ongoing research (Hall et al. 2012; Petit et al. 2020a; Petit et al. 2020b). Despite advancements in genomic analysis, questions persist regarding the genetic and epigenetic factors influencing sex expression in *C. sativa*, highlighting the complexity of sex determination in this species (Campbell et al. 2019).

As previously mentioned, because the current *C. sativa* genome assemblies lack a complete Y chromosome or monoecious chromosomes, it is challenging to accurately estimate their sizes (van Bakel et al. 2011; Laverty et al. 2019; Grassa et al. 2021). While the X chromosome is about 104 million base pairs (Mbp) in size (Grassa et al. 2021), the size of the Y chromosome is harder to determine but is estimated to be between 131 and151 Mbp based on existing data (Figure 13.2). The difference in genome size between males and females, mainly attributed to the larger Y chromosome in males, may also be influenced by repetitive genomic sequences, which make up about 64% of the *C. sativa* genome (Pisupati et al. 2018).

Various repetitive elements, such as copia-like retrotransposons and LINE transposable elements, are abundant on the Y chromosome and contribute to its larger size (Sakamoto et al. 2000; Moliterni et al. 2004; Sakamoto et al. 2005; Punja and Holmes 2020). Simple Sequence Repeats (SSRs) or microsatellites are also scattered throughout the genome, albeit at a lower density compared to the Y chromosome (Barcaccia et al. 2020). Moreover, gene copy number variations (Vergara et al. 2019) and gene rearrangements may further contribute to the Y chromosome's size variation, as observed in related species (Ming and Moore 2007; Ming et al. 2007; Bergero et al. 2008; Ming et al. 2011). Techniques like genomic in situ hybridization (GISH) confirm the abundance of repetitive DNA sequences on the Y chromosome (Razumova et al. 2023), shedding light on its composition and evolution.

13.5 SEX IDENTIFICATION DNA MARKERS

Despite the largely unknown mechanisms of sex determination and the role of the Y chromosome, DNA marker assays for sex determination have been developed, offering a tool to predict the sex of plants from seedlings before flowering. As noted previously, because male and female characteristics only become apparent during the reproductive stage in *C. sativa*, identifying sex early on is crucial through genetic markers. However, existing markers are often specific to certain varieties and not universally effective (Truta et al. 2007; Petit et al. 2020b).

One notable set of markers, MADC 1 through MADC 6 (Table 13.1), was discovered using techniques like Random Amplified Polymorphic DNA (RAPD) (Sakamoto et al. 1995; Mandolino et al. 1999). These markers, primarily specific to males, target male-associated DNA sequences on the Y chromosome. However, some markers, like MADC 5 and MADC 6, have been found in both sexes (Rahimi et al. 2019). Additionally, markers like CSP-1 have been identified in both male and female individuals (Toth et al. 2020). Female-specific markers, such as OPA-04 and OPA-05, are also available (Shao et al. 2003).

Despite progress, challenges remain, including the lack of reproducibility and universal applicability of these markers (Shao et al. 2003; Rahimi et al. 2019). Ongoing research aims to validate these markers across diverse populations to ensure their reliability in breeding programs. Efforts to develop more reliable markers are essential for effective use in breeding programs (Shao et al. 2003; Rahimi et al. 2019).

TABLE 13.1
Markers for Sex Determination, the Name of the Marker (Column 1), the Sex It Identifies (Column 2), the Primer Sequence (Column 3), and the Reference Where the Marker Was Published (Column 4)

Name	Sex	Primer Seq	Ref
MADC1	Female	ACGGCATATG	(Sakamoto et al. 1995)
MADC2	Male	GTGACGTAGGTAGAGTTGAA TGACGTAGGCTATGAGAG	(Mandolino et al. 1999)
OPA-04	Female	AATCGGGCTG	(Shao et al. 2003)
OPA-05	Female	CCGAATTCCC	
OPC08-2700	Male	GTGACGTAGG	(Mandolino et al. 2002)
OPD05-961/MADC5	Male	F: GAGCGGACATCATTGCCT R: ATCACCCACCGTTTAGG	(Törjék et al. 2002)
UBC354-151/MADC6	Male	F: TCAAACAACAACAAACCG R: GAGGCCGATAATTGACTC	
MADC3	Male		(Sakamoto et al. 2005)
MADC4	Male		
CSP1-FAM	Female	F: GAAGGTGACCAAGTTCATGCTA R: CTTGAAATGAGATGTCAAACC	(Toth et al. 2020) (Törjék et al. 2002)
MADC6	Male	F: GAAGGTCGGAGTCAACGGATTG R: AGCTTGAAATGAGATGTCAAACT	

There are several commercial tools available for growers to test for early sex determination. While many commercial labs accept samples from the growing sites to conduct DNA marker assays, there are other simple options. Companies such as MyFlora DNA and Leafworks accept samples to test whether a plant is a male. However, these tests cannot establish whether a plant is monoecious or has the possibility of producing male flowers. Growers may consider setting up sex determination tools with affordable investigative options including youPCR from Medicinal genomics (https://store.medicinalgenomics.com/youPCR-Solution-A-50-rxn-pre-aliquoted), and Y-spotter CannabisSexTest from MinipcrDX (https://dx.minipcr.com/product/y-spotter-60/). Though these kits require additional laboratory equipment such as a thermocycler and pipettes, they utilize simple PCR techniques with colorimetric reagents to indicate the presence of male DNA through a binding process that facilitates direct visualization of the result.

Another non-DNA marker base tool is also available, however, it is less known about the actual genetic mechanisms behind this technique (https://mariposatechnology.com/). The plant's sex can be distinguished by using a handheld Raman spectroscopy (RS) that scans leaf tissues. This device relies on the principle of inelastic light scattering, where incident photons interact with molecules in the sample, allowing for the direct chemical analysis of the sample by these scattered photons. This approach enables the detection of biochemical differences among female, male, and monoecious plants (Goff et al. 2022).

13.6 SEX MANIPULATION

While it's suggested that sex in *C. sativa* is determined by the X and Y chromosomes, it's influenced by genetic, hormonal, and environmental factors like photoperiod and temperature (Ram and Jaiswal 1970; Ram et al. 1972; Hall et al. 2012). For instance, low temperatures and short days tend to induce male flower formation, but long photoperiods can also affect sex expression (Faux et al. 2014). Dioecy usually occurs in equal ratios, but this balance can be disrupted under certain conditions, often favoring females (Hall et al. 2012). Because the exact role of the Y chromosome in sex determination is still unclear, and variations like XXY and XXXY individuals are female, it is suggested that there might be a dosage effect (Truta et al. 2007). Other factors like epigenetics may also play a role in sex determination, adding to the complexity of understanding how sex is determined in *C. sativa* (Campbell et al. 2019; Petit et al. 2020a).

Plant hormones like auxins, gibberellins, and ethylene play crucial roles in growth and development, including sex determination (Veen 1983; Kende and Zeevaart 1997). Exogenous hormone applications can manipulate sex expression, inducing male or female flowers on genetically opposite plants (Galoch 1978). For instance, gibberellic acid can promote male flower development on female plants, while ethephon can feminize males (Galoch 1978).

Masculinization of female plants can be achieved through treatments with compounds like silver nitrate or silver thiosulfate, which interfere with ethylene action (Veen 1983). These treatments induce the development of male flowers, allowing for controlled pollination and seed production (Mohan Ram and Sett 1982). These seeds are called feminized seeds as they lack a Y chromosome (Figure 13.3). However, the efficacy of these treatments can vary between plant strains and requires careful timing and application (DiMatteo et al. 2020).

Understanding these mechanisms of sex determination and manipulation is crucial for the cannabis industry, facilitating controlled breeding and the production of desirable traits.

To trigger the development of female flowers on male plants, a chemical called ethephon is commonly used (Ram and Jaiswal 1970). Ethephon induces the release of ethylene and manipulates auxin, resulting in the production of female-like flowers on male plants, capable of producing seeds upon pollination (Ram and Jaiswal 1970). Early application of ethephon after the formation of flower primordia enhances the formation of female flowers (Moon et al. 2020).

♀  ♂ 	X	X
X	XX	XX
Y	XY	XY

♀  ♂ 	X	X
X	XX	XX
X	XX	XX

FIGURE 13.3 Punnett squares illustrating the ratio of sex chromosomes in the progeny when the parents are a regular female and a regular male (left), and when the parents are a regular female and a female who was induced to produce pollen (right).

Different chemical treatments (Table 13.2) can lead to variations in the characteristics of treated plants. For instance, while male plants treated with gibberellic acid showed no change in sex expression, female plants exhibited elongated internodes and produced male flowers with normal stamens (Ram et al. 1972). However, the application of silver thiosulfate, colloidal silver, or gibberellic acid did not affect plant height (Flajšman et al. 2021). Pollen from reversed female flowers, induced to become male, showed irregularities and lower germination rates compared to pollen from naturally male flowers (DiMatteo et al. 2020; Flajšman et al. 2021).

Treatment with ethephon resulted in male plants producing smaller seeds compared to females, and showed differences in height but not in the number of branches (Moon et al. 2020). Additionally, the application of compounds like abscisic acid can influence THC levels and enzyme activity, with potential sex-dependent effects (Mansouri et al. 2009a; Mansouri et al. 2009b). Similarly, chloroethyl trimethylammonium chloride (Cycocel), a growth retardant, affects THC levels differently in male and female plants (Mansouri and Rohani 2014).

Studies on gene expression have identified differentially expressed genes in plants undergoing sex manipulation. These genes, related to sex expression and hormone signaling pathways like ethylene, auxin, cytokinin, and gibberellins, may control the masculinization or feminization of

TABLE 13.2
Chemical Compound, Concentration, and Effect for Feminization, Masculinization, or Other Alterations of *C. sativa* Plants

Compound	Effect
Ethephon	• 960 ppm showed the highest feminization of male plants
	• Increase carotenoids and THC at 10 and 100 μm
	• Increase THC content, highest effect at 1 and 100 μM. Decrease CBD
	• Female flowers on male plants had fewer stigmas, lighter seeds, with lower germination rate and were smaller
Silver nitrate	• Masculinization of female plants
Silver thiosulfate	• Apparently more effective at 150 μg
	• 0.3 mM produced more male flowers on genetically XX plants.
	• Masculinized female plants differentially expressed genes
	• Masculinized female plants differentially expressed genes
Gibberellic acid	• Increase THC and CBD

plants (Adal et al. 2021). Ethylene-related genes are distributed throughout the genome, with a significant portion located on the X chromosome (Monthony et al. 2023). Transcriptomic analyses have revealed differences in gene expression levels between male, female, and induced female flowers treated with silver thiosulfate, highlighting the complexity of sex determination in plants (Adal et al. 2021). Many of these genes are found on the X chromosome (Prentout et al. 2020).

13.7 CONCLUSIONS AND A POSSIBLE FUTURE

Sex determination in agriculture sparks debates across various sectors, influencing breeding strategies, crop management, and overall productivity. In the *C. sativa* world, understanding and controlling sex determination are critical for optimizing yield and quality.

When it comes to *C. sativa*, females are the golden ticket for cannabinoid content. Hence, early identification and removal of male plants are crucial to prevent pollination and seed production, ensuring the undisturbed growth of females and maximizing cannabinoid yield.

In grain production, monoecious plants, with both male and female organs, simplify pollination and seed formation. Similarly, having both dioecious males and females can be advantageous, promoting genetic diversity. Precision in sex determination maintains the right balance, enhancing pollination efficiency and boosting grain yields. For fiber hemp, a mix of males and females is essential, as males are believed to excel in fiber production. Striking this balance ensures optimal fiber quality and yield, meeting the demands of industrial applications.

The complexity of sex determination continues, and genes scattered across the genome may play a role, making trait modification challenging. And let's not forget the Y chromosome—large, enigmatic, and possibly harboring advantages like disease resistance, but the genetic mechanisms involved in sex determination are still largely unknown. Some findings suggest that the sex chromosomes in both *Cannabis* and *Humulus* may have evolved from a common ancestor, with dioecy emerging before their speciation event (Kovalchuk et al. 2020) and implying that this is, possibly, one of the oldest sex chromosome system in plants (Prentout et al. 2021). A detailed review on the sex mechanisms of the family Cannabaceae is soon to be available (Baek and Vergara in prep.).

Then there's the concept of sterile pollen individuals, offering a possible solution to prevent unwanted pollination and maintain crop integrity. Yet, concerns exist over genetic diversity loss and the resources needed to develop and maintain sterile lines. Triploid seeds in which a number of chromosomes are manipulated and fail to produce viable seeds are a hot topic in the marijuana market, promising more female plants with higher cannabinoid content. However, their use raises eyebrows due to potential environmental and health risks, lack of genetic diversity, and vulnerability to pests and diseases.

As research in sex determination continues, collaboration between scientists, breeders, and industry stakeholders will be crucial for ensuring sustainable and efficient production methods while preserving genetic diversity and environmental integrity. Further research into sex determination mechanisms, genomic analysis, and trait modification holds promise for revolutionizing *C. sativa* cultivation and product development, shaping the future of this versatile and economically significant plant species.

REFERENCES

Adal, A. M., K. Doshi, L. Holbrook, and S. S. Mahmoud. 2021. Comparative RNA-Seq analysis reveals genes associated with masculinization in female *Cannabis sativa*. *Planta* 253:1–17.

Akagi, T. and D. Charlesworth. 2019. Pleiotropic effects of sex-determining genes in the evolution of dioecy in two plant species. *Proceedings of the Royal Society B* 286:20191805.

Akagi, T., I. M. Henry, T. Kawai, L. Comai, and R. Tao. 2016. Epigenetic regulation of the sex determination gene MeGI in polyploid persimmon. *The Plant Cell* 28:2905–2915.

Balant, M., R. Rodríguez González, S. Garcia, T. Garnatje, J. Pellicer, J. Vallès, D. Vitales, and O. Hidalgo. 2022. Novel insights into the nature of intraspecific genome size diversity in *Cannabis sativa* L. *Plants* 11:2736.

Baránková, S., J. P. Pascual-Díaz, N. Sultana, M. P. Alonso-Lifante, M. Balant, K. Barros, U. D'Ambrosio, H. Malinská, V. Peska, and I. P. Lorenzo. 2020. Sex-chrom, a database on plant sex chromosomes. *The New Phytologist* 227:1594–1604.

Barcaccia, G., F. Palumbo, F. Scariolo, A. Vannozzi, M. Borin, and S. Bona. 2020. Potentials and challenges of genomics for breeding cannabis cultivars. *Frontiers in Plant Science* 11:573299.

Bergero, R., D. Charlesworth, D. Filatov, and R. Moore. 2008. Defining regions and rearrangements of the Silene latifolia Y chromosome. *Genetics* 178:2045–2053.

Campbell, B., Z. Dong, and J. K. McKay. 2019. Hemp genetics and genomics. In *Industrial hemp as a modern commodity crop*, D. W. Williams (Ed.), pp. 92–106. American Society of Agronomy Crop Science.

Campbell, L. G., K. Peach, and S. B. Wizenberg. 2021. Dioecious hemp (*Cannabis sativa* L.) plants do not express significant sexually dimorphic morphology in the seedling stage. *Scientific Reports* 11:16825.

Charlesworth, B. and D. Charlesworth. 1978. A model for the evolution of dioecy and gynodioecy. *American Naturalist* 112:975–997.

Charlesworth, D. 2002. Plant sex determination and sex chromosomes. *Heredity* 88:94–101.

Charlesworth, D. 2006. Evolution of plant breeding systems. *Current Biology* 16:R726–R735.

Charlesworth, D. 2019. Young sex chromosomes in plants and animals. *New Phytologist* 224:1095–1107.

Charlesworth, D., B. Charlesworth, and G. Marais. 2005. Steps in the evolution of heteromorphic sex chromosomes. *Heredity* 95:118–128.

DiMatteo, J., L. Kurtz, and J. D. Lubell-Brand. 2020. Pollen appearance and in vitro germination varies for five strains of female hemp masculinized using silver thiosulfate. *HortScience* 55:547–549.

Divashuk, M. G., O. S. Alexandrov, O. V. Razumova, I. V. Kirov, and G. I. Karlov. 2014. Molecular cytogenetic characterization of the dioecious *Cannabis sativa* with an XY chromosome sex determination system. *PloS One* 9:e85118.

Faux, A.-M., A. Berhin, N. Dauguet, and P. Bertin. 2014. Sex chromosomes and quantitative sex expression in monoecious hemp (*Cannabis sativa* L.). *Euphytica* 196:183–197.

Faux, A.-M., X. Draye, R. Lambert, R. d'Andrimont, P. Raulier, and P. Bertin. 2013. The relationship of stem and seed yields to flowering phenology and sex expression in monoecious hemp (*Cannabis sativa* L.). *European Journal of Agronomy* 47:11–22.

Flajšman, M., M. Slapnik, and J. Murovec. 2021. Production of feminized seeds of high CBD *Cannabis sativa* L. by manipulation of sex expression and its application to breeding. *Frontiers in Plant Science* 12:718092.

Gagne, S. J., J. M. Stout, E. Liu, Z. Boubakir, S. M. Clark, and J. E. Page. 2012. Identification of olivetolic acid cyclase from *Cannabis sativa* reveals a unique catalytic route to plant polyketides. *Proceedings of the National Academy of Sciences* 109:12811–12816.

Galoch, E. 1978. The hormonal control of sex differentiation in dioecious plants of hemp (Cannabis sativd). The influence of plant growth regulators on sex expression in male and female plants. *Acta Societatis Botanicorum Poloniae* 47:153–162.

Goff, N. K., J. F. Guenther, J. K. Roberts III, M. Adler, M. D. Molle, G. Mathews, and D. Kurouski. 2022. Non-invasive and confirmatory differentiation of hermaphrodite from both male and female cannabis plants using a Hand-Held Raman spectrometer. *Molecules* 27:4978.

Grassa, C. J., G. D. Weiblen, J. P. Wenger, C. Dabney, S. G. Poplawski, S. Timothy Motley, T. P. Michael, and C. Schwartz. 2021. A new Cannabis genome assembly associates elevated cannabidiol (CBD) with hemp introgressed into marijuana. *New Phytologist* 230:1665–1679.

Hall, J., S. P. Bhattarai, and D. J. Midmore. 2012. Review of flowering control in industrial hemp. *Journal of Natural Fibers* 9:23–36.

Harkess, A., K. Huang, R. van der Hulst, B. Tissen, J. L. Caplan, A. Koppula, M. Batish, B. C. Meyers, and J. Leebens-Mack. 2020. Sex determination by two Y-linked genes in garden asparagus. *The Plant Cell* 32:1790–1796.

Irish, E. E. and T. Nelson. 1989. Sex determination in monoecious and dioecious plants. *The Plant Cell* 1:737.

Kende, H. and J. Zeevaart. 1997. The Five" Classical" Plant Hormones. *The Plant Cell* 9:1197.

Kovalchuk, I., M. Pellino, P. Rigault, R. van Velzen, J. Ebersbach, J. R. Ashnest, M. Mau, M. Schranz, J. Alcorn, and R. Laprairie. 2020. The genomics ofcannabis and its close relatives. *Annual Review of Plant Biology* 71:713–739.

Lan, T., S. Zhang, B. Liu, X. Li, R. Chen, and W. Song. 2006. Differentiating sex chromosomes of the dioecious Spinacia oleracea L.(spinach) by FISH of 45S rDNA. *Cytogenetic and Genome Research* 114:175–177.

Laverty, K. U., J. M. Stout, M. J. Sullivan, H. Shah, N. Gill, L. Holbrook, G. Deikus, R. Sebra, T. R. Hughes, and J. E. Page. 2019. A physical and genetic map of *Cannabis sativa* identifies extensive rearrangements at the THC/CBD acid synthase loci. *Genome Research* 29:146–156.

Livingston, S. J., T. D. Quilichini, J. K. Booth, D. C. Wong, K. H. Rensing, J. Laflamme-Yonkman, S. D. Castellarin, J. Bohlmann, J. E. Page, and A. L. Samuels. 2020. Cannabis glandular trichomes alter morphology and metabolite content during flower maturation. *The Plant Journal* 101:37–56.

Mandolino, G., A. Carboni, M. Bagatta, V. C. Moliterni, and P. Ranalli. 2002. Occurrence and frequency of putatively Y chromosome linked DNA markers in *Cannabis sativa* L. *Euphytica* 126:211–218.

Mandolino, G., A. Carboni, S. Forapani, V. Faeti, and P. Ranalli. 1999. Identification of DNA markers linked to the male sex in dioecious hemp (*Cannabis sativa* L.). *Theoretical and Applied Genetics* 98:86–92.

Mansouri, H., Z. Asrar, and M. Mehrabani. 2009a. Effects of gibberellic acid on primary terpenoids and Δ9-tetrahydrocannabinol in *Cannabis sativa* at flowering stage. *Journal of Integrative Plant Biology* 51:553–561.

Mansouri, H., Z. Asrar, and J. Szopa. 2009b. Effects of ABA on primary terpenoids and Δ9-tetrahydrocannabinol in *Cannabis sativa* L. at flowering stage. *Plant Growth Regulation* 58:269–277.

Mansouri, H. and M. Rohani. 2014. Response of *Cannabis sativa* L. to foliar application of 2-chloro-ethyl-trimethyl-ammonium chloride. *Plant Physiology* 5:1225–1233.

McPartland, J. M., W. Hegman, and T. Long. 2019. Cannabis in Asia: its center of origin and early cultivation, based on a synthesis of subfossil pollen and archaeobotanical studies. *Vegetation History and Archaeobotany*, , *28*, 691–702.

Meier, C. and V. Mediavilla. 1998. Factors influencing the yield and the quality of hemp (*Cannabis sativa* L.) essential oil. *Journal of International Hemp Association* 5:16–20.

Ming, R., A. Bendahmane, and S. S. Renner. 2011. Sex chromosomes in land plants. *Annual Review of Plant Biology* 62:485–514.

Ming, R. and P. H. Moore. 2007. Genomics of sex chromosomes. *Current Opinion in Plant Biology* 10:123–130.

Ming, R., J. Wang, P. H. Moore, and A. H. Paterson. 2007. Sex chromosomes in flowering plants. *American Journal of Botany* 94:141–150.

Mohan Ram, H. and R. Sett. 1982. Induction of fertile male flowers in genetically female *Cannabis sativa* plants by silver nitrate and silver thiosulphate anionic complex. *Theoretical and Applied Genetics* 62:369–375.

Moliterni, V. M. C., L. Cattivelli, P. Ranalli, and G. Mandolino. 2004. The sexual differentiation of *Cannabis sativa* L.: a morphological and molecular study. *Euphytica* 140:95–106.

Monthony, A. S., M. de Ronne, and D. Torkamaneh. 2023. Exploring rthylene-related genes in *Cannabis sativa*: implications for sexual plasticity. *bioRxiv:2023.2004. 2028.538750.*

Moon, Y.-H., Y. J. Lee, S. C. Koo, M. Hur, Y. C. Huh, J.-K. Chang, and W. T. Park. 2020. Effect of timing of ethephon treatment on the formation of female flowers and seeds from male plant of hemp (*Cannabis sativa* L.). *Korean Journal of Plant Resources* 33:682–688.

Onofri, C. and G. Mandolino. 2017. Genomics and molecular markers in *Cannabis sativa* L. In *Cannabis sativa L.-botany and biotechnology*, S. Chandra, H. Lata, & M. ElSohly (Eds.), pp. 319–342. Springer. https://doi.org/10.1007/978-3-319-54564-6_15

Petit, J., E. M. Salentijn, M.-J. Paulo, C. Denneboom, and L. M. Trindade. 2020a. Genetic architecture of flowering time and sex determination in hemp (*Cannabis sativa* L.): a genome-wide association study. *Frontiers in Plant Science* 11:569958.

Petit, J., E. M. Salentijn, M.-J. Paulo, C. Thouminot, B. J. van Dinter, G. Magagnini, H.-J. Gusovius, K. Tang, S. Amaducci, and S. Wang. 2020b. Genetic variability of morphological, flowering, and biomass quality traits in hemp (*Cannabis sativa* L.). *Frontiers in Plant Science* 11:102.

Picq, S., S. Santoni, T. Lacombe, M. Latreille, A. Weber, M. Ardisson, S. Ivorra, D. Maghradze, R. Arroyo-Garcia, and P. Chatelet. 2014. A small XY chromosomal region explains sex determination in wild dioecious V. vinifera and the reversal to hermaphroditism in domesticated grapevines. *BMC Plant Biology* 14:1–17.

Pisupati, R., D. Vergara, and N. C. Kane. 2018. Diversity and evolution of the repetitive genomic content in *Cannabis sativa. BMC Genomics* 19:156.

Prentout, D., O. Razumova, B. Rhoné, H. Badouin, H. Henri, C. Feng, J. Käfer, G. Karlov, and G. A. Marais. 2020. An efficient RNA-seq-based segregation analysis identifies the sex chromosomes of *Cannabis sativa. Genome Research* 30:164–172.

Prentout, D., N. Stajner, A. Cerenak, T. Tricou, C. Brochier-Armanet, J. Jakse, J. Käfer, and G. A. Marais. 2021. Plant genera Cannabis and Humulus share the same pair of well-differentiated sex chromosomes. *New Phytologist* 231:1599–1611.

Punja, Z. K. and J. E. Holmes. 2020. Hermaphroditism in marijuana (*Cannabis sativa* L.) inflorescences–impact on floral morphology, seed formation, progeny sex ratios, and genetic variation. *Frontiers in Plant Science* 11:718.

Punja, Z. K., L. Ni, S. Lung, and L. Buirs. 2023a. Total yeast and mold levels in high THC-containing cannabis (*Cannabis sativa* L.) inflorescences are influenced by genotype, environment, and pre-and post-harvest handling practices. *Frontiers in Microbiology* 14:1192035.

Punja, Z. K., D. B. Sutton, and T. Kim. 2023b. Glandular trichome development, morphology, and maturation are influenced by plant age and genotype in high THC-containing cannabis (*Cannabis sativa* L.) inflorescences. *Journal of Cannabis Research* 5:12.

Rahimi, M., M. Shokrpour, S. A. Salami, and M. R. Taheri. 2019. Morphological and molecular study of sex determination in *Cannabis sativa* L. *Journal of Plant Physiology and Breeding* 9:137–146.

Ram, H. Y. M. and V. S. Jaiswal. 1970. Induction of female flowers on male plants of *Cannabis sativa* L. by 2-chloroethanephos-phonic acid. *Cellular and Molecular Life Sciences* 26:214–216.

Ram, H. Y. M. and V. S. Jaiswal. 1972. Induction of male flowers on female plants of *Cannabis sativa* by gibberellins and its inhibition by abscisic acid. *Planta* 105:263–266.

Razumova, O. V., M. G. Divashuk, O. S. Alexandrov, and G. I. Karlov. 2023. GISH painting of the Y chromosomes suggests advanced phases of sex chromosome evolution in three dioecious Cannabaceae species (Humulus lupulus, H. japonicus, and *Cannabis sativa*). *Protoplasma* 260:249–256.

Ren, G., X. Zhang, Y. Li, K. Ridout, M. L. Serrano-Serrano, Y. Yang, A. Liu, G. Ravikanth, M. A. Nawaz, and A. S. Mumtaz. 2021. Large-scale whole-genome resequencing unravels the domestication history of *Cannabis sativa. Science Advances* 7:eabg2286.

Renner, S. S. 2014. The relative and absolute frequencies of angiosperm sexual systems: dioecy, monoecy, gynodioecy, and an updated online database. *American Journal of Botany* 101:1588–1596.

Renner, S. S. and N. A. Müller. 2022. Sex determination and sex chromosome evolution in land plants. Pp.. *Philosophical Transactions of the Royal Society* 377:20210210.

Renner, S. S. and R. E. Ricklefs. 1995. Dioecy and its correlates in the flowering plants. *American Journal of Botany* 82:596–606.

Romanov, D. V., G. I. Karlov, and M. G. Divashuk. 2022. Developing oligo probes for chromosomes identification in Hemp (*Cannabis sativa* L.). *Plants* 11:1900.

Sakamoto, K., T. Abe, T. Matsuyama, S. Yoshida, N. Ohmido, K. Fukui, and S. Satoh. 2005. RAPD markers encoding retrotransposable elements are linked to the male sex in *Cannabis sativa* L. *Genome* 48:931–936.

Sakamoto, K., Y. Akiyama, K. Fukui, H. Kamada, and S. Satoh. 1998. Characterization; genome sizes and morphology of sex chromosomes in hemp (*Cannabis sativa* L.). *Cytologia* 63:459–464.

Sakamoto, K., N. Ohmido, K. Fukui, H. Kamada, and S. Satoh. 2000. Site-specific accumulation of a LINE-like retrotransposon in a sex chromosome of the dioecious plant *Cannabis sativa. Plant Molecular Biology* 44:723–732.

Sakamoto, K., K. Shimomura, Y. Komeda, H. Kamada, and S. Satoh. 1995. A male-associated DNA sequence in a dioecious plant, *Cannabis sativa* L. *Plant and Cell Physiology* 36:1549–1554.

Salentijn, E. M., J. Petit, and L. M. Trindade. 2019. The complex interactions between flowering behavior and fiber quality in hemp. *Frontiers in Plant Science* 10:614.

Shao, H., S.-J. Song, and R. C. Clarke. 2003. Female-associated DNA polymorphisms of hemp (*Cannabis sativa* L.). *Journal of Industrial Hemp* 8:5–9.

Shephard, H., J. Parker, P. Darby, and C. Ainsworth. 2000. Sexual development and sex chromosomes in hop. *New Phytologist* 148:397–411.

Shephard, H., J. Parker, P. Darby, and C. A. Charles. 1999. Sex expression in hop (Humulus lupulus L. and H. japonicus Sieb. et Zucc.) floral morphology and sex chromosomes. In C. A. Charles, ed. *Sex determination in plants* (pp. 139–150). BIOS Scientific Publishers Limited.

Sirikantaramas, S., F. Taura, Y. Tanaka, Y. Ishikawa, S. Morimoto, and Y. Shoyama. 2005. Tetrahydrocannabinolic acid synthase, the enzyme controlling marijuana psychoactivity, is secreted into the storage cavity of the glandular trichomes. *Plant and Cell Physiology* 46:1578–1582.

Small, E. 2016. *Cannabis: a complete guide.* CRC Press.

Small, E. and S. G. Naraine. 2016. Size matters: evolution of large drug-secreting resin glands in elite pharmaceutical strains of *Cannabis sativa* (marijuana). *Genetic Resources and Crop Evolution* 63:349–359.

Törjék, O., N. Bucherna, E. Kiss, H. Homoki, Z. Finta-Korpelová, I. Bócsa, I. Nagy, and L. E. Heszky. 2002. Novel male-specific molecular markers (MADC5, MADC6) in hemp. *Euphytica* 127:209–218.

Toth, J. A., G. M. Stack, A. R. Cala, C. H. Carlson, R. L. Wilk, J. L. Crawford, D. R. Viands, G. Philippe, C. D. Smart, and J. K. Rose. 2020. Development and validation of genetic markers for sex and cannabinoid chemotype in *Cannabis sativa* L. *Gcb Bioenergy* 12:213–222.

Truta, E., Z. Olteanu, S. Surdu, M.-M. Zamfirache, and L. Oprica. 2007. Some aspects of sex determinism in hemp. *Analele Stiintifice ale Universitatii "Alexandru Ioan Cuza" din Iasi Sec. II a. Genetica si Biologie Moleculara* 8(2):31–39.

van Bakel, H., J. M. Stout, A. G. Cote, C. M. Tallon, A. G. Sharpe, T. R. Hughes, and J. E. Page. 2011. The draft genome and transcriptome of *Cannabis sativa. Genome Biology* 12: 1–18.

Veen, H. 1983. Silver thiosulphate: an experimental tool in plant science. *Scientia Horticulturae* 20:211–224.

Vergara, D., E. L. Huscher, K. G. Keepers, R. M. Givens, C. G. Cizek, A. Torres, R. Gaudino, and N. C. Kane. 2019. Gene copy number is associated with phytochemistry in *Cannabis sativa. AoB Plants* 11:plz074.

Welling, M. T., M. A. Deseo, A. Bacic, and M. S. Doblin. 2021. Untargeted metabolomic analyses reveal chemical complexity of dioecious cannabis flowers. *Australian Journal of Chemistry* 74:463–479.

Yang, M.-Q., R. van Velzen, F. T. Bakker, A. Sattarian, D.-Z. Li, and T.-S. Yi. 2013. Molecular phylogenetics and character evolution of Cannabaceae. *Taxon* 62:473–485.

Zhou, Y., W. Sun, H. Cai, H. Bao, Y. Zhang, G. Qian, and C. Ge. 2019. The role of anti-Müllerian hormone in testis differentiation reveals the significance of the TGF-β pathway in reptilian sex determination. *Genetics* 213:1317–1327.

14 A Successful Cannabis Business (Glass House Farms)

Graham Farrar and Josh Karchmer

14.1 INTRODUCTION

Graham Farrar fell in love with cannabis back when he was in high school in Santa Barbara.

> "I smoked weed, and I sold weed," he says. "I saw firsthand that DARE and the War on Drugs were bullshit. I realized that not only were the people who smoke cannabis good people—often times they were better versions of themselves."

Still, Farrar could never imagine that his early passion for the plant would blossom into a $750 million publicly traded business. Farrar is the co-founder and President of Glass House Brands. Along with his business partner Kyle Kazan, he has built the largest vertically integrated cannabis company in California, a sprawling operation that includes cultivation, branded products, distribution, and retail operations.

Glass House has three large greenhouse grow facilities in California. Their crown jewel is their Camarillo farm spanning over 5.5 million square feet. "As far as I know, it's the largest cannabis greenhouse in the universe," Farrar says (Figure 14.1).

On a single day, their current operations can produce over 1,000 pounds of cannabis flower. Their Glass House Farms and Allswell brands are the most consumed cannabis flower in the state of California, the largest weed market in the world, beating out the second biggest competitor by more than 20%.

But the company's massive size is not the only story of its success. In an uncertain and highly competitive business that has sunk many others, Glass House has implemented a winning strategy of agricultural innovation, an integrated business model, and a dedication to providing quality and consistent products at the best possible value to consumers.

Not bad for a company whose earliest days can be traced back to Farrar's college dorm room in Boulder, Colorado, where he set up an automated hydroponic grow room in his 2x3 closet, complete with an automated flood and drain system, a high-pressure sodium lamp, and CO_2 supplementation.

14.2 ORIGIN STORY

In college at Boulder, Farrar's passion for cannabis merged with his interests in biology and technology. Growing his own high-quality cannabis sparked his entrepreneurial spirit, selling weed to students for $4,500 a pound.

After college, Farrar continued growing cannabis in California. Prop 215 had just kicked in, allowing Farrar to get his doctor's recommendation to gradually set up a small network of grow houses around Santa Barbara. From time to time, he would leave town to take long sailing adventures, leaving the grow houses to his friend, Jason Downs, who would eventually become Glass House's Director of Cannabis Operations.

 DOI: 10.1201/9781003386438-14

FIGURE 14.1 GlassHouse Farm Camarillo exterior.

FIGURE 14.2 Graham and Kyle observing the plant material.

During this time, Farrar met his future business partner, Glass House Co-founder, Chairman, and CEO Kyle Kazan (Figure 14.2). This fateful meeting would change both their lives—and the cannabis business. Kazan was an advocate for cannabis legalization. In his second career, he ran a successful real estate management company.

In 2015, Kazan and Farrar purchased a 150,000-square-foot greenhouse in Santa Barbara county. Kazan provided the initial funding, and Farrar set up the growing operations and systems. This meeting of the minds would ultimately become Glass House 1, the proof of concept that set the stage for their much bigger ambitions.

With one flourishing greenhouse on the ledger, Farrar and Kazan were able to convince investors to buy into their vision. Just a few years later, Glass House 2 was built two minutes away from their first grow. Farrar also secured the first adult-use license in Santa Barbara, and he and Kazan opened their first brick-and-mortar dispensary together, The Farmacy, on West Mission Street.

Building Glass House 3 proved more challenging. The company had its sights on a massive tomato and cucumber farm in Camarillo, but there were two problems—cannabis wasn't legal in the country, and there was also the little matter of a $250 million price tag. "For some reason, that didn't stop us," says Farrar.

Eventually, they were able to get an ordinance passed by voters to grow cannabis at the farm. They also reduced the price tag to $93 million cash. But in order to buy the property, the company needed to go public. So Glass House Brands was born in June of 2021.

14.3 STRATEGIC VISION & MISSION

From the very start, Farrar and Kazan had the vision to build a cannabis company for the normalized, legal market they believed was inevitable. Unlike many of their competitors, who set their sights on growing cannabis in remote desert locations where regulations were more lax, Glass House chose locations based on the fundamentals of agriculture—good climate, available labor, abundant water, and proximity to the massive California consumer market.

Behind every decision was the belief that what they were doing was worthwhile and that cannabis made the world a better place. With that as their core principle, they set out to produce high-quality yet affordable cannabis and create a sustainable company that cares for its employees.

Says Farrar, "We wanted to make something that everybody could have access to both in terms of cost and scale. A brand that could eventually sell on shelves at 7-Eleven, Vons, and Circle K around the country."

Another crucial part of their strategy was being vertically integrated across cultivation, brands, and retail. This allowed them to ensure quality control from cultivation all the way through the end user experience.

Glass House's bread and butter has always been cultivation. In addition to their own in-house brands, they also supply flower to many of the largest brands in California.

14.4 CHALLENGES, ADAPTABILITY, AND RESILIENCE

The cannabis industry is rife with uncertainties, unpredictable regulatory changes, and aggravating challenges that most other industries never have to deal with. Basic financial operations like banking, payroll, and insurance are major headaches for the cannabis industry, and have forced Glass House to get creative and resilient.

In the early days, Glass House was forced to make its employee payroll entirely in cash. A tedious process that took two people two full days to count and distribute in envelopes. Not to mention, transporting and storing large sums of cash was an ever-present security risk. Obstacles like these nearly ground their business to a halt. "It's like trying to ride a bike with the brake on," Farrar says.

Every local municipality has its own Byzantine cannabis rules and regulations that are constantly evolving. Glass House had to spend enormous effort developing zoning ordinances, getting approval across multiple commissions and appeals processes for each new facility.

What kept them motivated? An unwavering belief in the moral rightness of ending cannabis prohibition. As Farrar says, "We believe the truth always happens eventually, and no view of the future I can see has cannabis remaining a Schedule 1 banned substance. So we keep on keeping on while the moral arc of the universe bends towards justice."

14.5 INNOVATION AND CREATIVITY

Rather than reinventing the greenhouse from scratch, Glass House has taken an innovative approach, retrofitting existing greenhouses to optimize them for cannabis cultivation.

For example, the company's state-of-the-art facilities use efficient ebb-and-flood floors and positive pressure, which are standard for large-scale food producers but are new in the world of cannabis. Even using greenhouses instead of warehouses is a novel approach. While it's easier to grow

FIGURE 14.3 Camarillo Farm, ebb-and-flood floors.

cannabis in a warehouse, it's also more expensive and hurts the environment by using tremendous amounts of energy.

"We've chosen to take the harder road, but the one that we think has more long-term value," explains Farrar.

Glass House's technicians have even designed and fabricated new pieces of equipment, like specialized climate-controlled dry rooms, which allow for better airflow around the plants. Farrar estimates that this generates 20% more capacity by saving time in the drying process.

On the genetics side, Glass House has teamed up with top breeders and developed an extensive genetic library of over 100 strains that are constantly evaluated. Glass House 1 has essentially morphed into an R&D facility to find the most desirable cultivars for commercialization (Figure 14.3). A key innovation has been pivoting to tissue culture propagation to ensure a disease-free, vigorous source of clean "super mom" plants that can be cloned.

14.6 CASE STUDY: IDENTIFYING AND MITIGATING HPLVD

In fact, Glass House's large-scale cultivation allowed them to be one of the first to identify and characterize a new cannabis virus, HPLVd (Hop Latent Viroid), which wreaks havoc on potency and yields. Cannabis growers are fortunate, so far, because they only have to combat one virus like this, while other crops have to deal with many. But HPLVd is a serious issue both economically and for its impact on the quality of the end product. One study estimates as much as 90% of California cannabis is infected with HLVd, costing nearly $4 billion in lost yields.

Six years ago, the Glass House team noticed a worrisome problem—some of their plants exhibited stunted growth, brittle stems, smaller flowers, and other undesirable characteristics. Because their operations were so consistent with environmental control, standardized nutrients, and consistent cloning/growing practices, the team knew this was not due to random variances. Something was wrong, but what was it? Suddenly, Glass House found themselves on the front lines of a battle they never saw coming. They needed to strike back before it was too late.

Glass House reached out to a top plant virology expert at UC Davis who normally would have jumped at the opportunity. But this was a cannabis problem, and like many academics and institutions, he was fearful that he might lose his federal funding. However, he put them in touch with a graduate student who was more than willing to collaborate. A crack team of researchers, including Dark Heart Nursery Research and Phylos Bioscience, was assembled. By providing DNA/RNA samples from affected plants, Glass House enabled research that genetically sequenced a new pathogenic virus, eventually identified "Hop Latent Viroid" by matching its RNA to the existing know viral genomic sequences.

The scientists determined that Hop transfers primarily mechanically via tools and equipment that have had contact with infected plants. While their work did not cure the disease, it laid the foundation for PCR testing protocols to detect HPLVd before it's too late.

Initially, Glass House sent suspicious samples to a lab for testing, but it would take weeks to get the results back, during which millions of potentially infected clones had already been cut. A decision was made to partner with a biotech firm called MyFloraDNA to implement diagnostic testing onsite at the greenhouse. This decision resulted in a much more efficient and less expensive defense against HPLVd, which kept the greenhouse safer from infection.

14.7 SOCIAL JUSTICE

Glass House's partnerships extend beyond the cultivation and business of the plant. They have actively sought out collaborations to address the injustices and harms caused by cannabis criminalization.

One organization they have allied with is the non-profit Mission Green, led by Weldon Angelos, which is committed to obtaining the freedom of people imprisoned for nonviolent marijuana-related crimes. Kazan serves on the board of Mission Green and has made substantial contributions to support the organization's efforts.

When Mission Green secured the release of Luke Scarmazzo, who received a 22-year jail sentence in federal court for running a state-legal medical cannabis dispensary in Modesto, Glass House gave him a job as soon as he walked out of prison.

"We make sure that we don't forget that there are people sitting in cages for doing a lot less than what we do every day," Farrar says, pointing to another case in Georgia where a nonviolent criminal is being held in prison for possessing eight pounds of weed. "I mean, we produce that in the first six minutes of the day."

Glass House maintains an active philanthropy program. From donating product to veterans through its SB-34 program to providing Seniors with cannabis education and dispensary transportation, they have worked toward universal access for a range of groups.

14.8 EXPANSION & GROWTH

One of Glass House's main objectives is laying the foundations for rapid future expansion. To date, they have occupied three of the six greenhouses at their sprawling Camarillo location. However, they plan on converting the rest of the greenhouses in the next few years.

Says Farrar, "Our goal is to make brands and products consumers love and to become the most consumed cannabis brands on the planet, to add our label to everything we grow and eventually have our products distributed legally across the country and around the world."

14.9 CULTIVATING SUCCESS

A 2022 poll found that less than 25% of cannabis-touching businesses are profitable. Glass House has not only defied the odds and become a trailblazer in the cannabis industry, but it's also among the most successful cannabis brands in the country. A few key factors contributed to the company's remarkable success.

Forward-thinking leadership: Co-founders Graham Farrar and Kyle Kazan have guided the company with a clear vision of a brand that can thrive in a normalized, legal cannabis market. Their foresight has allowed Glass House to make strategic decisions that set the stage for success today and in a fully legalized future.

Innovative cultivation practices: Glass House has made an art form out of taking traditional ag greenhouses and converting them to maximize efficiency and quality. From ebb-and-flood floors to tissue culture propagation, the company has consistently pushed the boundaries of cannabis cultivation.

Proactive problem-solving: When faced with the very serious threat of the HPLVd virus, Glass House collaborated with top scientists to identify and mitigate the issue, demonstrating their ability to adapt and innovate.

Vertical integration: By controlling every aspect of the supply chain, from seed to sale, Glass House has been able to capture more profits and maintain a high level of quality control.

Commitment to decriminalization and legalization: Glass House has not forgotten the people who pushed cannabis forward during prohibition. The company has actively partnered with organizations like Mission Green to address the injustices caused by cannabis criminalization.

As Glass House continues to expand its operations, the company is anchored by the same philosophy that inspired Farrar in his earliest days as a cannabis entrepreneur.

"Cannabis is stress relief and medicine for so many people," he says. "It makes the world a better place."

15 Cannabis Policy and Legalization

Hirsh Jain

15.1 HISTORY OF STATE-LEVEL CANNABIS LEGALIZATION IN THE UNITED STATES

15.1.1 Medical Cannabis Legalization

The modern cannabis legalization movement in the United States began in 1996, when the voters of California passed Proposition 215, a state ballot initiative legalizing cannabis for medical purposes, in November 1996. California was the first state in the country to adopt a medical cannabis law. However, medical cannabis legalization quickly became a trend that swept the country. In the years that followed, many other states followed California's example and adopted medical cannabis laws of their own.

In 1998, voters in Alaska, Oregon and Washington legalized medical cannabis through ballot measures. In 1999, voters in Maine legalized medical cannabis through a ballot measure. In 2000, the Hawaii state legislature passed a law legalizing medical cannabis. Also in 2000, voters in Nevada and Colorado legalized medical cannabis through ballot measures. Thus, in the four years following California's legalization of medical cannabis in 1996, 7 other states legalized medical cannabis.

In 2003, the Maryland state legislature passed a law legalizing medical cannabis. In 2004, the Vermont state legislature passed a law legalizing medical cannabis. In 2004, voters in Montana legalized medical cannabis through a ballot measure. In 2006, the Rhode Island state legislature passed a law legalizing medical cannabis. Thus, in the decade following California's legalization of medical cannabis in 1996, 11 other states legalized medical cannabis.

The medical cannabis legalization trend would accelerate over the next 5 years, with New Mexico, Michigan, Massachusetts, New Jersey, Arizona, Delaware, and Connecticut adopting medical cannabis laws. By 2012, nearly half of the states in the country had legalized medical cannabis.

15.1.2 Adult Use Legalization

In 2012, voters in Colorado and Washington legalized "adult use" cannabis, that is, cannabis for use by adults 21 years of age. Colorado and Washington were the first two states to adopt regulatory approaches to cannabis akin to alcohol, in which access to cannabis would not be predicated on having a qualifying "medical" condition, but instead would be a privilege afforded to all adults 21 years of age or older. Colorado and Washington began adult use sales in January 2014.

As with California's decision to legalize medical cannabis in 1996, this trend began sweeping the country. In 2014, voters in Alaska and Oregon legalized adult use of cannabis by ballot measure. In 2016, voters in California, Nevada, Maine, and Massachusetts legalized adult use of cannabis

 DOI: 10.1201/9781003386438-15

by ballot measure. Thus, within 4 years of the first adult use cannabis laws passing, 8 states had legalized adult use cannabis.

In 2018, voters in Michigan legalized adult use of cannabis by ballot measure. In 2019, the Illinois state legislature passed a law legalizing adult use of cannabis. In 2020, voters in Arizona, Montana, and New Jersey legalized adult use of cannabis by ballot measure. Also in 2020, the Vermont state legislature passed a law legalizing adult use of cannabis. Thus, within 8 years of the first adult use cannabis law passing, 14 states had legalized adult use cannabis.

In 2021, voters in Connecticut, New Mexico, New York, and Virginia legalized adult use cannabis by ballot measure. In 2022, voters in Maryland and Missouri legalized adult use cannabis by ballot measure. Also in 2022, the Rhode Island state legislature passed a law legalizing adult use cannabis. In 2023, the Delaware and Minnesota state legislatures passed a law legalizing adult use cannabis. Also in 2023, voters in Ohio legalized adult use cannabis by ballot measure.

Thus, a little more than 10 years following the first adult use cannabis law passing, 24 states have legalized adult use cannabis.

15.1.3 State-Level Cannabis Legalization Efforts in 2024 & 2025

As of June 2024, there are 24 states that have legalized adult use cannabis. In addition, there are 38 states that have legalized cannabis for medical purposes.

There are six additional states that appear poised to legalize cannabis for adult use later in 2024, or in 2025. By the end of 2025, there thus may be up to 30 states that have legalized cannabis for adult use, well over half of the country.

- **New Hampshire**. As of June 2024, the New Hampshire state legislature is considering a cannabis legalization bill. If this bill passes, New Hampshire would be the 25th state to legalize adult use cannabis. It would also be the first state with unified Republican control of the state legislature and Governor's Office to pass an adult use cannabis law. If New Hampshire fails to pass an adult use cannabis legalization bill in June 2024, it will likely consider the issue again in 2025. There is no ballot initiative process in New Hampshire, so cannabis legalization must pass via the state legislature.
- **Florida.** In November 2024, voters in Florida will vote on an adult use cannabis ballot initiative. In order to pass, the initiative must receive 60% of the vote, the only state to have such a high threshold for the passage of state ballot initiatives.
- **South Dakota.** In November 2024, voters in South Dakota will vote on an adult use cannabis ballot initiative. In order to pass, the initiative must receive 50% of the vote.
- **North Dakota**. In November 2024, voters in North Dakota may vote on an adult use cannabis ballot initiative. In order to appear on the ballot, the initiative must receive enough signatures to qualify, which will be determined by August 2024. If the initiative appears on the ballot, in order to pass it must receive 50% of the vote.
- **Pennsylvania.** As of June 2024, the Pennsylvania state legislature is considering a cannabis legalization bill. If Pennsylvania fails to pass a cannabis legalization bill in 2024, it will likely consider the issue again in 2025. There is no ballot initiative process in Pennsylvania, so cannabis legalization must pass via the state legislature.
- **Hawaii**. In 2024, Hawaii considered an adult use cannabis legalization bill, but it did not pass. This issue will be taken up again in the 2025 legislative session. There is no ballot initiative process in Hawaii, so cannabis legalization must pass via the state legislature.

Thus, by the end of 2025, there may be up to 30 states that have legalized cannabis for adult use, well over half of the country. However, that would still leave at least 20 states that have failed to legalize adult use cannabis. Most analysts believe it will likely take several more years and perhaps until the end of the decade for every state in the country to legalize adult use cannabis.

It is important to consider that many states that continue to restrict cannabis access do not have a ballot initiative process, and have conservative state legislatures that have historically been hostile to cannabis reform. Despite the fact that there is widespread support for adult use cannabis legalization in many states (e.g. Texas), it seems unlikely that an adult use cannabis law will be passed in the next several years. The grassroots work of reforming cannabis laws at the state level is thus likely to persist until the end of this decade, at least.

15.2 FEDERAL POLICY ISSUES IN 2024 & 2025

15.2.1 Rescheduling

In May 2024, President Joe Biden publicly announced his Administration's support for the "rescheduling" of cannabis, from a Schedule I substance to a Schedule III substance under The Controlled Substances Act. Schedule I substances are those that are considered to have no acceptable medical value and a high potential for abuse, such as heroin. Schedule III substances are considered to have some medical value and a lower potential for abuse. Examples of Schedule III substances include Tylenol and codeine.

Shortly after President Biden's public announcement, the Department of Justice officially published a "proposed rule" rescheduling cannabis to Schedule III in the Federal Register. The publishing of this rule initiated a 60-day comment period that will conclude in July 2024.

After the comments are reviewed, a final rule will be issued. It is possible that the rescheduling of cannabis to Schedule III could be finalized in late 2024. While legal challenges and opposition from anti-cannabis organizations has the potential to delay its implementation, it seems very likely that cannabis will soon be a Schedule III substance.

If cannabis becomes a Schedule III substance, it would mark the most significant change to federal cannabis policy since The Controlled Substances Act was passed in 1970 and cannabis was placed on Schedule I, the most dangerous class of drugs. It would represent the United States federal government acknowledging for the first time that cannabis has medical value, nearly 30 years after California first affirmed this concept when Proposition 215 passed in 1996.

In addition to the immense symbolic value of cannabis being rescheduled, there would be two significant implications rescheduling would have for how cannabis is treated:

- **Research**. Research into cannabis has been heavily restricted because of its status as a Schedule I substance. If cannabis were rescheduled to Schedule III, it would become significantly easier for scientists and research institutions in the United States to study cannabis.
- **Taxation**. Section 280E of the IRS Tax Code prohibits companies that are trafficking in Schedule I or Schedule II substances from claiming ordinary business deductions on their federal taxes. As a result, cannabis companies in the United States pay an effective tax rate of 70%. This makes it nearly impossible to run a profitable cannabis business. The rescheduling of cannabis would normalize the tax treatment of cannabis businesses, and make it possible for a much larger universe of cannabis companies to succeed.

It is important to note, however, that even if cannabis is rescheduled, cannabis will still remain federally illegal. Rescheduling does not resolve the federal-state conflict that defined the cannabis space, in which an increasing number of states have legalized cannabis for medical or adult use purposes but cannabis remains illegal on the federal level. As such, the criminal justice implications of rescheduling would be quite modest. Rescheduling would not result in individuals being released from prison or in the expungement of cannabis records. In summary, though rescheduling would enable greater research into cannabis and normalize the tax treatment of cannabis businesses, cannabis would remain federally illegal.

Still, it is possible that rescheduling may make certain institutions that have previously been hesitant to serve or work with cannabis businesses to begin engaging with them. While these institutions may still be hesitant to work with cannabis businesses on account of cannabis being federally illegal, it may slowly make them more comfortable with serving such businesses.

- **Banks**. Of the 12,000 banks and credit unions in the United States, less than 600 (5%) serve cannabis businesses. Cannabis being federally illegal deters many of these banks from working with cannabis operators. Moreover, the compliance obligations that banks that serve cannabis businesses must follow are so onerous that many banks find serving cannabis businesses very resource-intensive and thus do not actively seek to serve cannabis clients. As noted, rescheduling would not address these issues immediately. However, it may slowly increase the universe of banks that are willing to serve cannabis companies, giving cannabis operators access to a wider range of providers and more competitive rates.
- **Insurance Companies.** Cannabis, being federally illegal, deters many insurers from working with cannabis operators. In addition, the lack of detailed history in most cannabis operators makes it a very difficult businesses to underwrite. As a result, cannabis companies have access to a very limited pool of insurance options, and often pay very high rates for very limited coverage. As noted, rescheduling would not address these issues immediately. However, it may start to increase the number of insurance carriers that are willing to provide lines of coverage to cannabis companies, giving these companies access to a wider range of coverage and more competitive rates.
- **Stock Exchanges**. Major American stock exchanges, such as the New York Stock Exchange (NYSE) and the Nasdaq, do not list plant-touching cannabis companies on their exchanges. The legal departments of these major exchanges have made the discretionary decision to prevent businesses trafficking in a Schedule I substance and thus violating US federal law from listing on these exchanges. The irony of this decision is that these major US exchanges continue to list cannabis companies based in other countries (e.g. Canadian cannabis companies) on their exchange. As noted, rescheduling would still render cannabis federally illegal. However, since the decision of the NYSE and Nasdaq to list US cannabis companies is entirely discretionary, it may change the calculus of these exchanges. If NYSE and Nasdaq become comfortable with listing US plant-touching cannabis companies, it will expose these companies to a much wider pool of investors, including institutional capital. It thus has the potential to significantly boost the valuations of public companies.

15.2.2 Federal Cannabis Legalization

Amidst the rescheduling of cannabis from Schedule I to Schedule III initiated by the Executive Branch, efforts to legalize or deschedule cannabis entirely continue to be introduced in the legislative branch, to little success. Despite polls revealing that 70% of Americans are supportive of cannabis legalization, that support is not reflected in Congress. In addition, even amongst Congressional advocates for cannabis reform, there remains widespread disagreement on how cannabis should be legalized and what an appropriate regulatory framework would look like. When combined with the general legislative paralysis that characterizes Congress, this has meant that while many cannabis legalization bills have been introduced in recent years, few have gained any meaningful traction.

- **Marijuana Opportunity Reinvestment and Expungement Act** (MORE). The MORE Act is a House bill introduced by members of the Democratic Party that would deschedule cannabis, facilitate the expungement of cannabis records and use taxes generated by cannabis sales to assist individuals negatively impacted by The War on Drugs. The MORE Act passed the House of Representatives in 2020 and again in 2021, but failed to gain support in the Senate.

- **Cannabis Administration and Opportunity Act** (CAOA). The CAOA Act is a Senate bill introduced by members of the Democratic Party that would deschedule cannabis, facilitate the expungement of cannabis records and use taxes generated by cannabis sales to assist individuals negatively impacted by The War on Drugs. The CAOA does not have the support of 60 Senators and so is unlikely to move forward in the Senate.
- **STATES Act.** The STATES Act is a bill that has been introduced in the House and the Senate, and has supporters in both the Democratic Party and Republican Party. The bill would deschedule cannabis and facilitate the expungement of cannabis records. Despite the fact that it has support from members of both parties, the STATES Act has never been prioritized by leadership and so has never been able to move forward in either Chamber.

15.2.3 The SAFE Banking Act

The SAFE Banking Act is legislation that has been introduced in both the House and Senate, and that would allow banks to serve cannabis companies without any fear of federal reprisal. Though the SAFE Banking Act would not change the fact that cannabis is federally illegal, it would provide banks that serve these businesses with "safe harbor," ensuring they are not subject to any federal prosecution for doing so.

If the SAFE Banking Act were passed, it would significantly increase the number of banks serving cannabis companies, allowing cannabis operators to gain access to a wide range of services, from a greater number of providers, at lower rates. In addition, SAFE Banking would make it more likely that credit card companies and major stock exchanges would become comfortable serving cannabis businesses. SAFE Banking would thus allow the cannabis industry to take advantage of the US financial system, which it has in large part been cut off from since its inception.

The SAFE Banking Act has passed the House of Representatives numerous times since 2019. However, it has never been able to pass the Senate, and hence has not become law. In September 2023, the SAFE Banking Act passed the Senate Banking Committee, a key hurdle before it would be considered by the full Senate. The Senate is likely to examine SAFE Banking in early 2025, during the "lame duck" session following the Presidential election. However, it would still need to pass the House, which is uncertain given that it is in Republican control.

15.2.4 The Farm Bill

The Farm Bill is the primary agricultural and food policy bill for the United States federal government, passing every 5 to 6 years. The most recent Farm Bill passed in 2018. The 2018 Farm Bill legalized industrial hemp, that is, it legalized cannabis plants that have a tetrahydrocannabinol (THC) concentration of no more than 0.3%, by removing it from Schedule I of the Controlled Substances Act. The Farm Bill also legalized all derivatives of these hemp products.

The years following the passage of the Farm Bill in December 2018 saw an explosion in "intoxicating hemp" products, those legalized by the 2018 Farm Bill but modified to have an intoxicating effect. As these products have proliferated in states across the country, there have been increasing calls to revise the next Farm Bill to "close the loophole" that legalized these products, which critics allege are unregulated and violate the spirit of the 2018 Farm Bill. There have also been equally vigorous efforts from the hemp industry to defend and retain the status quo, creating a "civil war" within the cannabis community.

As of June 2024, Congress is currently debating the next Farm Bill, which will likely pass in early 2025. We are likely to see language introduced that would redefine hemp in the 2024 Farm Bill to effectively make the intoxicating hemp-derived cannabinoid market illegal, along with robust efforts to oppose further restrictions on these products. How this issue resolves itself will have significant implications for the future of the US cannabis space.

15.3 JUDICIAL CHALLENGES

In May 2024, a Massachusetts federal court heard opening arguments in Cannaprovisions v. Garland, a case brought by a coalition of state-licensed cannabis companies challenging the constitutionality of federal cannabis prohibition. The coalition was represented by David Boeis, chairman of the firm Boies Schiller and Flexner LLP, and one of the most famous litigators in the country with a successful track record arguing in front of the Supreme Court.

This case is notable because it is arguably the highest profile judicial challenge to federal cannabis prohibition in the nation's industry, not only because of the stature of Mr. Boies but also because it comes at a moment when half the states in the country have legalized cannabis in defiance of federal prohibition. The filing of the case also reflects the failure of the cannabis movement to make significant progress on federal reform through the legislative branch, leading the industry to pursue reform through the judicial system.

The argument that cannabis companies are making in Cannaprovisions v. Garland is that the Constitution's Commerce Clause should preclude the federal government from interfering in state-legal cannabis activity, because it is regulated within a state's borders. In order for this argument to be successful, the Supreme Court would have to overturn its 2005 precedent, Gonzales v. Raich, which upheld the federal government's power to enforce federal cannabis prohibition within a state's borders. If the Supreme Court were willing to take up Cannaprovisions v. Garland and consider overturning its precedent, the earliest it would likely hear the case is 2026. This case is likely to be the highest profile judicial challenge to federal cannabis policy over the next several years, and is worth watching closely.

15.4 INTERNATIONAL CANNABIS MARKET

Though the United States is the largest cannabis market in the world, the cannabis reform movement is an increasingly global phenomenon. There are numerous countries across the world that are on the path to cannabis normalization.

15.4.1 North America

15.4.1.1 Canada

In 2018, Canada federally legalized cannabis. By many metrics, Canada is the most advanced country in the world when it comes to cannabis, with a functional national market having existed for more than 5 years. The Canadian cannabis industry generates nearly $5 billion in annual sales.

15.4.1.2 Mexico

In June 2021, Mexico's Supreme Court declared that national cannabis prohibition is unconstitutional and decriminalized cannabis, putting Mexico on a path to cannabis reform. However, Mexico's legislature has yet to develop the framework for a commercial model that enables cannabis sales. Thus, while Mexico is on the path to cannabis legalization, it still lags far behind its counterparts in North America.

15.4.2 Europe

15.4.2.1 Malta

In 2021, Malta became the first country in the European Union to legalize cannabis for adults. Consumers in Malta are allowed to possess 7 grams of cannabis and can grow four plants at home. However, Malta lacks a dispensary model as exists in Canada and the United States, and instead requires consumers to obtain cannabis through non-profit cooperatives.

15.4.2.3 Germany

In April 2024, a landmark cannabis reform bill passed in Germany. This bill allows adults to possess up to 25 grams of cannabis and grow up to three plants for personal use. In addition, the legislation allows adults to join "social clubs" where they could buy up to 25 grams of cannabis, with a cap of 50 grams per month. In addition, the legislation specified that medical cannabis will no longer be classified as a narcotic in Germany and will be treated as a standard prescription.

These changes represent a significant liberalization of cannabis in Germany. Because of its status as the largest economy in Europe, these changes to cannabis policy are anticipated to impact cannabis policy in neighboring jurisdictions as well, spurring a "domino effect" of cannabis reform across the European continent. In addition, Germany plans to liberalize its cannabis laws even further in 2025, and develop a more full-scale commercial model that more closely resembles models in Canada and the United States.

15.4.3 Other European Countries

Several other countries in Europe have adopted or are in the process of adopting models similar to Germany. Switzerland, the Czech Republic, and the Netherlands are three such countries that have adopted similar "pilot programs" that enable sales on a small, non-commercial scale.

A number of other countries have opted instead for a model focused on decriminalization, in which possession for small amounts is immune from punishment, but there is no commercial market in which cannabis products can be purchased. Countries such as Austria, Belgium, Italy, Portugal, and Spain fit this model.

15.4.4 Asia

15.4.4.1 Thailand

In June 2022, Thailand decriminalized cannabis, representing a dramatic departure from its prior policy of cannabis prohibition. However, the Thai government did not establish a clear regulatory framework for cannabis sales, leading to a rapid proliferation of cannabis establishments across the country. While this dramatically expanded cannabis access in Thailand, the speed of the change inspired opposition from segments of Thai society. As of June 2024, Thailand is considering legislation that would impose significant restrictions on cannabis, preserving access to it for medicinal purposes but restricting its recreational use. This issue will be finalized in late 2024.

15.4.4.2 Other Asian Countries

Most other countries in Asia strictly enforce cannabis prohibition. Some countries, such as Japan, Malaysia, and the Philippines, allow for medical cannabis in very narrow circumstances, but the vast majority of Asian countries continue to heavily restrict access to any form of cannabis.

15.4.5 South America

15.4.5.1 Uruguay

In 2013, Uruguay became the first country in the world to fully legalize cannabis. However, Uruguay has deliberately rolled out a very limited commercial model, with few opportunities for cannabis to be legally purchased.

15.4.5.2 Colombia

In June 2023, the Colombian Senate voted against legislation that would have legalized adult use cannabis in the country. Colombia continues to have a robust medical cannabis market, and is seen as a potential major player in the global cannabis industry given its ideal climate for cannabis cultivation and potential low cost of production. However, it will likely be several years before an adult use measure is taken up again in Colombia.

15.4.5.3 Other South American Countries

Most other countries in South America continue to heavily restrict cannabis access and enforce a policy of cannabis prohibition.

16 Intellectual Property for Cannabis Plants[1]

Dale Hunt

16.1 INTRODUCTION

Intellectual property (IP) laws and systems make it possible to treat ideas, inventions, creations, and certain secrets as property that can be bought, sold, and used in ways that are similar to the handling of tangible property. Patents, trademarks, copyrights, and trade secrets are the forms of IP most people are familiar with. How these forms of IP are used and leveraged in mainstream agriculture corresponds with and demonstrates how IP can be used to protect *Cannabis* plants and related products.

Patenting plants, particularly cannabis, is controversial and has been the subject of some of the most contentious disputes in the industry. Many view the plant as sacred and, therefore, believe strongly that no person or commercial entity should have exclusive ownership of any part of it. Cannabis comes from nature, they argue, so how can any human or business entity claim its invention? Further, today's commercial varieties were all derived from plants nurtured by generations of small farmers—often in impoverished or war-ravaged countries—or bred by small-scale cultivators who risked their freedom to bring them to market. For many, this difficult history acts as a backdrop that should discourage ethical parties from participating in what these advocates see as aggressive and self-serving appropriations of a common good. Those who stand on the shoulders of giants should tread lightly and gratefully, if at all.

On the other hand, plant breeders would argue that their work is to produce new plant varieties that would not exist in nature. They cultivate, document, select, and cross plants, combining characteristics of many ancestor plants to develop unique traits through numerous generations of the breeding and selection process. This often requires a commitment of many years and significant expense. When this work results in a distinctive variety, IP protection allows them to bring it to market, recoup costs, and continue breeding as a profession. Those who embrace the IP system for protection of new plant varieties argue that all discovery and invention builds on prior generations of knowledge and do not agree that the laborious development of novel plant varieties should be treated any differently than other creative endeavors. They further argue that the patent laws already treat products of nature as unpatentable subject matter. They point out that those who say that all plants (or all forms of cannabis) are products of nature do not understand the process of plant breeding or the differences that must exist between starting material and final products that would justify a patent or commercial success, in the first place.

Most criticisms surrounding IP protection in the cannabis industry are directed at the patent system. The primary criticism is the potential for abuse; patenting and patent disputes are the arena of the well-funded. Patents grant significant, albeit temporary, monopoly powers to their owners. The monopoly power of patents on important medicines or highly productive crops carries with it the potential for artificially high prices that can be abusive and that may limit more widespread access to such basics as quality healthcare and food.

 DOI: 10.1201/9781003386438-16

TABLE 16.1
Types of IP

IP Type	Function	Example
Patents	Protect inventions	Patent protection for Edison's light bulb
Trademarks	Protect brands	Trademark "Apple" as a brand of computers
Copyrights	Protect specific expressions of an original work	Pixar's copyright of the movie "Toy Story"
Trade Secrets	Protect valuable commercial secrets	The secret formula for Coca Cola

The counterargument to this is the fact that, without patent protection, far fewer parties could afford to devote themselves to plant breeding, and far fewer new plant varieties would be developed. And, without patent protection, any truly special variety would have to be maintained and farmed under strict security, because even a small cutting from a single plant could be used to create unlimited copies of the plant. They argue, therefore, that patent protection actually increases public access because the patent provides permits the owner to license the variety into many different markets without losing control over the variety.

Regardless of personal ethical views surrounding plant patents, they are already a reality in agriculture, including for both hemp and non-hemp cannabis plants. This chapter explains the basic types of IP, how they are used in mainstream agriculture, and how this applies to the cannabis industry for both marijuana and hemp forms of cannabis. It concludes with a critical ethical discussion on patenting nature, lessons from mainstream agriculture, and how plant patenting is shaping the future of cannabis breeding. See Table 16.1.

The most common justification for a legal system to protect IP rights is establishing an incentive for creativity that rewards artists and inventors for sharing their creations with the public. IP systems make it possible for such sharing to enhance, instead of diminish, the value of the creation for its creator. This value enables creators to keep creating. For example, an author who can make a living writing books will be able to keep writing books instead of having to pursue other employment, which would limit her time and opportunity to write. Likewise, an inventive person or company that can gain financial benefits from researching and developing new products and devices will continue to invest in research and development, presumably resulting in benefits to the public through continued innovation.

This principle of IP rights establishing an incentive to continue creating is most consistently aligned with patents and copyrights. It does not align as well with trademarks. While trademark rights can undoubtedly benefit their owners, the policy justification for trademarks is rooted in protecting consumers. If a particular mark is associated with a consistent source of goods that bear the mark, consumers can have a reasonable expectation of consistency and quality that such a source has delivered in the past.

The justification for trade secrets is based upon principles of fair competition. The reasoning is that if someone has a valuable secret and takes reasonable steps to keep it secret, it should be wrong for another person to steal the secret. This is compared to the fact that it is accepted as wrong for someone to steal tangible property that the owner considers valuable and tries to keep reasonably secure.

16.2 TERMINOLOGY NOTE

It is essential to point out a terminology problem in the cannabis industry before going in-depth on available IP protection. People using the same words can mean different things, and some of these words directly connect with legal definitions and regulatory requirements. The industry's standard terminology is frequently confusing and is sometimes arguably incorrect, but colloquial use will continue to perpetuate it. The terms "hemp," "cannabis," "strain," "variety," and "cultivar" are among those most inconsistently and confusingly used.

16.2.1 Cannabis, Hemp, and Marijuana

The term "cannabis" can refer to an industry, to products derived from plants, or to any plant within the *Cannabis* genus. To be as correct as possible, one would write *Cannabis* when referring to plants of the genus and cannabis when referring to the industry or products other than basic plant parts, such as the flower.

Hemp is now legally defined in the United States as any *Cannabis sativa* plant that contains less than 0.3% Δ9-THC by dry weight. This definition does not align with the historical usage of the term "hemp," which refers to cannabis varieties primarily cultivated to harvest their fibrous stalks and nutritious seeds. This problematic bifurcation has resulted in separate "cannabis" and "hemp" industries despite there being a serious inconsistency in saying "cannabis and hemp." Saying "cannabis and hemp" is like saying "people and men." Men are a subset of people, much as hemp is a subset of the genus *Cannabis*.

To avoid confusion, the terms are used here as they are in common practice. "Marijuana" is used to refer to high-THC flowers, still federally illegal, and "hemp" refers to any part of the plant with less than 0.3% THC content by dry weight.

The term "marijuana" exemplifies how the industry can dramatically shift how it uses terminology. The Northern Mexican slang term "marijuana" was deliberately used in the propaganda campaigns that led to US Prohibition, and as a result, marijuana is the term written into law. The term was phased out of use in the legal industry to disassociate with the racist origins of its legislative use. A separate faction within the cannabis community and industry is now pushing for the re-acceptance of the term "marijuana" because the word itself is not racist, even if the people who used it to demonize the people who originated its use were. For simplicity and consistency with the law, "marijuana" refers to the flowers of the cannabis plant that test higher than 0.3% Δ9-THC by dry weight.

16.2.2 Strains, Varieties, and Cultivars

The terms "strain," "variety," and "cultivar" are often used interchangeably in cannabis, which is also incorrect.

"Strain" is a general term used in the industry to refer to anything far enough along in the breeding process to have an identity, even if it is not completely consistent or stable. Many plants referred to as strains are actually stable and consistent, but not all are. Giving a name to a strain may seem helpful, but in many cases, it only adds to the confusion because the same name or very similar variations of a name may be applied to numerous different strains that share few or no recent ancestors. Likewise, an inversion of this problem also exists: many examples of the same genetic selection, or its close relatives, have numerous different names. There is, as yet, no widely accepted convention or protocol to avoid this problem.

Further confusion arises from the fact that, in a strict sense, the term "strain" is not accepted as appropriate when applied to plants. Traditionally, "strain" applies to viruses and microorganisms such as bacteria and fungi. Purists, therefore, argue that the term "strain" should not be used at all in the industry and should be replaced with "variety" or "cultivar."

The term "variety" can refer to any group of plants of the same species that share some important traits that collectively distinguish such a group from others within that species. Varieties can either be the result of natural selection or selective breeding. In contrast, the term "cultivar" (a portmanteau of "cultivated variety") is limited to varieties that are bred and selected for cultivation.

Some purists argue against using the term "variety" because cultivated varieties should be referred to as cultivars. However, this position ignores the conventional use of the term "variety" in such authoritative and relevant sources as the USDA Plant Variety Protection Act, the International Union for the Protection of New Varieties of Plants (also known as UPOV), and the nearly universal usage of "variety" over "cultivar" by the US Patent and Trademark Office (USPTO) in the context of plant patents. For this reason, the term "variety" is used here to refer to a stable strain, and "strain" is used when it is unclear whether a breeder's selection is sufficiently stable and uniform to qualify as a variety.

My counterargument to the purists who disfavor the use of the term "strain" in all cases is that there needs to be a term to refer to selections that are the work of a plant breeder even before such selections may be stable and consistent enough to be called varieties or cultivars. If something a breeder or grower is working with cannot correctly be called a variety or cultivar, what should it be called? One could use the term "selection," but given the already widespread conventional use of the term "strain" within the industry, I see no harm in its continued use. My experience with breeders of other types of plants is that the term they typically use is "selection."

"Strain" can be thought of as a noncommittal kind of term. A strain may be a plant worth working with, cultivating, and using, but may or may not be stable enough to be considered a finished variety or variety.

16.3 HOW TO PROTECT A NEW VARIETY

This is a simplified step-by-step summary of the IP protection process and does not replace qualified legal advice.

1. Breeding and selection (phenohunt) complete. Collect and store relevant data about the variety, which will be required to file the patent.
2. Choose the type of protection. Is commercially relevant propagation possible with both seeds and clones, or exclusively with either seeds or clones, but not both?

If the plant cannot be consistently and stably propagated by seeds, then it should be protected with a plant patent (PP). Sexually reproduced varieties are not eligible for PP, and PP are not infringed by use of the plant in sexual propagation such as to produce seeds or to breed new varieties.

If a stable seed line exists or can be developed, a stronger form of protection would be a variety-specific utility patent (VSUP).

If the plant legally qualifies as hemp, it is also eligible for USDA plant variety protection (PVP). (See Table 16.2 for a comparison of PP, VSUP, and PVP.)

TABLE 16.2
Comparing Intellectual Property Options for Plants

Plant IP	THC Level	Seed Deposit?	Does NOT Cover	Relative Cost*	Months to Issue**
USDA PVP	< 0.3%	Yes	Breeding; saving seed for replanting	$	18–24
USPTO Plant Patent	Any	No	Seeds; breeding	$	18–24
USPTO Utility Patent (VSUP)	Any	Yes	n/a (coverage is flexible)	$$	24–36

* while cost can vary, typically a VSUP will cost close to 2x the cost of a PP or PVP
** typical timing—exceptions occur

3. Make an appropriate disclosure to an attorney. This disclosure will require comparisons of the new variety with parents and closest known patents. A plant patent requires a detailed botanical description. A PVP and VSUP require a seed deposit. DNA testing is not required but can be included in the disclosure if desired.
4. The attorney files an application with USDA or USPTO.

16.4 TRADEMARKS VS. PATENTS

There is a misconception that patenting is not available for any kind of marijuana-related invention because patents and trademarks are federal rights granted by the US government, and marijuana remains federally illegal. In fact, marijuana patents are approved very commonly at the federal level, although trademarks are not.

Patents and trademarks are both granted by the USPTO. When patent examiners review a patent application, they apply US patent law, which has no legal-use requirement. Meanwhile, trademark examiners apply US trademark law, which does require that the trademark be applied to products "in commerce," which has been interpreted as legal commerce between the states.

Although trademarks are not available for marijuana products, which cannot legally cross state lines under federal law, trademark protection can be obtained for hemp products. Obtaining state trademarks is a level of protection marijuana breeders and cultivators can utilize to protect their brands ahead of a change in federal law. In the absence of federal trademark protection for marijuana, state trademarks will strengthen common-law trademark protection, and may support federal trademark applications if they become available in the future.

Federal patent examiners grant patents for marijuana because the patent statute does not require the use (legal or otherwise) of an invention for it to be patentable. The USPTO has allowed cannabis-related patents at least since the 1940s, shortly after federal Prohibition began. The apparent lack of patents in this space before adult use legalization seems to be from a lack of applications rather than any meaningful change in policy.

16.5 IP PROTECTION FOR CANNABIS PLANTS

There are three different forms of protection cannabis breeders can obtain for plant varieties in the US, each of which has its limitations and benefits:

- US Department of Agriculture Plant Variety Protection (USDA PVP)
- USPTO Plant Patents
- USPTO Utility Patents

Another specialized form of IP established for the protection of plant varieties is known as plant breeders' rights (PBR), plant variety rights, plant variety protection, or UPOV rights. Since these all arise from the same international system set under the International Union for the Protection of New Varieties of Plants (also known as UPOV), they will all be called PBR for simplicity. Almost all other countries that offer IP protection for plant varieties use the PBR system, which corresponds to one of the forms of protection available in the US, the USDA PVP system.

16.6 WHAT MAKES A PLANT PATENTABLE?

A plant is patentable if it was either developed by intentional breeding and selection or was discovered in a cultivated area. In either case, it can be considered that such a plant originated from human activity, while a plant discovered in the wild would be considered an unpatentable product of nature. Any plant originated by human activity is thus eligible for a patent, but it must also meet a few other requirements for a patent to be granted. These include novelty and an adequate description of the plant.

Novelty is lost if the new variety is available to the public for more than a year before filing a patent application. So, the timing of the filing is critical. Likewise, a nominal novelty or obviousness problem can arise if the applicant cannot adequately distinguish the new variety from its parents or from other known varieties (as discussed in more detail below). This is usually not a problem, in practical terms, because a variety that cannot be distinguished in some way from other varieties would probably not be worth patenting. The breeding and selection process is based mainly upon the goal of identifying a new variety with a combination of traits that are not already present in other varieties.

Novelty rules for PVP are different and are based on prior commercial activity involving the new variety. For a *Cannabis* PVP application filed in the US, the variety cannot have been in commercial use for more than a year in the US or for more than four years anywhere in the world. These same rules apply in other UPOV countries: local (in the country where the application is filed) commercial activity cannot have begun more than one year before the filing date, and commercial activity anywhere cannot have begun more than four years before the filing date.

The requirement of adequate description in a plant patent is met with a very detailed botanical description of the plant. For a VSUP, the seed deposit is the primary way the subject variety is "described."

Many people who have sought patents for other subject matter are familiar with the fact that one of the standards for patentability under the patent statute is non-obviousness. While this is formally a requirement for patenting plant varieties, it is rarely a genuine issue during the examination process. This is because of the fact that any individual organism that results from sexual reproduction is, in terms of its complete genetic code, the first of its kind.

When one understands the vastness of any genome and the biology of genetic recombination in sexual reproduction, it becomes clear that it would be statistically impossible for two separate events of sexual reproduction, even involving the same parents (unless both parents are almost fully homozygous throughout their genome), to ever produce two individuals that are exactly alike. For this reason, if a plant variety is a product of sexual reproduction, followed by further asexual reproduction to establish a stable clonal line, or followed by further inbreeding and selection to establish a stable seed line (either of which properly could be referred to as a variety), the variety will be inherently new in terms of its overall genotype. This is why the usual novelty requirement for any patent is not a function of prior art as to other plants, but is instead only a function of possible prior art created by pre-filing commercial activity as to the variety itself. Some other variety, even if it is very similar in phenotype, but which originated from a different cross, cannot be literally the same as the new variety, and therefore it would not be considered good prior art in a novelty rejection.

However, while it is very rare, it would be possible to receive a meaningful obviousness rejection as to a prior-art plant variety. If there is no phenotypic difference between two genetically different plant varieties, it would be appropriate for the USPTO to reject the later-filed application as being obvious over the earlier-filed application.

In everyday practice, Examiners do not issue obviousness rejections in plant patent applications, while in VSUP applications such rejections are easily overcome by some combination of pointing out the differences between the prior-art variety and the new variety, and agreeing to make a seed deposit. Essentially, the seed deposit provides additional "information" about the variety that goes beyond any description that could be reduced to writing, which is typically considered sufficient to effectively distinguish it from a prior-art variety.

Likewise, because of the variability in sexual reproduction,

16.7 THE USDA PVP SYSTEM

The USDA PVP system is based on the international model legislative framework established by *L'Union internationale pour la protection des obtentions végétales* (UPOV, in English, the International Union for Protection of Plant Varieties). Under this framework, a country that wishes to become a member of UPOV must enact legislation that substantially conforms to UPOV standards as part of its national laws. These standards include a requirement that a proposed new plant variety be tested for "DUS," or distinctness, uniformity, and stability.

DUS testing is conducted as part of the examination of the application by growing plants to maturity from deposited seeds or plantlets.

The UPOV system has two important exemptions:

1. It is not an infringement for someone to save seed from one planting to replant the same amount of cultivated area in succeeding seasons and;
2. It is not an infringement for someone to develop new varieties by breeding with protected varieties.

This system also has a longer novelty period (defining the filing deadline) than the patent system. A variety is still considered "new," and therefore protectable, if the application for protection is filed within one year of commercialization within the country where the application is filed and within four years of commercialization anywhere in the world. Note that trees and vines have an even longer period before losing novelty, but that longer period does not apply to cannabis varieties.

The USDA PVP system is the US implementation of UPOV and conforms to the same standards as other the UPOV/PRB laws in other countries. However, it is currently only available for varieties that qualify as hemp because part of a complete application for PVP is a seed deposit, and the USDA will not accept a deposit of seeds from a variety that would, in its mature form, qualify as illegal marijuana.

Although many in the cannabis industry have come to rely on a position that all *Cannabis* seeds qualify as hemp because they do not contain any significant amounts of THC, seeds deposited with the USDA may be germinated and grown to maturity to perform DUS testing. Thus, USDA PVP protection is not presently an option for marijuana varieties.

Likewise, USDA PVP may also not be an ideal option even for hemp varieties because of the exemptions that make it permissible to breed with a protected variety and to save seed for replanting. While it is possible to avoid these exemptions with contractual provisions that make it a breach of contract to breed or save seed, such limitations only apply to someone who is party to such a contract. There would be no recourse against a third party who legitimately obtained the protected variety and used it for breeding.

The procedure for obtaining a USDA PVP certificate is relatively straightforward. The application primarily consists of a questionnaire that addresses key features of the plant in comparative terms.

16.8 USPTO PLANT PATENT

Plant patents can provide adequate protection for some *Cannabis* varieties but are not ideal in most cases because a plant patent is only infringed by clonal propagation. Some varieties are only valuable in clone form and are not especially attractive to use in breeding. There are varieties that, despite the breeder's best efforts, do not seem capable of producing enough seeds for the development or preservation of stable seed-propagation of the variety.

When a plant patent is appropriate, the applicant must provide a complete botanical description of the variety. There are numerous—and some obscure—features that USPTO examiners want the applicant to describe.

Many question whether a chemical or DNA description is necessary for a plant patent. The answer is a very clear "no." The plant patent system was created in the 1930s and still reflects the botanical "whole plant" approach to describing new plant varieties. If the applicant has DNA or biochemical information, it is possible and may be helpful to include such information in the application, but it is not a requirement.

What is required is the name of the variety, a description of how it was obtained (such as the parents and steps in the breeding and selection process), and a description of how the new variety can be distinguished from its parents and the closest known varieties. Photographs of the variety are also required and should show its distinctive features (if such features are visible).

A very common mistake in plant patent applications is using, as the name of the new variety, a potentially distinctive or valuable brand name. To avoid this mistake, it is extremely important to appreciate the interplay of plant patents and trademarks. This derives from a fundamental doctrine of trademark law that generic terms cannot be treated as trademarks belonging to and controlled by a trademark registrant.

In this context, a generic term is a word whose meaning and association with a particular thing is already in the public lexicon. For example, the word "ball" has a specific meaning in English that is associated with "a solid or hollow spherical or egg-shaped object that is kicked, thrown, or hit in a game."

Imagine a business that makes and sells balls. It would be improper to grant a trademark for the term BALL for exclusive use by that business, because it would remove the generic word "ball" from free use by everyone. A trademark registration would make the mark BALL the exclusive domain of that one business. However, if the business had a name not automatically associated with balls, such as "Deep River," then registering the mark DEEP RIVER as a brand of balls would be uncontroversial because such a registration would not interfere with established use of a generic term.

Returning to plant patents, it is critical to know that whatever name is used in a patent application is, by definition, the generic term for that variety. As such, it is the word designated for free use by the public to refer to the variety. And therefore, since it is considered just as generic as "ball," it is inherently unavailable for trademark protection.

So, if the owner of a new cannabis variety wants to use a name, such as RIBEYE as a trade name for commercial purposes, establishing common-law trademark rights from its correct use, and keeping the door open for eventual federal trademark registration, it would be extremely important not to list "Ribeye" as the name of the new variety in the patent application.

The correct approach here is always to assign two names to any variety worth protecting. One is the generic name for the patent application, and the other is the potential commercial name. Failure to keep these two separate will make it impossible to establish any meaningful trademark value in the name. Using the example in which RIBEYE is the desired commercial name, a corresponding generic name could be something like RE-1 as the generic name. Packaging could designate RIBEYE as a brand of patent-pending *Cannabis* variety RE-1.

16.9 USPTO UTILITY PATENT

The strongest and most flexible form of protection for a cannabis variety is a USPTO utility patent because a utility patent permits claiming the variety as an invention in multiple ways. Examples of this flexibility include claiming:

- Seeds of the variety as exemplified by the seeds on deposit;
- Plants or parts of plants grown from such seeds;
- Progeny of the seeds or the plants (within certain limits to be negotiated during the examination process);

- Products made from the seeds or the plants;
- Using the plants in a breeding program; etc.

This flexibility of claiming the invention avoids the critical limitations of USDA PVP and USPTO plant patents by "closing the loopholes" that define certain essential activities, such as breeding and propagating with seeds, which are acts that do not infringe a PVP or plant patent.

It is important to note that some utility patents are broad enough to cover more than one variety and instead cover groups of plants of any origin that have certain characteristics in common. For convenience and clarity, in *Cannabis*, these can be referred to as "chemotype patents." Such chemotype patents are extremely difficult to obtain and may be subject to significant validity concerns. That approach to patenting is not common and is beyond the scope of this chapter.

The type of utility patent that is more reasonable and much more feasible for *Cannabis* varieties is limited to one specific variety per patent. The specific variety is exemplified and defined by the seeds deposited as part of the application process. To distinguish this narrower kind of utility patent from chemotype patents, I use the term "variety-specific utility patent" (VSUP).

So why doesn't everyone choose utility patents? There are two main reasons:

1. The requirement of a seed deposit and;
2. The greater expense associated with the application and the examination process. If a variety simply cannot produce stable seeds, it will not be eligible for a utility patent.

16.10 MATERIAL TRANSFER AGREEMENTS

Some commercial breeders license their seeds, clones, and tissue culture explants to commercial cultivators as their business model. They enter into material transfer agreements with their licensees to restrict and profit from the use of their valuable germplasm. Even the best material transfer agreements do not give the breeder rights against someone who didn't sign them. Therefore, plant variety patents enhance the protections of material transfer agreements.

These agreements are essential because they connect royalty obligations with rights to access and use. They include details such as how germplasm is obtained, if and how plants can be propagated, and how much can be propagated. However, just like a trade secret, once the limited access granted under an MTA fails—the germplasm gets out—there is little or no recourse against those not in privity of contract.

In mainstream agriculture, it is common and profitable to protect a unique plant variety with plant IP and then enter into licensing agreements for cultivators to access the germplasm. If a variety were to become so popular that there was a demand for it in many countries, it would often be more profitable to license germplasm to local cultivators rather than establish infrastructure in each market.

NOTE

1 Adapted from the author's work, which appears in *Cannabis Breeding* by James Loud (2024).

17 The CoNNect Community
Cultivating Collaboration and Innovation in the Cannabis Industry

Sergio Picazo

17.1 INTRODUCTION

In the dynamic world of cannabis cultivation, staying ahead requires not just keeping up with the latest genetics & advancements, but actively engaging in the dialogue that shapes the future of the industry. The CoNNect Community serves this critical role, offering a platform for breeders, cultivators, extractors, and brands with manufacturers to come together and forge connections that go beyond basic networking. Founded in early 2021 by Sergio "The Custodian" Picazo of CLTVTD Genetics/Validated Curations, The CoNNect Community has rapidly become a pivotal space for industry leaders to share knowledge, discuss hot topics, and drive innovation from seed to sale. The community quickly gained support from local hydro store juggernaut, Kevin Vu, who has propelled the CoNNect to where it is today. Kevin believes in fostering collaboration and making sure the cultivators were fed & comfortable while they learned after a long day of work.

17.2 PURPOSE AND PROCESS

The primary mission of the CoNNect Community is to elevate industry standards through innovative solution trials and discussions on important industry topics. Initially, these gatherings took place monthly at Big Sexy Brewing, located in the Sacramento Green Zone, where cultivators and industry experts could share techniques and learn from supplier partners in a private, collaborative environment. The aim was not just to educate, but to build a community that supports and elevates each other's business practices to grow better cannabis.

As the community grows, the vision expands. The CoNNect Community plans to develop an ecosystem that delivers scalable education and networking opportunities worldwide. This will include a network of affiliates capable of hosting local community events, thus fostering a global community of cannabis professionals.

17.3 INNOVATION AND EDUCATION

At the heart of the CoNNect Community's events are the industry-relevant topics that spur conversation and innovation. These in-person gatherings feature booth activations by manufacturing partners and brands that utilize cutting-edge products and services essential for operating successful cannabis businesses. Collaborative panel discussions bring a unique perspective to the space while incorporating subject matter experts with focused objectives for each event. A recent collaboration with Grow World integrates a live streaming platform that not only offers commercial trial applications but also helps convert these trials into purchases through a unified ecosystem of hydrostores. These

DOI: 10.1201/9781003386438-17

affiliate stores lean into the education provided by Th CoNNect as a training tool to become not only just a local shop, but a wealth of information for their cultivation community.

17.4 PAYOFF AND COMMUNITY IMPACT

The ultimate goal of the CoNNect Community is to strengthen long-term community development. This includes enhancing the ability to cultivate high-quality cannabis and educating retailers on achieving these standards. By promoting transparency and trust, the community helps consumers understand and maximize the medicinal and recreational benefits of cannabis. The benefits of such a community are manifold:

- **Networking and Collaboration:** Building strong relationships within the industry to foster support and shared growth.
- **Innovation and Exploration:** Introducing and implementing new solutions that advance the industry.
- **Education:** Sharing successful cultivation strategies and business practices to set industry benchmarks.

17.5 SPECIAL EVENTS: THE BATTLE ROYALE

A standout event within the CoNNect Community is the Battle Royale circuit. This competition allows cultivators to showcase their best cultivars based on aroma, smoothness, and taste. It is judged by guest judges who are recognized leaders within the cannabis industry and each contender must have a representative from their farm. This event serves a dual purpose: it provides a platform for cultivators to demonstrate their expertise while passionately sharing tech to improve quality and leveling up, in addition to understanding the genetic similarities from each submission. Each submission receives a genomic fingerprint, Certificate of Cultivar Identity (CCI), powered by MyFloraDNA, which provides a unique overview of how the contenders' genetics are from one another. The Battle Royale has created curiosity amongst the industry to learn more about this service while waiting for other markets to host the circuit. There will be an annual global Battle Royale where the former champions compete for the title within their respective category. The market demand of specific flavors allows the Battle Royale to bring awareness to cultivators who may not have a strong brand presence that can propel them into a position for retail demand.

Through having the ability to utilize a genomics tool, breeders & cultivators can help position themselves for trust and transparency with genetics while ensuring consistency across their gardens. One of the biggest challenges in this industry is the renaming of genetics. With a tool like the CCI, cultivation brands provide a seal of authenticity driving trust and transparency with public health & safety.

17.6 CONCLUSION

The CoNNect Community is more than just a networking platform; it is a catalyst for innovation and quality improvement in the cannabis industry. By fostering a collaborative environment that connects all verticals of the industry, the community helps pave the way for a healthier, more sustainable approach to cannabis cultivation and use. As the community continues to expand and evolve, it promises to remain at the forefront of the cannabis industry's most pressing discussions and developments.

18 The Evolving Landscape of Cannabis and Its Future Prospects

Angel Fernandez Marti and Jaime Prohens

18.1 INTRODUCTION

Cannabis, once relegated to the fringes of industrial and medicinal applications due to legal and social constraints, is now emerging as a pivotal crop with a vast array of uses. The surge in research and cultivation of cannabis is indicative of its growing significance. This book aims to bridge the existing knowledge gap by providing comprehensive and updated insights into cannabis genetics, breeding approaches, genomics, and other "omics," contributed by esteemed researchers and important players in the field.

18.1.1 The Importance of Cannabis in Modern Research

Cannabis is increasingly recognized for its multifaceted utility. Industrially, it offers applications ranging from biofuels to biodegradable plastics. The plant's fibers are used in textiles, construction materials, and even in the automotive industry. Medically, its compounds, particularly cannabinoids like THC and CBD, have demonstrated efficacy in treating conditions such as chronic pain, epilepsy, and multiple sclerosis. This dual importance has spurred a boom in cannabis research, necessitating a resource that consolidates current findings and methodologies.

18.1.2 Economic and Social Impact

The economic potential of cannabis is vast. With the legalization of both medical and recreational cannabis in various regions and countries, the market has seen exponential growth. This has led to the creation of numerous jobs, from cultivation and processing to retail and research. The social implications are equally significant. Legal cannabis has reduced the burden on the criminal justice system and provided patients with access to alternative treatments that were previously unavailable.

18.2 CHAPTER OVERVIEW

This book is structured into eighteen chapters, each addressing crucial aspects of cannabis research:

- Origin and Taxonomy of Cannabis: This chapter delves into the evolutionary history and classification of cannabis, providing a foundation for understanding its genetic diversity and potential. It covers the geographic origins of cannabis and its domestication history, shedding light on how different cultivars have evolved.

DOI: 10.1201/9781003386438-18

- Intellectual Property and Protection for Cannabis Plants: Here, the types of intellectual property (IP) protections available for cannabis plants and the best strategies to protect these valuable genetic resources are explored. This chapter covers patents, plant breeders' rights, and trademarks.
- Legislation Aspects: Legal barriers have historically hindered Cannabis research and breeding. This chapter discusses current laws, regulatory changes, and their implications on research and industry. It provides an overview of the legal landscape in various countries and the impact of legalization on research opportunities.
- Genetic Resources: Focusing on breeding techniques, this section covers the genetic resources available for cannabis cultivation, including traditional breeding methods and modern genetic approaches. It discusses the importance of preserving genetic diversity and the role of gene banks.
- Breeding Techniques: Detailed descriptions of breeding techniques are provided to enhance crop quality and resilience. This chapter includes case studies of successful breeding programs and the development of high-yield cultivars.
- Tissue Culture: The use of in vitro culture techniques in cannabis breeding are explored to facilitate precise and efficient breeding programs. This chapter covers tissue culture, micropropagation, and the development of genetically stable clones.
- Omics Technologies: A set of chapters dedicated to genomics, transcriptomics, and metabolomics, offering insights into the molecular underpinnings of cannabis traits and functions. These chapters delve into the genetic makeup of cannabis, gene expression profiles, and metabolic pathways that contribute to its unique properties.
- Integrated Pest Management: Effective pest management is critical for sustainable cannabis cultivation. This chapter discusses integrated pest management strategies, including biological control, cultural practices, and the use of biopesticides. Successful case studies are provided to illustrate practical applications.
- Gene Editing Prospects: Cutting-edge gene editing technologies, such as CRISPR, and their potential applications in Cannabis research and breeding are examined in this chapter. It discusses the ethical considerations, regulatory challenges, and the future potential of gene-edited cannabis.
- Future Perspectives: The concluding chapter looks ahead at the future of Cannabis research, identifying emerging trends and potential breakthroughs that could shape the industry. Topics include the potential for cannabis in personalized medicine, the development of novel cannabis-based products, and the integration of AI and big data in Cannabis research.

18.3 CONCLUSION

The rapid advancement in Cannabis research underscores the necessity of a comprehensive resource that collates the latest developments in genetics, genomics, and breeding. By addressing the most pertinent topics and challenges in Cannabis research, this book aspires to be an indispensable reference for scientists, breeders, and industry stakeholders. Through collaborative efforts and continued innovation, we can unlock the full potential of cannabis, driving forward both industrial applications and medical advancements.

The future of Cannabis is bright, with ongoing research promising to unveil even more of its hidden potential. As legal barriers continue to fall and public perception shifts, cannabis stands poised to become one of the most important crops of the 21st century. Its contributions to medicine, industry, and agriculture are only beginning to be realized, and this book serves as a testament to the remarkable journey of cannabis from a stigmatized plant to a cornerstone of modern research and innovation.

Index

P

R

S

T

V

W

9781032478975